Modern Masonry: Building Techniques, Tools and Materials

Modern Masonry: Building Techniques, Tools and Materials

Editor: Patrick Walker

NY RESEARCH PRESS
New York

Published by NY Research Press
118-35 Queens Blvd., Suite 400,
Forest Hills, NY 11375, USA
www.nyresearchpress.com

Modern Masonry: Building Techniques, Tools and Materials
Edited by Patrick Walker

International Standard Book Number: 978-1-64725-470-4 (Hardback)

Cataloging-in-Publication Data

Modern masonry : building techniques, tools and materials / edited by Patrick Walker.
p. cm.
Includes bibliographical references and index.
ISBN 978-1-64725-470-4
1. Masonry. 2. Building. 3. Building--Equipment and supplies. 4. Building materials. I. Walker, Patrick.
TA670 .M63 2023
693.1--dc23

Contents

Preface

Every book is a source of knowledge and this one is no exception. The idea that led to the conceptualization of this book was the fact that the world is advancing rapidly; which makes it crucial to document the progress in every field. I am aware that a lot of data is already available, yet, there is a lot more to learn. Hence, I accepted the responsibility of editing this book and contributing my knowledge to the community.

Masonry refers to the construction of structures from distinct units placed and bound together. It can be characterized as the arrangement of various materials in various ways with or without mortar. Modern masonry involves constructing buildings using modern materials such as concrete-block joists, concrete slabs, and walls made of pressed bricks, cement blocks, perlite, tiles, stones, and glass blocks. The most popular masonry materials used in industrialized countries are concrete block and brick, which can be either load bearing or non-load-bearing. Concrete blocks, particularly those with hollow cores, provide compressive strength and are appropriate for constructing buildings with light transverse loading. This book unravels the recent studies in the field of modern masonry. It presents researches on the building techniques, tools and materials which are used within this field. This book is a vital tool for all researching or studying masonry as it gives incredible insights into emerging trends and concepts.

While editing this book, I had multiple visions for it. Then I finally narrowed down to make every chapter a sole standing text explaining a particular topic, so that they can be used independently. However, the umbrella subject sinews them into a common theme. This makes the book a unique platform of knowledge.

I would like to give the major credit of this book to the experts from every corner of the world, who took the time to share their expertise with us. Also, I owe the completion of this book to the never-ending support of my family, who supported me throughout the project.

Editor

1

Improvement of the Durability of Recycled Masonry Aggregate Concrete

Tereza Pavlů [1,2,*], Kristina Fořtová [2], Jakub Řepka [1,2], Diana Mariaková [1,2] and Jiří Pazderka [1]

1 Department of Building Structures, Faculty of Civil Engineering, Czech Technical University in Prague, Thakurova 7, 160 00 Prague, Czech Republic; jakub.repka@cvut.cz (J.Ř.); diana.mariakova@fsv.cvut.cz (D.M.); jiri.pazderka@fsv.cvut.cz (J.P.)

2 Research Team Architecture and the Environment, University Centre for Energy Efficient Buildings, Czech Technical University in Prague, Trinecka 1024, 273 43 Bustehrad, Czech Republic; kristina.fortova@cvut.cz

* Correspondence: tereza.pavlu@cvut.cz

Abstract: The use of recycled masonry aggregate for concrete is mostly limited by the worse properties in comparison with natural aggregate. For these reasons it is necessary to find ways to improve the quality of recycled masonry aggregate concrete and make it more durable. One possibility is utilization of crystalline admixture which was verified in this study by laboratory measurements of key material properties and durability. The positive influence of mineral admixture was proved for freeze-thaw resistance. The positive impact to carbonation resistance was not unambiguous. In conclusion, the laboratory evaluation shows how to improve the durability of recycled masonry aggregate concrete, however, it is necessary to investigate more about this topic.

Keywords: recycled masonry aggregate concrete; crystalline admixture; durability of concrete

---

1. Introduction

The main advantages and disadvantages related to recycled aggregate (RA) used as aggregate for concrete are generally known and have been described many times. On the one hand, the replacement of natural aggregate (NA) leads to the reduction of primary source consumption. On the other hand, the utilization of recycled aggregate mostly negatively affects the properties of concrete, especially its water absorption, mechanical properties and durability. For this reason, the utilization of RA from construction and demolition waste (CDW) as aggregate for concrete has limited structural applications worldwide. Furthermore, the possible utilization of RA depends on its quality, composition, and properties. Generally, there are three basic types of RA known: recycled concrete aggregate (RCA), masonry aggregate (RMA), and mixed recycled aggregate (MRA) [1,2]. Each type of aggregate has its own specifications and limits of utilization.

The RCA mostly contains natural aggregate with attached mortar, unbound natural aggregate, and cement paste. The attached mortar causes two main problems related with RCA utilization as aggregate for concrete: (1) transition zones between old and new mortar which negatively influence the mechanical properties of recycled aggregate concrete (RAC) [3–6]; (2) the higher porosity and water absorption of RCA than the companion NA. The water absorbability of RCA, which ranges up to 15% [2], is necessary to know due to the mixture design for further workability of the concrete mix. These aspects lead to the higher water absorption, decline of freeze-thaw resistance and reduction in the carbonation resistance of recycled aggregate concrete (RAC) [7–9]. Generally, the replacement rate up to 30% of RCA has no significant influence on total porosity and compressive strength of RAC.

When the replacement rate is higher than 30% of RCA in the mixture, the total porosity increases and compressive strength decreases significantly [5].

The freeze-thaw resistance of concrete mostly defines the possible future utilization of concrete structural elements in the exposition environment. As expected, the freeze-thaw resistance of RAC decreases with the increasing replacement ratio of NA by RA in the concrete mixture [10], as a result of the increasing porosity of concrete containing RCA [11]. Moreover, it is linearly correlated with its water absorption capacity [10]. In general, the freeze-thaw resistance of concrete is evaluated by comparison of several physical properties and mechanical properties before and after a defined number of freezing and thawing cycles [12,13]. During freezing, the water expands throughout the pores and makes pressure inside the concrete, which could lead to local cracks. Furthermore, the freeze-thaw resistance is also influenced by a water-cement (w/c) ratio, where increasing w/c ratio leads to decreasing freeze-thaw resistance due to the higher amount of mixing water.

The knowledge of carbonation resistance of concrete is essential due to the future utilization of reinforced concrete structural elements in which the corrosion of reinforcement would be affected. The reinforcing bars are endangered by corrosion when the passive coating is destroyed by carbonation and chloride ingress. Corrosion of steel bars starts earlier in RAC than in NAC and depends on the amount of RA in the RAC. Furthermore, the steel corrosion is also influenced by the electrical resistivity of concrete which decreases with the increasing replacement rate of aggregate and porosity of concrete. Carbonation of concrete, which can be described as a physical-chemical process, is influenced by the permeability of concrete, its moisture content, cement content and water/cement ratio, mineral additions, aggregate type, porosity and the environment's CO_2 content, relative humidity and temperature [9,14]. Generally, according to the previous studies [1,14–24] the carbonation depth of RAC also increases with an increasing replacement level of aggregate in concrete. The significant decline of carbonation resistance is shown for concrete with a higher replacement rate than 30% [17,23]. This is caused by the higher water absorption of RA than NA, which leads to a greater porosity of RAC than NAC the same w/c ratio [9,13]. In previous studies [9,25], the greater negative influence of the fine fraction of RA in comparison with the coarse fraction of RA was evaluated. This result was probably caused by the higher water absorption of fine RCA in comparison with coarse RCA, which leads to worse permeability of RAC. The carbonation depth of RAC tested during time periods is approximately two times higher than the carbonation depth of NAC [26]. Generally, it could be said that the results of limitation of the RCA utilization as aggregate for concrete in the case of carbonation resistance corresponds with the limitation in general, whereas the maximal replacement rate is defined as 30% for the coarse fraction (4–8 mm and 8–16 mm) and 20% for the fine fraction (0–4 mm) of RCA. On the contrary, the opposite result was also observed in which the RAC carbonation depth decreased. In this case, the replacement ratio of NA by RCA was greater than 70% and RCA had about 40% of adhered mortar. This phenomenon could be caused by the cement content in the adhered mortar, which could slow down the carbonation rate [27].

In spite of the negative aspects of worse carbonation resistance of RAC, there is one positive point of view. Concrete in general absorbs CO_2 from the atmosphere. RAC with higher porosity and consequently greater carbonation depth absorbs more CO_2 then conventional concrete, which brings environmental benefits [28,29].

The masonry aggregate (RMA), with a high content of waste masonry such as red bricks, calcium silicate bricks, ceramic blocks, aerated concrete, plaster, etc., and mixed recycled aggregate (MRA) which combines RMA and RCA, has not found satisfactory use yet. According to the Czech standards [30], the RA with a lower content of waste concrete than 90% is not efficiently used as aggregate for concrete, due to its negative impact on mechanical properties and durability. The properties of RMA, such as porosity, water absorption, density, and resistance—for instance, resistance to wear, abrasion resistance, or freeze (thaw resistance), etc. are different in comparison with NA, which depends on the composition which is connected with the recycling process and consequently influences the properties of recycled masonry aggregate concrete (RMAC) [31–56].

In the case of RMA, higher porosity and water absorption are caused by porous materials contained in RMA, such as red bricks, ceramic blocks, aerated concrete, and concrete particles with adhered mortar. It follows that the consequences of high porosity and water absorption of RMAC are slightly different than that of RAC; however, the impact on the mechanical properties and durability is similar. For better properties of fresh and hardened concrete, the additional water compensating water absorbability of RMA needs to be added to the concrete mixture during mixing [43] or the RMA must be soaked in water for 24 h before mixing [49]. Generally, the mechanical properties and durability of recycled aggregate concrete decrease with an increasing replacement rate of aggregate in concrete by RMA [43,49,57]. A maximal replacement rate without significant decline of mechanical properties of RMAC comparable with NAC has been found in 15% coarse RMA. On the contrary, the mechanical properties of RMAC with 100% replacement rate decreases up to 65% [49]. The partial replacement of natural sand by fine RMA has had no significant impact on the compressive strength of recycled aggregate concrete, probably due to the high amount of fine particles and silica and alumina contents in crushed bricks, which could lead to pozzolanic reactions [57]. Moreover, it was found that the carbonation depth of concrete with RMA is greater in comparison with RCA for the same replacement ratio of aggregate. Furthermore, the carbonation depth of concrete with MRA, the combination of RMA and RCA, increases with increasing amount of RMA in the mixture. It follows that the masonry content worsens the carbonation resistance [9].

Durability of concrete is the ability of concrete to resist various types of effects during its service period of exposure to its surrounding environment. The durability of RMAC is usually weaker than NAC and RAC due to its higher porosity and water absorption. In previous studies, the possibilities of durability improvement were verified. At first, there were few verified possible ways to improve the freeze-thaw resistance of RAC. When RCA is presoaked during the two-stage mixing approach [58] and the water absorption of RCA is compensated (the use of additional water resulting from the absorbability of RCA), the concrete mix achieves a higher compressive strength and durability [59–61], due to the water contained in the porous RA affecting the internal curing effect, in which the water is gradually released for further cement hydration [61–63]. Furthermore, the addition of a suitable amount of fly ash or metakaolin [64–66] to the concrete mixture is due to the ability of the mineral admixtures to react with $Ca(OH)_2$ to form an additional C–S–H gel to increase the density and strength of concrete. Secondly, it was found [67] that the carbonation depth of RAC at early ages could be positively influenced by water-reducing superplasticizers. However, the effectivity of this solution decreases with time. The use of superplasticizers leads to the crystal growing, which causes the structure of concrete to be denser. Moreover, the other way to improve the carbonation resistance of RAC is the addition of mineral admixtures [13] such as high volumes of fly ash, due to its ability to fill the pores, which consequently improves the microstructure. [68]. On the contrary, the utilization the mineral additions as a partial replacement of cement causes a lowering of pH of the concrete and leads to worse carbonation resistance [9]. Finally, the other way to reduce the carbonation depth is the lowering of the w/c ratio [9].

The main aim of this study is the improvement of the RMAC mixture durability by the crystalline admixture. This approach is completely new due to the fact that there are no studies where the durability properties of RMAC concrete were improved and furthermore, no studies where the properties of concrete with RA are improved by crystalline admixture. The combination of these two topics is the main aim of this study. The verification of improvement in RMAC properties was proven by the laboratory measurements of the physical and mechanical properties and durability performance, especially freeze-thaw resistance and carbonation resistance.

2. Materials and Methods

In total, 8 concrete mixtures were prepared and tested to verify the possibility of improvement of the RMAC properties by crystalline admixture. Crystalline admixtures (CA) are permeability reducing admixtures, which are inherently hydrophilic due to their easy reaction with water. The chemical reaction between the crystalline admixture with cement and water leads to the increasing density of calcium silicate hydrate (CSH) and resistance of water penetration. The proper function of the waterproofing effect of crystalline material in concrete is achieved by the reaction of various chemical components necessary when the porous system of concrete reaches a sufficient level of moisture [69]. It was found that the crystalline admixture could improve the durability of reinforced concrete [70]. For example, the depth of penetration could be reduced by almost 50% in selected cases [71]. The previous experimental measurements verify the positive impact of crystalline admixture on the durability of concrete [72–75].

There were two types of NAC mixtures—one was a reference mixture with a natural aggregate without crystalline admixture, and the other mixtures contained two types of RMA as a full replacement of NA and different amounts of crystalline admixture. The durability properties were tested in two steps. The first four mixtures (NAC 1 and RMACs 1) were tested at age 28 days. The second four mixtures (NAC 2 and RMACs 2) were tested at age 60 days to show the influence of the crystalline admixture over time [74,76].

2.1. *Recycled Aggregate*

In the previous studies, it was found that the dry density of coarse RMA ranges between 1800 and 2700 kg/m^3 and fine RMA between 2000 and 2500 kg/m3, which is generally lower than natural gravel and sand [32,40,43,49,52,53]. The range of water absorption of coarse RMA has been established from 10% to 19%, which is up to 25 times higher than natural gravel [32,40,49,52], and fine RMA from 12% to 15%, which is more than 10 times higher than natural sand [43,52,57].

In this study, one type of NA and two types of RMA were used. Both types of RMA were prepared from construction and demolition waste by a Czech recycling center, which were separated into fractions of 0–4, 4–8, and 8–16 mm (see Figure 1). Both types of RMA contained more than 70% of the waste masonry (red brick, aerated concrete, and plaster). Furthermore, the waste concrete and unbound aggregates were contained. All tested properties of RMA differed from NA, especially the water absorption capacity, which was more than ten times higher and ranged from 7.8 to 12.4% for coarse fraction and was 3.7 and 13.3 for fine fraction of RMA. This evaluation shows slightly lower water absorption of RMA in comparison with results of previous studies [32,40,43,49,52,57]. The dry density of RMA was lower in comparison with NA with a decline up to 25% which corresponds with the results of previous studies [32,40,43,49,52,53]. Furthermore, the RMA contains more fine particles and has different granulometry in comparison with NA and does not meet the requirements in Standard [77] (see Figure 2). Therefore, the basic properties of aggregates (see Table 1) are presented to show the differences in the materials used for the preparation of the concrete mixtures.

(**a**) RMA 1 fraction 0–4 mm (**b**) RMA 1 fraction 4–8 mm (**c**) RMA 1 fraction 8–16 mm

Figure 1. Recycled masonry aggregate (RMA 1).

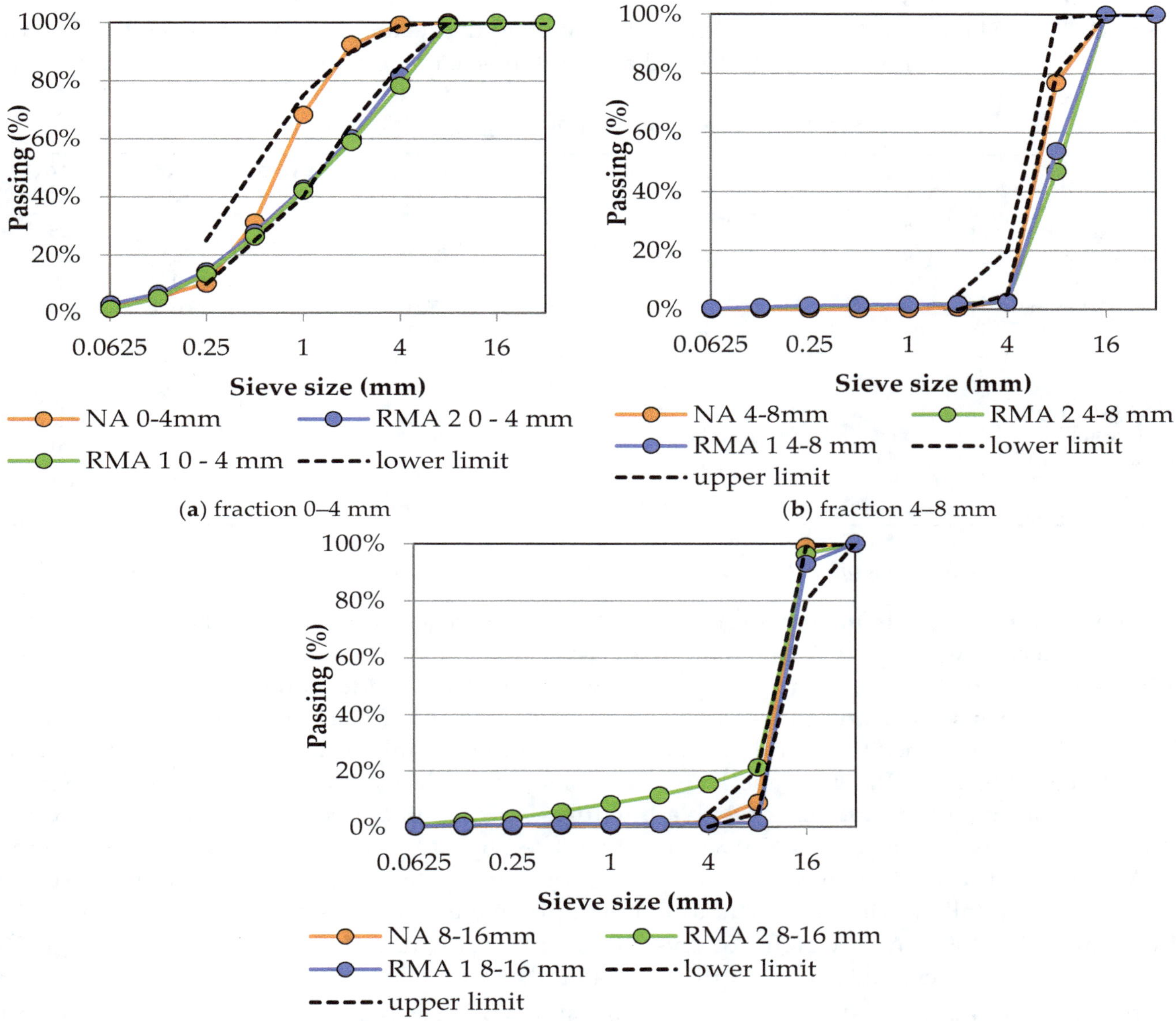

(**a**) fraction 0–4 mm (**b**) fraction 4–8 mm

(**c**) fraction 8–16 mm

Figure 2. Sieving curves for natural aggregate, recycled masonry aggregate with limits defined in the standard [77] used in concrete mixtures.

Table 1. Physical properties of particular fractions of used aggregates.

Types of Recycled Aggregate	Grading (mm)	Content of Finest Particles	Oven-Dried Particle Density		Water Absorption Capacity (%)	
		f (%)	ρ_{RD} (kg/m^3)	σ	WA_{24} (%)	σ
Natural aggregate (NA)	0–4	2.0	2570	81	1.0	0.0
	4–8	0.1	2530	12	1.7	0.3
	8–16	0.2	2540	12	1.9	0.2
Recycled masonry aggregate 1 (RMA 1)	0–4	2.9	2340	108	3.7	0.6
	4–8	0.4	1920	62	12.4	0.7
	8–16	0.7	2130	72	7.8	0.5
Recycled masonry aggregate 2 (RMA 2)	0–4	1.3	1950	166	13.3	2.1
	4–8	0.4	2050	41	10.6	0.4
	8–16	0.3	1990	33	10.6	1.1

Selected properties of RMA were tested according to the requirements of Czech European standards [77]. The properties with the highest impact to the recipe design were tested. The basic properties of RMA are shown in Table 2 and the granulometry is shown in graphs in Figure 2.

Table 2. Concrete mix proportion, per cubic meter.

Designation	NAC 1 C0	RMAC 1 C0	RMAC 1 C1	RMAC 1 C3	NAC 2 C0	RMAC 2 C0	RMAC 2 C1	RMAC 2 C3
Cement (kg/m)	260	260	260	260	260	260	260	260
Water (kg/m^3)	169	219	219	219	169	284	284	284
Sand (kg/m^3)	710	0	0	0	710	0	0	0
NA 4/8 (kg/m^3)	520	0	0	0	520	0	0	0
NA 8/16 (kg/m^3)	609	0	0	0	609	0	0	0
RMA 0/4 (kg/m^3)	0	807	807	807	0	949	949	949
RMA 4/8 (kg/m^3)	0	54	54	54	0	32	32	32
RMA 8/16 (kg/m^3)	0	653	653	653	0	500	500	500
Crystalline admixture (kg/m^3)	0	0	5	10	0	0	5	10
w/c eff (-)	0.65	0.65	0.65	0.65	0.65	0.65	0.65	0.65
w/c (-)	0.65	1.09	1.09	1.09	0.65	0.84	0.84	0.84

2.2. Recycled Aggregate Concrete Mixtures

The laboratory measurements were carried out on eight concrete mixtures with the same exposition class XC1, effective w/c ratio 0.65, and amount of cement CEM I 42.5 R 260 kg/m^3. The mixtures were optimized using the Bolomey particle size distribution curve. The additional water was calculated for RMAC mixtures according to the water absorption capacity of RMA and moreover the higher effective w/c ratio 0.65 positively influenced the effect of the crystalline admixture. Two mixtures of conventional concrete (NAC 1 C0 and NAC 2 C0) of strength class C25/30 only with NA of particle size up to 16 mm and without crystalline admixture was manufactured as a reference to compare with the other mixtures in which NA was fully replaced by RMA. Two mixtures of RMA concrete (RMAC 1 C0 and RMAC 2 C0) were prepared without the crystalline admixture. In the other four mixtures, the crystalline admixture was added to improve the properties of RMAC. Mixture RMAC C1 contains 1.5% (of cement weight) of crystalline admixture and mixture RMAC C3 contains 3% (of cement weight) of crystalline admixture (see Table 2).

Physical and mechanical properties and durability were tested at age 28 (NAC 1 and RMAC 1) and 60 days (NAC 2 and RMAC 2) according to valid Czech standards. Samples of dimensions $100 \times 100 \times 400$ mm^3, $150 \times 150 \times 150$ mm^3, and $100 \times 100 \times 100$ mm^3 were used for testing.

2.3. Evaluation Methodology

The mechanical properties were tested on Controls MCC8 50-C8422/M (Controls Group, Milan, Italy) according to the following standards to obtain the target values: compressive strength EN 12390-3 (2003); flexural strength EN 12390-5 (2009); static modulus of elasticity EN 12390-13 (2014); dynamic modulus of elasticity EN 12504-4 (2005). Samples for testing of mechanical properties were stored and cured in a stable laboratory environment during solidification and maturation, and after 28 and 60 days were determined by laboratory tests. Compressive strength was tested on cubic specimens $150 \times 150 \times 150$ mm^3 and other properties were tested on prismatic specimens $100 \times 100 \times 400$ mm^3.

Water absorption capacity by immersion was tested on cubic specimens $100 \times 100 \times 100$ mm^3. Specimens were treated by water, and after stabilization of weight, dried in an oven at a temperature of 105 ± 2 °C until stabilization of weight. The saturated surface-dried density and dry density were measured on these samples. Samples NAC 1 and RMAC 1 for testing were stored and cured in a stable laboratory environment, and samples NAC 2 and RMAC 2 were stored and cured in water during solidification and maturation for 60 days to verify the influence of activation of the crystalline admixture by water. Capillary water absorption was determined by measuring the rate of water absorption by the capillaries. The ends of the fractured prismatic specimens of 100 × 100 × approx. 150 mm^3, which were tested after the flexural strength test, were immersed in water up to a maximum height of 5 mm and measured by periodically weighing the surface-dried sample until their weight stabilized.

Prismatic specimens 100 × 100 × 400 mm^3 for examination of thefreeze-thaw resistance were stored and cured in water during solidification and maturation for 28 and 60 days. Subsequently, the samples were placed in freezing-thawing equipment KD 20 developed by the Ecofrost company (Czech Republic, Olomouc) for testing frost resistance according to the Czech standard CSN 73 1322 (1969). Test specimens occurred in freezing cycles in which the temperature of the freezer must be between −15 °C and +20 °C. One freezing cycle consists of four hours of freezing and two hours of thawing. During freezing, the test specimens are stored in water of +20 °C. At the end of each freezing stage (25 cycles), one set of surface-dried test beams (3 pieces) shall be tested to determine their dimensions, weight, bulk density, and dynamic modulus of elasticity according to EN 12504-4 (2005). After the frost resistance test (typically 100 cycles), the test beams are tested for flexural strength according to EN 12390-5 (2009).

Prismatic specimens 100 × 100 × 400 mm^3 for examination of the carbonation resistance of mixtures NAC 1 and RMAC 1 were stored and cured in a stable laboratory environment, and samples NAC 2 and RMAC 2 were stored and cured in water during solidification and maturation for 60 days to verify the influence of activation of the crystalline admixture by water. Afterwards, the specimens were stored in a laboratory incubator with air circulation with CO_2 atmosphere CO2CELL (MMM group, Germany, Munich). Samples were placed in the environment with 3.0 ± 0.2% CO_2 concentration for 28 days, which was inspired by the Standard CSN EN 12390-12. However, the evaluating of the carbonation resistance was modified to compare the influence of the same environment to the different mixtures with different amounts of the crystalline admixture. Furthermore, the flexural strength and carbonation depth were measured. The carbonation depth was evaluated on the prismatic specimens split in halves by the use of the phenolphthalein indication method, due to the reduction of pH in concrete by the effect of CO_2. The solution of 0.8 g of phenolphthalein powder was dissolved in a solution of 70 mL ethanol and 30 mL of deionized water.

3. Results and Discussion

3.1. Physical Properties

The water absorption of concrete influences its durability [78,79]. The values of density, water absorption by immersion, and capillary water absorption are shown in Table 3. The density of RMAC is lower in comparison with NAC, with a maximal decline of 25%. Test results of the density showed no significant relationship depending on the amount of crystalline admixture in the RMAC mixture. The water absorption by immersion was more than three times higher for the RMAC in comparison with NAC. In addition, the results of the capillary water absorption of RMAC showed higher values in comparison with NAC (see Figure 3). Furthermore, the capillary water absorption was higher for NAC 1 and RMAC 1 mixtures due to the different concrete treatment during solidification and maturation. However, both types of water absorption of RMAC 2 mixtures increased with higher amounts of crystalline admixture. On the contrary, both types of water absorptions of RMAC 1 mixtures decreased with higher amounts of crystalline admixture. This was probably caused by the different concrete treatment during solidification and maturation and shows the impact of the activating of crystalline admixture during curing in water. Examined results of the density and water absorption of RMAC mixtures confirm the results reported in previous studies [34,78–81].

Table 3. Average values of results of physical properties of concrete, including standard deviation.

Recycled Concrete Mixture	Dry Density		Water Absorption by Immersion		Capillary Water Absorption	
Designation	(kg/m^3)	σ	(%)	σ	(kg/m^2)	σ
NAC 1 C0	2240	21	5.3	0.1	6.45 (1)	0.68
RMAC 1 C0	1726	22	17.3	0.7	24.31 (1)	2.08
RMAC 1 C1	1765	65	17.0	1.0	18.19 (1)	1.59
RMAC 1 C3	1755	26	16.6	0.3	17.05 (1)	1.18
NAC 2 C0	2141	32	5.5	0.2	0.93 (2)	0.25
RMAC 2 C0	1708	21	17.0	0.8	4.45 (2)	0.69
RMAC 2 C1	1673	20	18.1	0.3	4.74 (2)	0.61
RMAC 2 C3	1659	7	18.3	0.4	5.03 (2)	0.11

(1) Weight stabilization after 120 h; (2) Weight stabilization after 72 h.

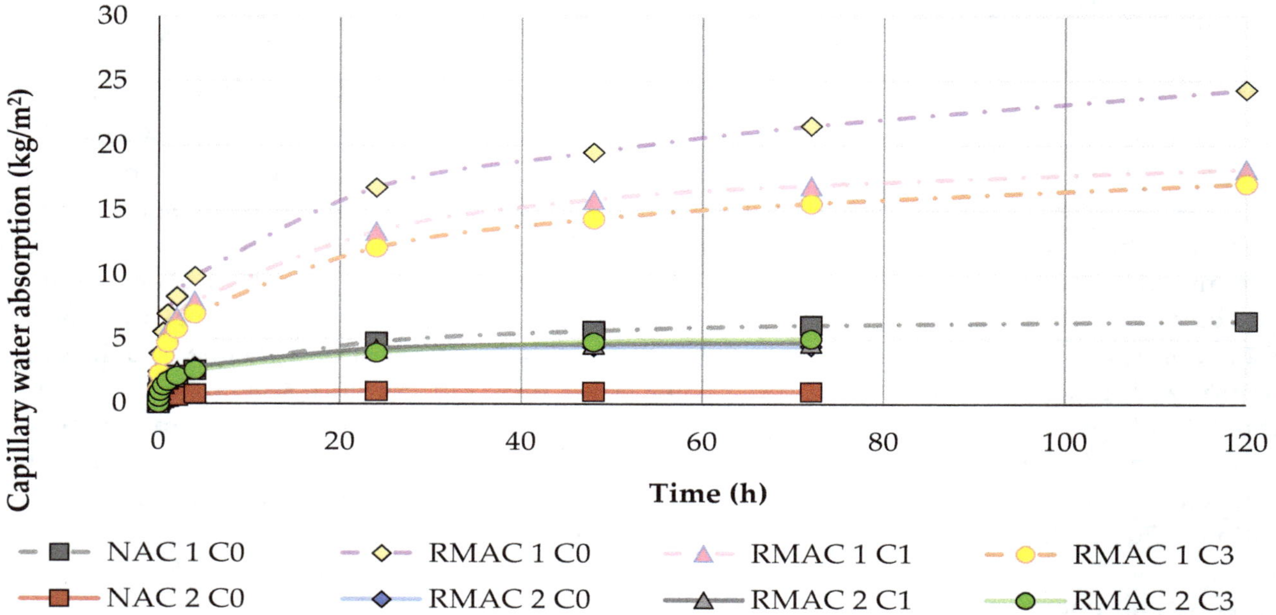

Figure 3. Comparison and progression of capillary water absorption of NAC and RMAC.

3.2. Mechanical Properties

The compressive and flexural strength and static and dynamic modulus of elasticity were tested for all concrete mixtures due to the knowledge of the material properties before durability examination (see Table 4). The compressive strength is one of the most important properties of concrete, which determines its use. This property is mostly negatively influenced by using recycled aggregate, especially from recycled masonry [30–56]. The compressive strength was evaluated on the cubic specimens before storing prismatic specimens in the freeze-thaw chamber. The testing age of samples NAC 1 and RMAC 1 was 28 days and 60 days for the samples NAC 2 and RMAC 2. The samples NAC1 and RMAC1 were also tested after 60 days for comparison with samples NAC2 and RMAC 2. Test results of the compressive strength showed lower values for all tested RMACs in comparison with the NAC, with a maximal decline of 54% (see Figure 4). There are also differences between the compressive strength of RMAC 1 and RMAC 2 mixtures, which were probably caused by the different quality of RMA. Test results also verify the increasing compressive strength with time, however, without significant influence of the crystalline admixture. The flexural strength of RMAC mixtures was also lower in comparison with NAC. The lowest decline was measured for the RMAC 2 mixture which was up to 50%, while the maximal decline of RMAC 1 mixtures was up to 30%. The addition of the crystalline admixture shows the opposite influence for RMAC 1 and RMAC 2 mixtures. As reported in previous studies [82–84], the static modulus of elasticity decreases more than other mechanical properties while replacing RMA in concrete mixture. This decline for full replacement rates was more than 50%, which applied in this case too. The decline of dynamic modulus of elasticity shows the highly negative influence of replacing NA in concrete mixture, with a maximal decline more than 50%. For this property, a negative impact of crystalline admixture was also shown, which decreased the dynamic modulus between 5 and 10%.

Table 4. Average values of results of mechanical properties of concrete, including standard deviation.

Recycled Concrete Mixture	Compressive Strength				Flexural Strength		Static Modulus of Elasticity		Dynamic Modulus of Elasticity	
	28 Days		60 Days							
Designation	(MPa)	σ	(MPa)	σ	(MPa)	σ	(GPa)	σ	(GPa)	σ
NAC 1 C0	34.0	0.7	38.2	0.4	6.6	0.4	38.2	4.8	34.8	4.1
RMAC 1 C0	20.9	0.5	23.1	1.5	4.2	0.2	15.9	1.7	21.4	1.1
RMAC 1 C1	24.7	0.4	25.3	0.9	4.7	0.3	15.9	0.2	21.2	1.0
RMAC 1 C3	24.2	0.3	26.5	0.3	4.7	0.1	15.2	0.5	19.9	2.4
NAC 2 C0	-	-	35.3	1.4	6.7 (1)	0.7	33.8 (1)	0.4	39.1	5.1
RMAC 2 C0	-	-	18.2	0.4	4.1 (1)	0.2	16.9 (1)	0.7	22.0	1.4
RMAC 2 C1	-	-	18.2	0.7	3.4 (1)	0.0	15.4 (1)	0.1	21.9	1.5
RMAC 2 C3	-	-	16.2	0.6	3.4 (1)	0.1	15.9 (1)	0.4	18.9	1.0

(1) Examined in 60 days.

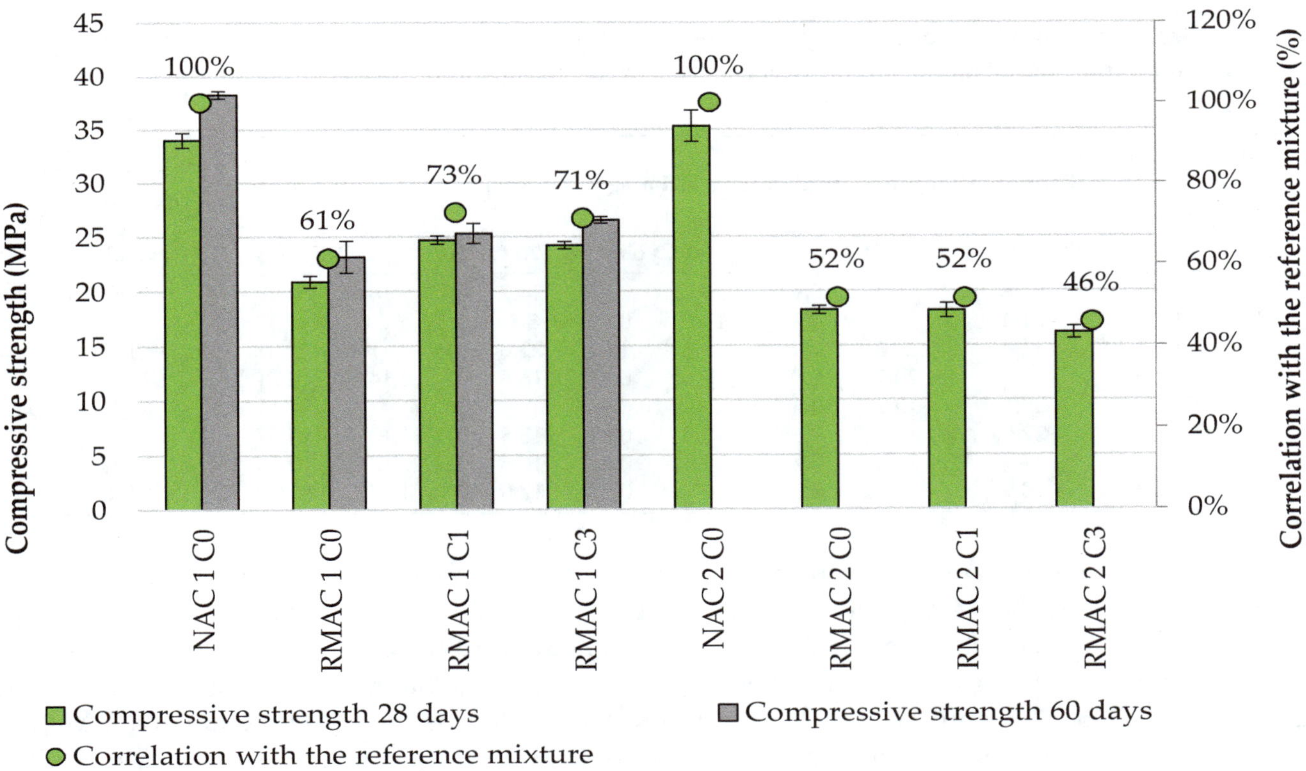

Figure 4. Comparison of compressive strength of NAC and RMAC at age 28 and 60 days.

3.3. Durability Properties

The durability of concrete is essential due to the desire to maintain performance of concrete throughout the service life of a structure. This study deals with two durability properties in which the positive influence of the addition of crystalline admixture was predicted. The evaluated properties were freeze-thaw resistance, which is essential for future utilization in the outdoor environment, and carbonation depth, which it is necessary to know due to the reinforcement design.

3.3.1. Freeze-Thaw Resistance

Freezing and thawing are often used as one possible way to evaluate concrete durability. The freeze-thaw resistance of concrete is mostly evaluated by measuring the dynamic elastic modulus, weight loss rate, and flexural strength loss rate after exposure to freeze-thaw cycles. Generally, this property is mainly affected by porosity, water content, environmental conditions, and aggregate types [13].

Two properties were evaluated for samples exposed to freezing and thawing cycles: dynamic modulus of elasticity and flexural strength. The dynamic modulus of elasticity was evaluated after 0, 25, 50, 75, and 100 cycles (Table 5) The resulting values were compared to the reference values, which were measured on the samples after 28 days, then they were placed in freezing-thawing chamber. The frost resistance coefficient was determined from these values as the quotient of two values measured before and after freezing and thawing. The frost resistance coefficient and its linear developing trend are shown in Figure 5. The evaluation of flexural strength after 0 and 100 cycles was also performed. The frost resistance coefficient was determined from the flexural strength before and after freezing and thawing cycles the same way as for the dynamic modulus of elasticity (see Table 6 and Figure 6). Concrete mixture is frost resistant when the frost resistance coefficient does not decrease under the value 0.75.

Table 5. Dynamic modulus of elasticity measured by ultrasonic method and frost resistance coefficient determined from the dynamic modulus of elasticity after freezing and thawing cycles.

Recycled Concrete Mixture	Dynamic Modulus of Elasticity (GPa) + Frost Resistance Coefficient (-)									Freeze-Thaw Resistance
Designation	0 cycles	25 Cycles		50 Cycles		75 Cycles		100 Cycles		Cycles
NAC 1 C0	34.8	32.8	0.94	26.7	0.77	24.7	0.71	16.5	0.48	50
RMAC 1 C0	21.4	14.9	0.70	12.8	0.59	8.2	0.38	6.8	0.32	0
RMAC 1 C1	21.2	22.3	1.05	22.1	1.04	22.0	1.04	21.4	1.01	100
RMAC 1 C3	19.9	20.1	1.01	17.4	0.87	15.6	0.78	14.2	0.71	75
NAC 2 C0	39.1	15.7	0.40	6.2	0.16	-	-	-	-	0
RMAC 2 C0	22.0	15.0	0.68	13.5	0.61	13.8	0.63	12.6	0.57	0
RMAC 2 C1	21.9	16.4	0.75	14.8	0.68	16.5	0.75	17.4	0.79	100
RMAC 2 C3	18.9	15.0	0.79	14.5	0.77	17.0	0.90	17.1	0.90	100

The results of the dynamic modulus of elasticity tested during the freezing and thawing cycles by the ultrasonic method show a significant influence of crystalline admixture. All four mixtures without crystalline admixture have the lowest freeze-thaw resistance according to the frost resistance coefficient determined from the dynamic modulus of elasticity, which means that the frost resistance coefficient was lower than 0.75. Three mixtures (NAC 1, 2 and RMAC 2 C0) was not frost resistant at all. Mixture RMAC 1 C0 was frost resistant until 50 cycles. On the contrary, the majority of mixtures with crystalline admixture were frost resistant for the whole test period, which was 100 cycles. Only the mixture RMAC 1 C3 was frost resistant until 75 cycles. Furthermore, the results of freeze-thaw resistance of concrete mixtures with crystalline admixture determined from the dynamic modulus of elasticity verify the positive impact of crystalline admixture in time due to the testing age of the mixtures, when the mixtures RMAC 2 were stored in the freezing-thawing chamber at age 60 days, unlike mixtures RMAC 1, which were stored in the freezing-thawing chamber at age 28 days.

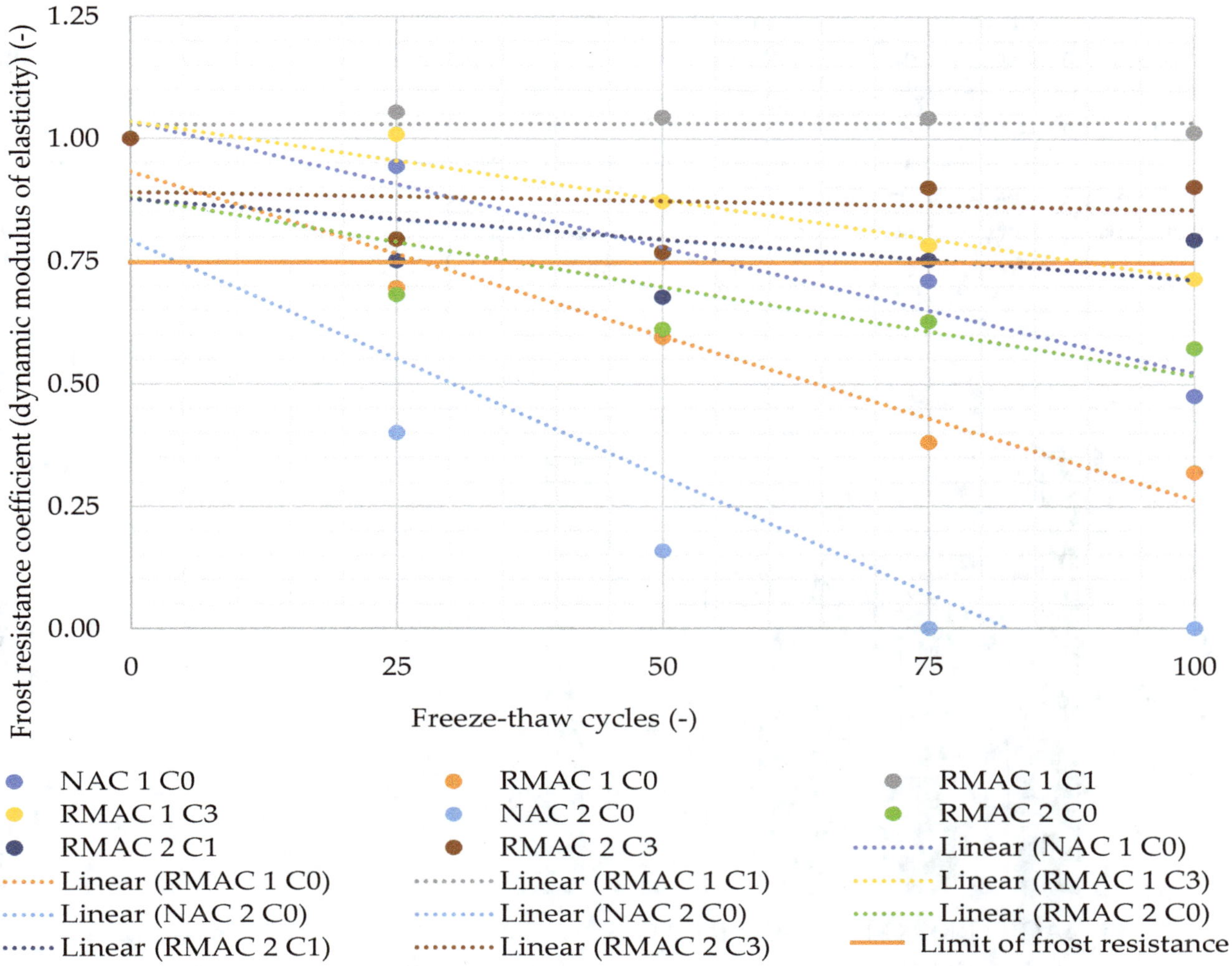

Figure 5. Frost resistance coefficient determined from the dynamic modulus of elasticity.

Table 6. Flexural strength and frost resistance coefficient determined from flexural strength after freezing and thawing cycles.

Recycled Concrete Mixture	Flexural Strength				Frost Resistance Coefficient
	(MPa)		σ		(-)
Designation	0	100	0	100	
NAC 1 C0	6.6	2.4	0.4	0.4	0.36
RMAC 1 C0	4.2	1.8	0.2	2.5	0.42
RMAC 1 C1	4.7	4.4	0.3	0.3	0.93
RMAC 1 C3	4.7	2.5	0.1	1.1	0.52
NAC 2 C0	6.7	1.5	0.7	0.3	0.23
RMAC 2 C0	4.1	2.6	0.2	0.4	0.63
RMAC 2 C1	3.4	2.8	0.0	0.4	0.81
RMAC 2 C3	3.4	3.0	0.1	0.3	0.89

The second possibility of evaluating the frost resistance according to the Czech standard is the frost resistance coefficient determined from flexural strength. This test could be measured only on samples which were not damaged during freeze-thaw cycles. In the case of this study, the flexural strength could be tested after 100 freeze-thaw cycles for the majority of concrete mixtures, however, the mixture NAC 2 C0 was tested after only 50 cycles and for mixture RMAC 1 C0 stayed testable in only one sample after 100 cycles. In general, the results of the flexural strength correspond with the results of dynamic modulus of elasticity and show a significant influence of crystalline admixture (see Figure 7). All four mixtures without crystalline admixture have the lowest freeze-thaw resistance according to the frost resistance coefficient determined from the flexural strength, which means that the frost resistance coefficient was lower than 0.75. On the contrary, the majority of mixtures with crystalline admixture were frost resistant for 100 cycles. Only the mixture RMAC 1 C3 had a lower frost resistance coefficient than 0.75, which was 0.52. Furthermore, the results of freeze-thaw resistance of concrete mixtures with crystalline admixture determined from the flexural strength show the same positive impact of crystalline admixture in time due to the testing age of the mixtures as was described for results of frost resistance determined from dynamic modulus of elasticity.

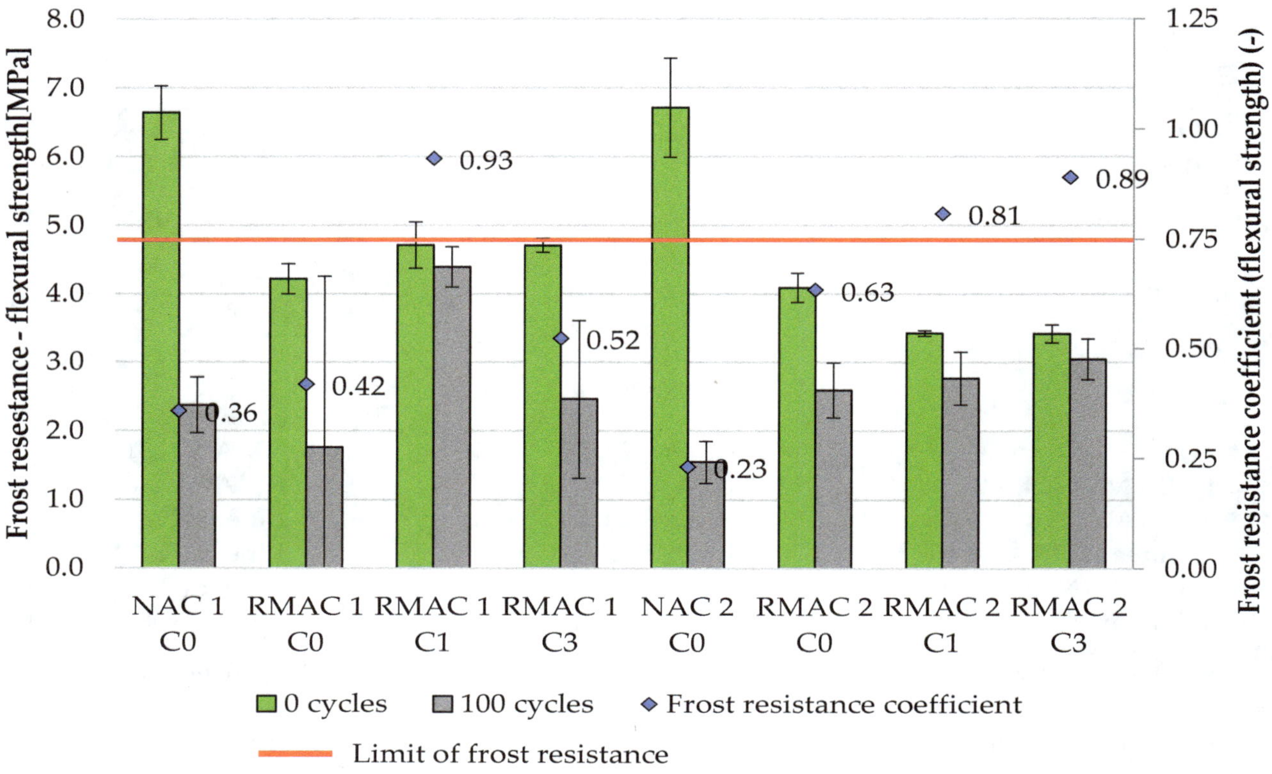

Figure 6. Flexural strength for 28 days (before freezing), after 100 freeze-thaw cycles and frost resistance coefficient determined from flexural strength.

3.3.2. Carbonation Resistance

Carbonation of concrete can be described as a physicochemical process in which a number of chemical reactions take place in the presence of carbon dioxide (CO_2), which promotes the reduction of pH in concrete. There are many studies about the positive or negative influence of CO_2 to the microstructure and properties of concrete. On one hand, carbonation can cause important changes in porosity, pore size distribution, connectivity, specific surface and also significant changes in transport properties, including water permeability, capillary retention, ion and gas diffusion. Furthermore, the decalcification of the C-S-H phase was found [85]. On the other hand, carbon dioxide, which penetrates the concrete mainly through a diffusion mechanism, slowly progresses from the surface of the concrete. In the presence of moisture, CO_2 forms carbonic acid, which reacts with calcium hydroxide (Ca $(OH)_2$) to form calcium carbonate ($CaCO_3$) and can slightly increase strength and reduce permeability due to the deposition of $CaCO_3$ in the cavities of the cement matrix [9,86].

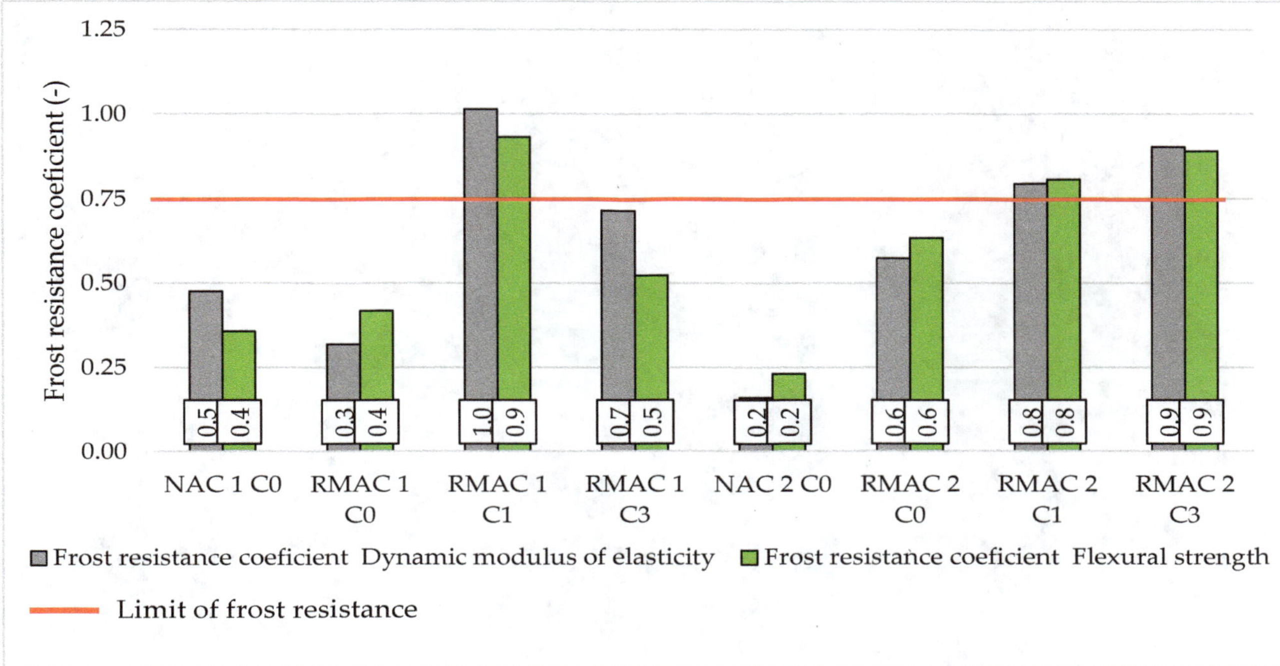

Figure 7. The comparison of the frost resistance coefficients evaluated from dynamic modulus of elasticity and flexural strength.

Three properties were evaluated for samples exposed to the high concentration of CO_2. The depth of decrease of pH of concrete was measured by the phenolphthalein method (see Figure 8). Furthermore, the dynamic modulus of elasticity and flexural strength were evaluated. These properties were tested on the prismatic specimens which were placed in the laboratory incubator with air circulation with a CO_2 atmosphere. The dynamic modulus of elasticity and flexural strength of concrete mixtures was also evaluated before placing the samples to the laboratory incubator with air circulation with CO_2 atmosphere and after 28 days effecting 3.0 ± 0.2% of CO_2 (see Table 7). The resulting values were compared to the reference values, which were measured on the samples after 28 days before then they were placed in an environment with a higher value of CO_2.

In general, the results of the carbonation show the deep effect of CO_2 for all RMAC mixtures, which shows no significant improvement in the carbonation resistance of RMAC with the addition of crystalline admixture. The results are evaluated by the indicator of increase the depth of carbonation compared to the NAC. This indicator of RMAC mixtures was more than twice as high as in NAC mixtures. This result slightly corresponds with the results presented in the previous study [9], where carbonation depth of RAC with 100% replacement rate of coarse aggregate was 2.5 times higher in comparison with NAC; however, the carbonation depth of RAC with the same replacement rate of fine aggregate was 8.7 times higher. In this study, both fractions of RMA were used for concrete. In addition, the indicator of RMAC mixtures without crystalline admixture is slightly higher in comparison with RMAC with crystalline admixture, which is more visible for mixture RMAC 1, in which the indicator decreases from 2.5 to approximately 1.7. On the contrary, the clearly positive effect was not shown for mixture RMAC 2. Furthermore, the influence of water curing and longer curing of concrete mixtures with crystalline admixture, where RMAC 2 mixtures were 60 days treated in water, was not shown to have a significant effect on the carbonation depth (see Figure 9).

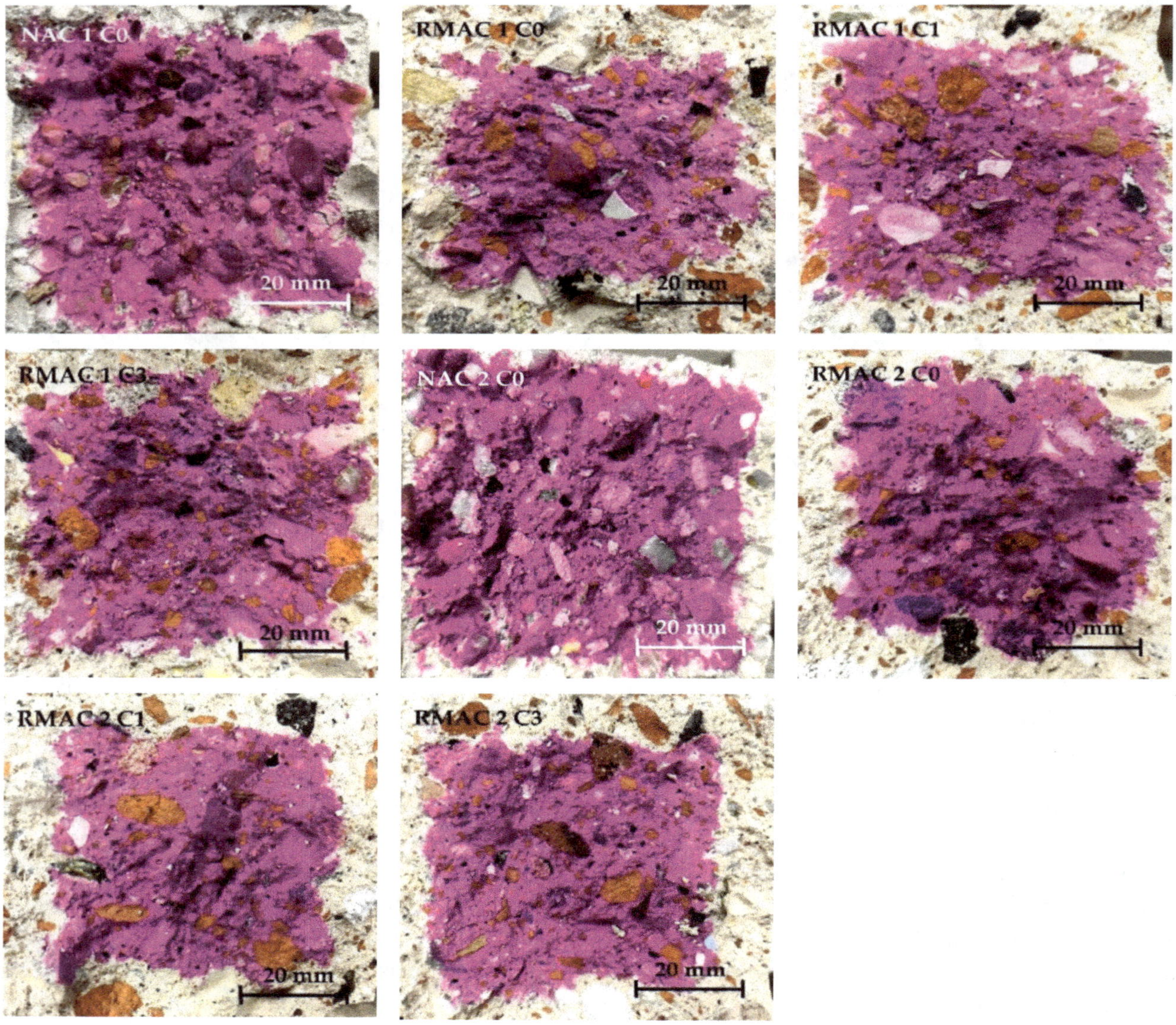

Figure 8. Carbonation depth of NAC and RMAC.

Table 7. Flexural strength, dynamic modulus of elasticity and carbonation depth after 28 days in laboratory incubator with air circulation with CO_2 atmosphere.

Recycled Concrete Mixture	Flexural Strength				Dynamic Modulus of Elasticity				Indicator of Increase of Carbonation Depth Compared to NAC
	No Exposition to CO_2		Exposition to CO_2		No Exposition to CO_2		Exposition to CO_2		
Designation	(MPa)	σ	(MPa)	σ	(GPa)	σ	(GPa)	σ	(mm)
NAC 1 C0	6.6	0.4	6.4	0.6	34.8	4.1	30.5	0.7	1.00
RMAC 1 C0	4.2	0.2	3.9	0.1	21.4	1.1	15.7	0.8	2.54
RMAC 1 C1	4.7	0.3	4.0	0.2	21.2	1.0	15.8	0.9	1.70
RMAC 1 C3	4.7	0.1	4.1	0.2	19.9	2.4	17.5	0.9	1.74
NAC 2 C0	6.7	0.7	7.4	0.4	39.1	5.1	29.4	4.2	1.00
RMAC 2 C0	4.1	0.2	3.6	0.3	22.0	1.4	17.2	1.0	2.28
RMAC 2 C1	3.4	0.0	3.4	0.2	21.9	1.5	15.9	0.9	2.23
RMAC 2 C3	3.4	0.1	3.4	0.1	18.9	1.0	15.5	1.4	2.09

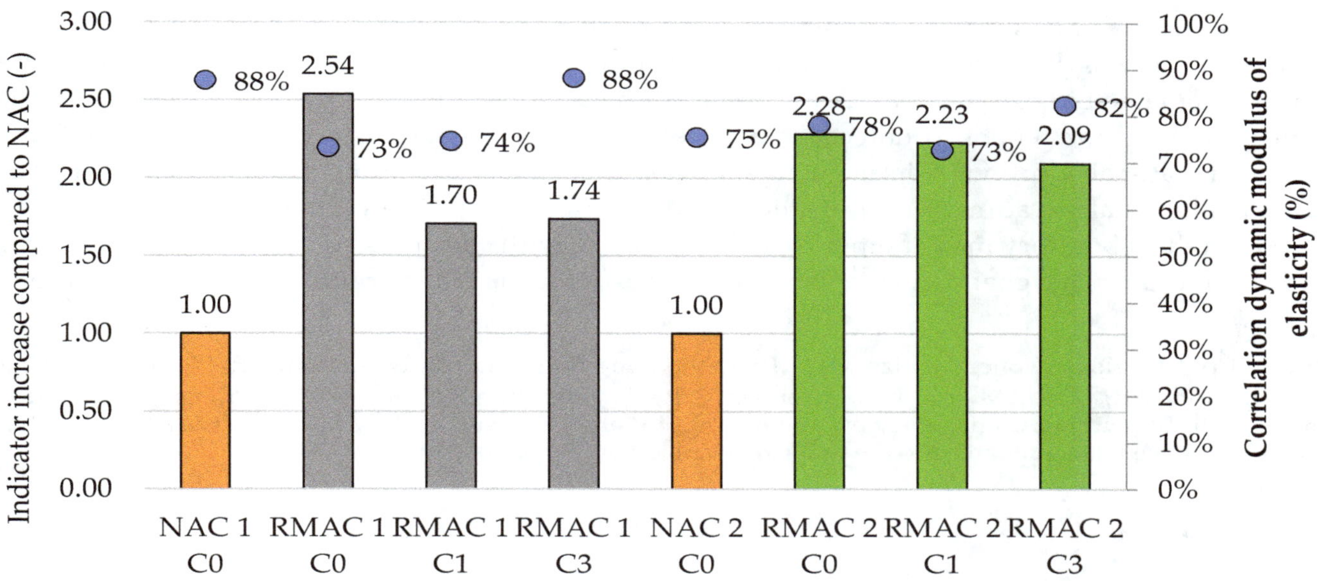

Figure 9. Carbonation resistance—the indicator increases of measured carbonation depth compared to NAC and correlation of dynamic modulus of elasticity of NAC and RMAC samples no exposed to CO_2 and exposed to CO_2.

The flexural strength of all RMAC mixtures after exposure to CO_2 is slightly lower in comparison with the sample which was not exposed to CO_2. However, there was no significant decline of flexural strength of RMAC 2 mixtures with the crystalline admixture, which could show the positive impact of longer water curing of samples (see Table 7). In addition, the dynamic modulus of elasticity of RMAC mixtures with crystalline admixture had also shown better results and a lower decline of properties for samples with crystalline admixture. The correlations of the dynamic modulus of elasticity before and after the exposition of CO_2 are shown in Figure 8.

4. Conclusions

In this study, the experimental verification of the improvement of the physical, mechanical and durability properties of concrete containing recycled masonry aggregate by crystalline admixture was examined and discussed. It is generally known that the mechanical properties and durability decrease with the replacement of natural aggregate by recycled aggregate. The durability of recycled masonry aggregate concrete is negatively influenced by higher porosity and water absorption of concrete, which could be improved by adding suitable admixture. In the previous studies, a few possibilities of improvement of properties using admixtures have been tested. In this case, the improvement of the durability of recycled masonry aggregate concrete by the crystalline admixture was verified, which has been carried out for the first time in this study. The final conclusions that have been reached can be summarized in the following points:

1. The improvement of mineral admixture utilization was verified. However, it was found it that has no significant positive impact on the evaluated mechanical properties.
2. The water absorption by immersion was approximately three times higher without the positive effect of crystalline admixture. On the contrary, capillary water absorption verified the positive impact of crystalline admixture, however, it was still more than two times higher.
3. The utilization of crystalline admixture leads to better freeze-thaw resistance of recycled masonry aggregate concrete, which meets the requirement of frost resistance according to the Czech standard.
4. The carbonation depth of RMAC was more than two times higher in comparison with conventional concrete. For one set of samples (mixtures RMAC 1), the carbonation depth was clearly improved by the crystalline admixture.

The novelty of this study was the utilization of crystalline admixture to improve the durability of concrete. It was predicted that this admixture could fill the pores due to its reaction with water, which is contained in recycled masonry aggregate. The positive impact of crystalline admixture is not clearly shown for water absorption capacity of recycled masonry aggregate concrete due to different results of water absorption by immersion and capillary water absorption. However, there are clear benefits of crystalline admixture utilization on the durability, especially on freeze-thaw resistance. Although the laboratory measurements have been done with emphasis on different types of curing and concrete ages, the goal of further laboratory measurements will be confirmation of these results.

Author Contributions: Conceptualization, T.P.; methodology, T.P. and J.P.; validation, T.P., J.Ř., D.M. and K.F.; investigation, T.P., J.Ř., D.M. and K.F.; resources, T.P.; data curation K.F. and D.M.; writing—original draft preparation, T.P. and J.P.; writing—review and editing, T.P. and J.P.; visualization, T.P.; project administration, J.P. All authors have read and agreed to the published version of the manuscript.

References

1. Gomes, M.; De Brito, J. Structural concrete with incorporation of coarse recycled concrete and ceramic aggregates: Durability performance. *Mater. Struct.* **2008**, *42*, 663–675. [CrossRef]
2. Pavlu, T. The Utilization of Recycled Materials for Concrete and Cement Production—A Review. *IOP Conf. Ser. Mater. Sci. Eng.* **2018**, *442*, 012014. [CrossRef]
3. Otsuki, N.; Miyazato, S.-I.; Yodsudjai, W. Influence of Recycled Aggregate on Interfacial Transition Zone, Strength, Chloride Penetration and Carbonation of Concrete. *J. Mater. Civ. Eng.* **2003**, *15*, 443–451. [CrossRef]
4. Kong, D.; Lei, T.; Zheng, J.; Ma, C.; Jiang, J.; Jiang, J. Effect and mechanism of surface-coating pozzalanics materials around aggregate on properties and ITZ microstructure of recycled aggregate concrete. *Constr. Build. Mater.* **2010**, *24*, 701–708. [CrossRef]
5. Arredondo-Rea, S.P.; Corral-Higuera, R.; Soberón, J.M.V.G.; Gámez-García, D.C.; Bernal-Camacho, J.M.; Rosas, C.A.; Ungsson-Nieblas, M.J. Durability Parameters of Reinforced Recycled Aggregate Concrete: Case Study. *Appl. Sci.* **2019**, *9*, 617. [CrossRef]
6. Chinchillas-Chinchillas, M.J.; Rosas, C.A.; Arredondo-Rea, S.P.; Soberón, J.M.V.G.; Corral-Higuera, R. SEM Image Analysis in Permeable Recycled Concretes with Silica Fume. A Quantitative Comparison of Porosity and the ITZ. *Materials* **2019**, *12*, 2201. [CrossRef]
7. Kim, N.; Kim, J.; Yang, S. Mechanical Strength Properties of RCA Concrete Made by a Modified EMV Method. *Sustainability* **2016**, *8*, 924. [CrossRef]
8. Montero, J.; Laserna, S. Influence of effective mixing water in recycled concrete. *Constr. Build. Mater.* **2017**, *132*, 343–352. [CrossRef]
9. Silva, R.; Neves, R.; De Brito, J.; Dhir, R. Carbonation behaviour of recycled aggregate concrete. *Cem. Concr. Compos.* **2015**, *62*, 22–32. [CrossRef]
10. Tuyan, M.; Mardani-Aghabaglou, A.; Ramyar, K. Freeze–thaw resistance, mechanical and transport properties of self-consolidating concrete incorporating coarse recycled concrete aggregate. *Mater. Des.* **2014**, *53*, 983–991. [CrossRef]
11. Rangel, C.S.; Amario, M.; Pepe, M.; Martinelli, E.; Filho, R.T. Durability of Structural Recycled Aggregate Concrete Subjected to Freeze-Thaw Cycles. *Sustainability* **2020**, *12*, 6475. [CrossRef]
12. Cai, G.; Liu, L.; Yan, S.; Zhang, M.; Xia, J.; Xie, Y. Effect of freeze-thaw cycles on mechanical and porosity properties of recycled construction waste mixtures. *Constr. Build. Mater.* **2019**, *210*, 347–363. [CrossRef]
13. Guo, H.; Shi, C.; Guan, X.; Zhu, J.; Ding, Y.; Ling, T.-C.; Zhang, H.; Wang, Y. Durability of recycled aggregate concrete—A review. *Cem. Concr. Compos.* **2018**, *89*, 251–259. [CrossRef]
14. Del Bosque, I.S.; Heede, P.V.D.; De Belie, N.; De Rojas, M.S.; Medina, C. Carbonation of concrete with construction and demolition waste based recycled aggregates and cement with recycled content. *Constr. Build. Mater.* **2020**, *234*, 117336. [CrossRef]

15. Amorim, P.; De Brito, J.; Evangelista, L. Concrete Made with Coarse Concrete Aggregate: Influence of Curing on Durability. *ACI Mater. J.* **2012**, *109*, 195–204. [CrossRef]
16. Buyle-Bodin, F.; Hadjieva-Zaharieva, R. Influence of industrially produced recycled aggregates on flow properties of concrete. *Mater. Struct.* **2002**, *35*, 504–509. [CrossRef]
17. Evangelista, L.; De Brito, J. Durability performance of concrete made with fine recycled concrete aggregates. *Cem. Concr. Compos.* **2010**, *32*, 9–14. [CrossRef]
18. Sagoe-Crentsil, K.K.; Brown, T.; Taylor, A.H. Performance of concrete made with commercially produced coarse recycled concrete aggregate. *Cem. Concr. Res.* **2001**, *31*, 707–712. [CrossRef]
19. Leelawat, T.; Dhir, R.K.; Limbachiya, M.C. Suitability of recycled concrete aggregate for Use in BS 5328 Designated mixes. *Proc. Inst. Civ. Eng. Struct. Build.* **1999**, *134*, 257–274. [CrossRef]
20. Rao, A.; Jha, K.N.; Misra, S. Use of aggregates from recycled construction and demolition waste in concrete. *Resour. Conserv. Recycl.* **2007**, *50*, 71–81. [CrossRef]
21. Xiao, J.; Lei, B.; Zhang, C. On carbonation behavior of recycled aggregate concrete. *Sci. China Ser. E Technol. Sci.* **2012**, *55*, 2609–2616. [CrossRef]
22. Leemann, A.; Loser, R. Carbonation resistance of recycled aggregate concrete. *Constr. Build. Mater.* **2019**, *204*, 335–341. [CrossRef]
23. Zega, C.J.; Di Maio, Á.A. Use of recycled fine aggregate in concretes with durable requirements. *Waste Manag.* **2011**, *31*, 2336–2340. [CrossRef] [PubMed]
24. Cartuxo, F.; Jorge, D.B.; Luís, E.; Jiménez, J.R.; Ledesma, E.F. Increased Durability of Concrete Made with Fine Recycled Concrete Aggregates Using Superplasticizers. *Materials* **2016**, *9*, 98. [CrossRef]
25. Torgal, F.M.A.S.P.; Jalali, S. Compressive strength and durability properties of ceramic wastes based concrete. *Mater. Struct.* **2010**, *44*, 155–167. [CrossRef]
26. Zhu, Y.-G.; Kou, S.-C.; Poon, C.-S.; Dai, J.-G.; Li, Q.-Y. Influence of silane-based water repellent on the durability properties of recycled aggregate concrete. *Cem. Concr. Compos.* **2013**, *35*, 32–38. [CrossRef]
27. Lei, B.; Xiao, J.-Z. Research on carbonation resistance of recycled aggregate concrete. *Jianzhu Cailiao Xuebao/J. Build. Mater.* **2008**, *11*, 605–611.
28. Kjellsen, K.O.; Guimaraes, M.; Nilsson, Å. *The CO_2 Balance of Concrete in a Life Cycle Perspective*; Danish Technological Institute: Taastrup, Denmark, 2005.
29. Kikuchi, T.; Kuroda, Y. Carbon Dioxide Uptake in Demolished and Crushed Concrete. *J. Adv. Concr. Technol.* **2011**, *9*, 115–124. [CrossRef]
30. *CSN EN 206+A1 Concrete: Specification, Performance, Production and Conformity*; CEN: Prague, Czech Republic, 2018. (In Czech)
31. Senthamarai, R.; Manoharan, P.D. Concrete with ceramic waste aggregate. *Cem. Concr. Compos.* **2005**, *27*, 910–913. [CrossRef]
32. Yang, J.; Du, Q.; Bao, Y. Concrete with recycled concrete aggregate and crushed clay bricks. *Constr. Build. Mater.* **2011**, *25*, 1935–1945. [CrossRef]
33. Crushed Bricks as Coarse Aggregate for Concrete. *ACI Mater. J.* **1999**, *96*, 478–484. [CrossRef]
34. Medina, C.; De Rojas, M.I.S.; Thomas, C.; Polanco, J.A.; Frias, M.F. Durability of recycled concrete made with recycled ceramic sanitary ware aggregate. Inter-indicator relationships. *Constr. Build. Mater.* **2016**, *105*, 480–486. [CrossRef]
35. Vieira, T.; Alves, A.; Jorge, D.B.; Correia, J.; Silva, R. Durability-related performance of concrete containing fine recycled aggregates from crushed bricks and sanitary ware. *Mater. Des.* **2016**, *90*, 767–776. [CrossRef]
36. Correia, J.R.; De Brito, J.; Pereira, A.S. Effects on concrete durability of using recycled ceramic aggregates. *Mater. Struct.* **2006**, *39*, 169–177. [CrossRef]
37. Rashid, K.; Razzaq, A.; Ahmad, M.; Rashid, T.; Tariq, S. Experimental and analytical selection of sustainable recycled concrete with ceramic waste aggregate. *Constr. Build. Mater.* **2017**, *154*, 829–840. [CrossRef]
38. González, J.S.; Gayarre, F.L.; Pérez, C.L.-C.; Ros, P.S.; López, M.A.S. Influence of recycled brick aggregates on properties of structural concrete for manufacturing precast prestressed beams. *Constr. Build. Mater.* **2017**, *149*, 507–514. [CrossRef]
39. Nepomuceno, M.C.; Isidoro, R.A.; Catarino, J.P. Mechanical performance evaluation of concrete made with recycled ceramic coarse aggregates from industrial brick waste. *Constr. Build. Mater.* **2018**, *165*, 284–294. [CrossRef]

40. Cachim, P.B. Mechanical properties of brick aggregate concrete. *Constr. Build. Mater.* **2009**, *23*, 1292–1297. [CrossRef]
41. Anderson, D.J.; Smith, S.T.; Au, F.T. Mechanical properties of concrete utilising waste ceramic as coarse aggregate. *Constr. Build. Mater.* **2016**, *117*, 20–28. [CrossRef]
42. Zheng, C.; Lou, C.; Du, G.; Li, X.; Liu, Z.; Li, L. Mechanical properties of recycled concrete with demolished waste concrete aggregate and clay brick aggregate. *Results Phys.* **2018**, *9*, 1317–1322. [CrossRef]
43. Alves, A.; Vieira, T.; Jorge, D.B.; Correia, J. Mechanical properties of structural concrete with fine recycled ceramic aggregates. *Constr. Build. Mater.* **2014**, *64*, 103–113. [CrossRef]
44. Nematzadeh, M.; Dashti, J.; Ganjavi, B. Optimizing compressive behavior of concrete containing fine recycled refractory brick aggregate together with calcium aluminate cement and polyvinyl alcohol fibers exposed to acidic environment. *Constr. Build. Mater.* **2018**, *164*, 837–849. [CrossRef]
45. Corominas, A.G.; Etxeberria, M. Properties of high performance concrete made with recycled fine ceramic and coarse mixed aggregates. *Constr. Build. Mater.* **2014**, *68*, 618–626. [CrossRef]
46. Khalaf, F.M.; Devenny, A.S. Properties of New and Recycled Clay Brick Aggregates for Use in Concrete. *J. Mater. Civ. Eng.* **2005**, *17*, 456–464. [CrossRef]
47. Khalaf, F.M.; Devenny, A.S. Recycling of Demolished Masonry Rubble as Coarse Aggregate in Concrete: Review. *J. Mater. Civ. Eng.* **2004**, *16*, 331–340. [CrossRef]
48. Pacheco-Torgal, F.; Tam, V.; Labrincha, J.; Ding, Y.; de Brito, J. *Handbook of Recycled Concrete and Demolition Waste*; Woodhead Publishing Limited: Sawston, UK, 2013; ISBN 978-0-85709-690-6.
49. Debieb, F.; Kenai, S. The use of coarse and fine crushed bricks as aggregate in concrete. *Constr. Build. Mater.* **2008**, *22*, 886–893. [CrossRef]
50. Devenny, A.; Khalaf, F.M. The use of crushed brick as coarse aggregate in concrete. *MI* **1999**, *12*, 81–84.
51. Verian, K.P.; Ashraf, W.; Cao, Y. Properties of recycled concrete aggregate and their influence in new concrete production. *Resour. Conserv. Recycl.* **2018**, *133*, 30–49. [CrossRef]
52. Uddin, M.T.; Mahmood, A.H.; Kamal, R.I.; Yashin, S.; Zihan, Z.U.A. Effects of maximum size of brick aggregate on properties of concrete. *Constr. Build. Mater.* **2017**, *134*, 713–726. [CrossRef]
53. Chen, H.-J.; Yen, T.; Chen, K.-H. Use of building rubbles as recycled aggregates. *Cem. Concr. Res.* **2003**, *33*, 125–132. [CrossRef]
54. Nili, M.; Sasanipour, H.; Aslani, F. The Effect of Fine and Coarse Recycled Aggregates on Fresh and Mechanical Properties of Self-Compacting Concrete. *Mater.* **2019**, *12*, 1120. [CrossRef] [PubMed]
55. Desmyter, J.; Van Dessel, J.; Blockmans, S. The use of recycled concrete and masonry aggregates in concrete: Improving the quality and purity of the aggregates. In *Exploiting Wastes in Concrete*; Thomas Telford Publishing: London, UK, 1999; pp. 139–149. ISBN 978-0-7277-4724-2.
56. Cavalline, T.L.; Weggel, D.C. Recycled brick masonry aggregate concrete. *Struct. Surv.* **2013**, *31*, 160–180. [CrossRef]
57. Khatib, J.M. Properties of concrete incorporating fine recycled aggregate. *Cem. Concr. Res.* **2005**, *35*, 763–769. [CrossRef]
58. Tam, V.W.; Gao, X.; Tam, C. Microstructural analysis of recycled aggregate concrete produced from two-stage mixing approach. *Cem. Concr. Res.* **2005**, *35*, 1195–1203. [CrossRef]
59. Grabiec, A.M.; Zawal, D.; Rasaq, W.A. The Effect of Curing Conditions on Selected Properties of Recycled Aggregate Concrete. *Appl. Sci.* **2020**, *10*, 4441. [CrossRef]
60. Bogas, J.A.; De Brito, J.; Ramos, D. Freeze–thaw resistance of concrete produced with fine recycled concrete aggregates. *J. Clean. Prod.* **2016**, *115*, 294–306. [CrossRef]
61. Yildirim, S.T.; Meyer, C.; Herfellner, S. Effects of internal curing on the strength, drying shrinkage and freeze–thaw resistance of concrete containing recycled concrete aggregates. *Constr. Build. Mater.* **2015**, *91*, 288–296. [CrossRef]
62. Bendimerad, A.Z.; Roziere, E.; Loukili, A. Plastic shrinkage and cracking risk of recycled aggregates concrete. *Constr. Build. Mater.* **2016**, *121*, 733–745. [CrossRef]
63. Seara-Paz, S.; González-Fonteboa, B.; Martínez-Abella, F.; González-Taboada, I. Time-dependent behaviour of structural concrete made with recycled coarse aggregates. Creep and shrinkage. *Constr. Build. Mater.* **2016**, *122*, 95–109. [CrossRef]
64. Role of Chemical and Mineral Admixtures on the Physical Properties and Frost-Resistance of Recycled Aggre-gate Concrete. *ACI Mater. J.* **1998**, *95*, 558–563. [CrossRef]

65. Liu, Q.; Cen, G.; Cai, L.; Wu, H. Frost-resistant performance and mechanism of recycled concrete for airport pavement. *Huazhong Keji Daxue Xuebao (Ziran Kexue Ban)/J. Huazhong Univ. Sci. Technol. (Nat. Sci. Ed.)* **2011**, *39*, 128–132.
66. Sun, J.-Y.; Geng, J. Effect of particle size and content of recycled fine aggregate on frost resistance of concrete. *Jianzhu Cailiao Xuebao/J. Build. Mater.* **2012**, *15*, 382–385. [CrossRef]
67. Matias, D.; De Brito, J.; Rosa, A.D.; Pedro, D. Mechanical properties of concrete produced with recycled coarse aggregates—Influence of the use of superplasticizers. *Constr. Build. Mater.* **2013**, *44*, 101–109. [CrossRef]
68. Corinaldesi, V.; Donnini, J.; Giosuã, C.; Mobili, A.; Tittarelli, F. Durability Assessment of Recycled Aggregate HVFA Concrete. *Appl. Sci.* **2020**, *10*, 6454. [CrossRef]
69. Rahhal, V.; Bonavetti, V.; Delgado, A.; Pedrajas, C.; Talero, R. Scheme of the Portland cement hydration with crystalline mineral admixtures and other aspects. *Silic. Ind.* **2009**, *74*, 347–352.
70. Escoffres, P.; Desmettre, C.; Charron, J.-P. Effect of a crystalline admixture on the self-healing capability of high-performance fiber reinforced concretes in service conditions. *Constr. Build. Mater.* **2018**, *173*, 763–774. [CrossRef]
71. Azarsa, P.; Gupta, R.; Biparva, A. Crystalline Waterproofing Admixtures Effects on Self-healing and Permeability of Concrete. In Proceedings of the 1st International Conference on New Horizons in Green Civil Engineering (NHICE-01), Victoria, BC, Canada, 25–27 April 2018.
72. Bohus, S.; Drochytka, R. Cement Based Material with Crystal-Growth Ability under Long Term Aggressive Medium Impact. Available online: https://www.scientific.net/AMM.166-169.1773 (accessed on 1 November 2020).
73. Jianxin, W.; Guangda, L.; Liping, L. Influence of Shenkebao permeable crystalline admixture on concrete resistance to chemical corrosion and freeze-thaw cycles. *Jiangsu Build. Mater.* **2010**, *2*, 18–21.
74. Pazderka, J.; Hájková, E. Crystalline admixtures and their effect on selected properties of concrete. *Acta Polytech.* **2016**, *56*, 306–311. [CrossRef]
75. Reiterman, P.; Pazderka, J. Crystalline Coating and Its Influence on the Water Transport in Concrete. Available online: https://www.hindawi.com/journals/ace/2016/2513514/ (accessed on 30 October 2020).
76. Weng, T.-L.; Cheng, A. Influence of curing environment on concrete with crystalline admixture. *Monatshefte für Chemie—Chem. Mon.* **2013**, *145*, 195–200. [CrossRef]
77. *CSN EN 12620+A1 Aggregate for Concrete*; CEN: Prague, Czech Republic, 2008. (In Czech)
78. De Brito, J.; Saikia, N. *Recycled Aggregate in Concrete*; Green Energy and Technology; Springer: London, UK, 2013; ISBN 978-1-4471-4539-4.
79. Dhir, R.K.; de Brito, J.; Silva, R.V.; Lye, C.Q. 10—Recycled Aggregate Concrete: Durability Properties. In *Sustainable Construction Materials*; Dhir, R.K., de Brito, J., Silva, R.V., Lye, C.Q., Eds.; Woodhead Publishing Series in Civil and Structural Engineering; Woodhead Publishing: Duxford, UK, 2019; pp. 365–418. ISBN 978-0-08-100985-7.
80. Pavlů, T.; Šefflová, M. Study of the Freeze-Thaw resistance of the fine-aggregate concrete. In Proceedings of the EAN 2016—54th International Conference on Experimental Stress Analysis, Srni, Czech Republic, 30 May–2 June 2016.
81. Bravo, M.; De Brito, J.; Pontes, J.; Evangelista, L. Durability performance of concrete with recycled aggregates from construction and demolition waste plants. *Constr. Build. Mater.* **2015**, *77*, 357–369. [CrossRef]
82. Pavlu, T.; Šefflová, M. The Static and the Dynamic Modulus of Elasticity of Recycled Aggregate Concrete. *Adv. Mater. Res.* **2014**, *1054*, 221–226. [CrossRef]
83. Silva, R.V.; De Brito, J.; Dhir, R.K. Establishing a relationship between modulus of elasticity and compressive strength of recycled aggregate concrete. *J. Clean. Prod.* **2016**, *112*, 2171–2186. [CrossRef]
84. Dhir, R.K.; de Brito, J.; Silva, R.V.; Lye, C.Q. 9—Deformation of Concrete Containing Recycled Concrete Aggregate. In *Sustainable Construction Materials*; Dhir, R.K., de Brito, J., Silva, R.V., Lye, C.Q., Eds.; Woodhead Publishing Series in Civil and Structural Engineering; Woodhead Publishing: Duxford, UK, 2019; pp. 283–363. ISBN 978-0-08-100985-7.
85. Morandeau, A.; Thiery, M.; Dangla, P. Investigation of the carbonation mechanism of CH and C-S-H in terms of kinetics, microstructure changes and moisture properties. *Cem. Concr. Res.* **2014**, *56*, 153–170. [CrossRef]
86. Neville, A.M. *Properties of Concrete*, 5th ed.; Pearson: Harlow, UK, 2011; ISBN 978-0-273-75580-7.

2

Selected Aspects Regarding the Restoration/ Conservation of Traditional Wood and Masonry Building Materials: A Short Overview of the Last Decade Findings

Radu Claudiu Fierascu, Mihaela Doni and Irina Fierascu *

National Institute for Research & Development in Chemistry and Petrochemistry—ICECHIM Bucharest, Emerging Nanotechnologies Group, 202 Spl. Independentei, 060021 Bucharest, Romania; radu_claudiu_fierascu@yahoo.com (R.C.F.); mihaela.doni@icechim.ro (M.D.)

* Correspondence: dumitriu.irina@yahoo.com

Abstract: Vernacular buildings are usually constructed using materials at hand, including wood, natural stone and bricks (either clay or mud bricks). All those materials are exposed to a series of environmental factors, affecting their structure and integrity. The literature review was conducted using different databases (Scopus, Web of Science, ScienceDirect, SpringerLink) using as keywords the historical material, "heritage" and the terms regarding the desired effect, within the envisaged time period (2010–2019). The assessment of the results was performed by manual inspection (reading the entire article) and the selection of the works to be inserted in the current review was made by evaluating the contribution to the field. This review summarizes different aspects related to the restoration and conservation of wooden and masonry elements of traditional buildings, including materials used for biocidal interventions, protection against abiotic factors, cleaning and consolidation agents. Finally, a critical discussion regarding the current limitations and future perspectives concludes the review work, envisaging the role of researchers specialized in materials science in the context of cultural heritage conservation.

Keywords: vernacular constructions; wood; stone; restoration; conservation

1. Introduction

Traditional buildings represent an important heritage of each civilization. These constructions are specific to each nation and region, considering a multitude of factors, including climate and availability of materials [1], without any one of them being decisive. Traditional buildings were usually constructed using the materials at hand, including wood, natural stone and bricks (either clay or mud bricks) [2,3]. All those materials are exposed to a series of environmental factors, affecting their structure and integrity. A good example of different construction materials can be observed in the Romanian traditional buildings (Figure 1) which are usually consisting of a mixture of wood and masonry (either mud or clay bricks and stones). As the traditional buildings are considered to be a proof of continuity of a civilization [4], great attention must be given to the development of appropriate materials for their restoration and conservation.

Figure 1. Examples of Romanian traditional buildings incorporating different construction materials (stone, bricks and wood)—pictures belonging to the authors' private collection.

Cultural heritage of a nation is formed not only of well-known monuments. Traditional houses and their construction methods represents an important legacy, which is transmitted from one generation to another, which was adapted to the times and needs, but permanently reflecting the environmental, cultural, technological, economic and historical conditions of the local context.

The present review focuses on the recent developments regarding the materials used for the restoration and conservation of traditional buildings (masonry, stone and wood) and represents a critical discussion of aspects related to their utilization; very useful in the transdisciplinary studies related to further development of new materials or for the use of the current ones.

The literature review was conducted using different databases (Scopus, Web of Science, ScienceDirect, SpringerLink) using as keywords the historical material ("wood", "masonry", "brick", "stone" "building", "mortar"), "heritage" and the terms regarding the desired effect ("restoration", "conservation", "biocid*" or "consolidation") within the envisaged time period (2010–2019). The assessment of the results was performed by manual inspection (reading the entire article) and the selection of the works to be inserted in the current review was made by evaluating the contribution to the field (the entire process is described in Figure 2).

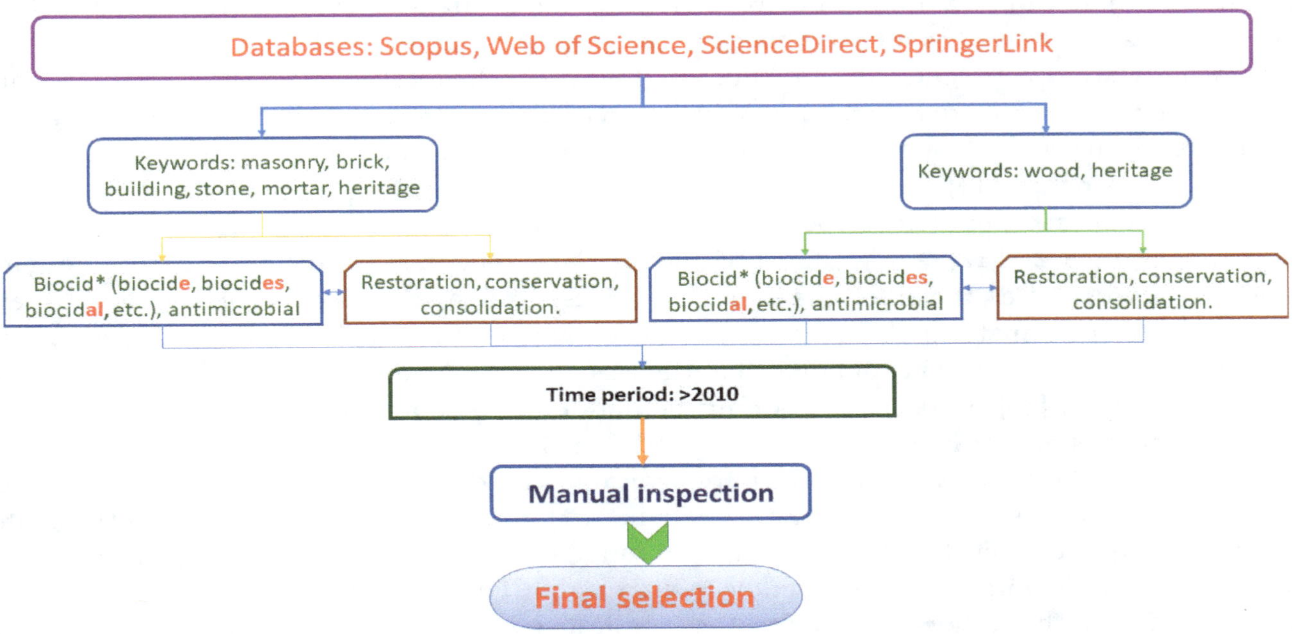

Figure 2. General procedure followed for the selection of articles to be included in the present review.

2. Materials Used for Restoration and Conservation of Wooden Elements

2.1. *Biocidal Materials Used for Wooden Elements*

Wood is considered to be one of the most common construction materials, with traditional buildings made of wood being encountered all over the world [5]. Due to its biodegradable nature, wood is exposed to a wide variety of biotic and abiotic agents, including fungi, insects, termites, moisture fluctuation or weathering [6]. Thus, the approach for restoration or conservation of the wood artifacts should comprise two aspects: the biocidal treatment and a chemical consolidation approach (Figure 3).

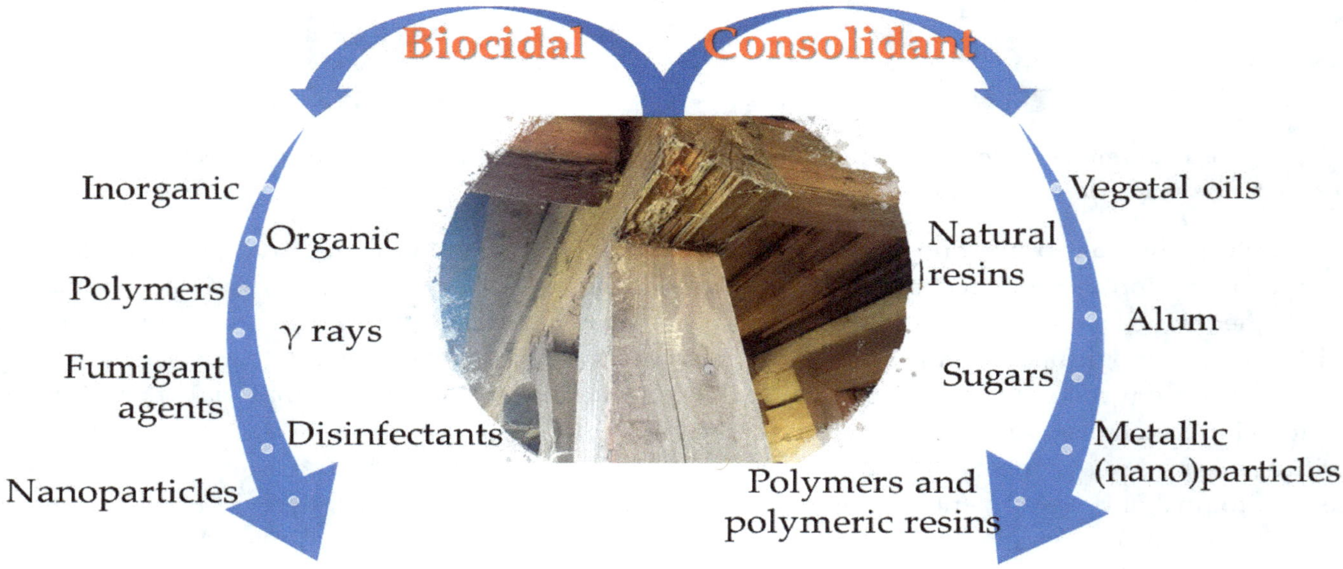

Figure 3. Aspects covered by the present review regarding the conservation and restoration of wooden building materials.

Another important biotic factor affecting all the objects belonging to the cultural heritage is represented by the human interventions. If we speak, either about interventions with negative effects (through the use of inappropriate materials or techniques), vandalism or about the lack of any intervention, these can be avoided only by increasing the awareness level, both at the level of the general population, as well as decision makers [7,8].

Wood itself can be categorized using different physical or chemical characteristics, one of the most important being the nature of the wood—softwood (gymnosperm trees—conifers, cycads, Ginkgo and gnetophytes) or hardwood (angiosperm trees) [9,10] and the conservation/restoration strategy should be adapted to its nature [11].

One of the most encountered problems in the preservation of wood artifacts is represented by its biodegradation. Unger [12] reviewed in 2012 the problems related to the historical use of inorganic (fluorides, fluorosilicates, metallic sulfates and chlorides, alkali arsenates or boron compounds) and organic biocides (chlorinated hydrocarbons, cyclodiene compounds, phenol derivatives, organometallic compounds or organophosphates), both on the artifacts themselves and to human health. The European standard EN 15003 [13] proposes the hot air methodology to eliminate fungal and insect attacks. However, the method cannot be always applied (considering the state of degradation and the characteristics of the materials). With the understanding of those negative effects, the search of new biocides represented an important research area in the last decades. In 2011, Clausi et al. [14] evaluated the antifungal effect (against white and brown rot fungi) of two widely used polymeric consolidants (Paraloid B72 and Regalrez 1126) applied on White poplar (*Populus alba*) and Norway spruce (*Picea abies* (L.) H. Karst). The authors observed that the application of individual consolidants led to different inhibition of the fungi (10% Paraloid being effective against white-rot fungus, while 5% Regalrez was effective against brown-rot fungus; the consolidant mixture inhibited both types of fungi). The results suggested that the combined application of the consolidants could slow the fungal growth in treated samples (behavior depended on wood species and treatment type); on the long-term, no treatment was proven to be effective in completely inhibiting fungal growth. Stejskal et al. [15] proposed the use of the fumigating agent hydrogen cyanide that proved to be effective against pinewood nematodes, Asian long-horned beetles and house longhorned beetles, important pests of the construction wood and historical buildings. The authors recorded a 100% mortality against the cerambycids (after 1-h exposure) and nematodes (after 18-h exposure), at a 20 g/m^3 concentration of fumigant agent. Although effective, the chemical compound was prohibited for use due to its high toxicity (together with another effective pesticide, methyl bromide) [16]. This led to the search of alternative chemical and non-chemical pesticidal agents.

Koziróg et al. [17] evaluated several active ingredients against a series of bacteria and molds isolated from the historical wooden surfaces at the former Auschwitz II-Birkenau concentration and extermination camp. The authors noticed that the most promising active ingredients were didecyldimethylammonium chloride and *N*-(3-aminopropyl)-*N*-dodecylpropane-1,3-diamine. As a next step of the study, the effect of commercially available biocides based on those active ingredients, applied by spraying or fogging, were evaluated in terms of bacterial and fungal inhibition, the best results being obtained for Boramon and Rocima 101 (12 months effectiveness, respectively 3 months).

Goffredo et al. [18] evaluated the potential of inorganic nanoparticles (consisting of 1% TiO_2 nanoparticles solutions, supplemented with silver or copper nanoparticles, at concentrations of 0.60% and 3.15%) as antifungal agents against *Aspergillus niger*, applied on pine (*Pinus sylvestris* L.—softwood) and beech (*Fagus sylvatica* L.—hardwood). The best results were obtained for the solutions containing 3.15% metallic nanoparticles, without any significant visual alteration of the visual aspect of wooden surfaces.

Recent advances in the conservation of wood artifacts were recently reviewed by Walsh-Korb and Averous [19]. Thus, the current review will only aim to briefly present the common conservation approaches (Table 1), and to complete the cited work with the most recent developments.

2.2. *Consolidants Used for Wooden Elements*

Historically speaking, one of the first used consolidants were the vegetal oils (such as linseed or Tung oil) and natural resins (such as colophony) [19]. Having advantages such as being easy to obtain and apply, the main disadvantages (such as the aesthetic changes induced by the colophony, extensive drying periods of the linseed oil leading to wood structure softening or inhomogeneous film structure obtained using Tung oil) led to the search for alternative treatments [20–22]. A similar approach, is represented by the use of sugars or sugar alcohols, which, through controlled drying, give rise to crystals consolidating the material. However, development of large crystals could damage the wood materials; in the same time, sugars can be a very good source of nutrients for the development of microbial growth [23,24]. One of the first inorganic treatments of the historical wooden elements is represented by the use of potassium aluminum sulfate dodecahydrate, developed in the middle of the XIXth century. However, its major drawbacks (among which the acid degradation of cellulose is one of the most important) led to abandoning the treatment, but not the general use of inorganic consolidants [25]. More recently, several types of inorganic materials (including nanoparticles) were applied for the consolidation of wood materials. As some such types of particles also possesses antimicrobial properties, their application can also cover the biocidal need in restoration procedures [19,26–28].

With the development of polymer science, several polymeric materials or polymeric resins were successfully tested for the conservation of wood materials. Among those, several types of commercially available materials are currently applied in our days [29–34]. Finally, in recent years, the use of biobased materials (such as keratin, cellulose or chitosan) was supported by several literature studies [35–37].

Besides the materials previously presented, Cataldi et al. [38] proposed the addition of microcrystalline cellulose (5%–20%) into a UV-light curable methacrylic-siloxane resin formulation, with potential application as historical wood coating. The addition led to the increase of the dynamic moduli, flexure stiffness, and hydrophobicity, accompanied by a decrease of the thermal expansion coefficient, which suggest that the photo-curable micro-composites could be used to recover the mechanical and physical properties of damaged wood, being a good alternative to the traditional resins currently used.

Another interesting approach was presented by Moise et al. [39]. Using polyester acrylate styrene free resin and gamma curing, the authors observed the increase of thermal, chemical and photochemical stability. In the same time, the gamma curing offers the advantage of disinfection of the material. However, great attention must be payed upon gamma treatment, as the ionizing radiation can damage the cellulose structure [40]. Additionally, considering the radioprotection issues, the method cannot be usually applied for in situ interventions [19].

In our opinion, the previously described materials present advantages such as slowing the fungal growth in treated samples, in situ application (for most of the proposed solutions), recovery of damaged properties of wooden materials, etc. The main disadvantage of the current approaches is represented by the lack of long-term protection (the need for repeating the treatment). Additionally, the future research should be focused on the development of multi-phase treatments, which could adapt to environmental changes in real-time, besides offering a protection for multiple degradation causes [19].

Table 1. Classic and modern conservation agents for wood artifacts (references presented in chronological order).

Procedure	Agent	Characteristics	Application Method	Ref.
Inhibition of fungal growth	Paraloid B72 and Regalrez 1126	Slowing of fungal growth in treated samples; short-term effect	Surface application	[14]
Consolidant	Vegetable oil (linseed, Tung), natural resins (colophony)	Water repellent, crack-filling, non-toxic, easy to obtain and apply	Surface application	[20–22]
Consolidant	Polymers and polymeric resins (melamine- or urea-formaldehyde, Paraloid B72, Regalrez 1162, Poly(ethylene glycol), Acrylic resins, Silanes, Epoxy resins)	Wide-spread treatments, easy to apply, good stability; last generation treatments (acrylic or epoxy resins) although have better properties, require vacuum application	Surface application, immersion	[29–34]
Consolidant	Sugars and sugar alcohols (sucrose, lactilol, trehalose)	Reversible, non-toxic, increase stability upon crystallization of sugars	Immersion of the wood artifacts	[23,24]
Consolidant	Inorganic particles (calcium hydroxide, magnesium hydroxide, titanium dioxide, alkaline carbonate)	Ability to neutralize acids within the wood (in some cases even continuous deacidification), reduce cellulose hydrolysis; some have biocidal action	Spraying, surface applications	[26–28]
Consolidant	Biobased solutions (keratin, cellulose, chitosan)	Natural resource, easy to apply, good compatibility; can undergo the same degradation issues as the wood	Immersion	[35–37]
Biocidal	didecyldimethylammonium chloride, *N*-(3-aminopropyl)-*N*-dodecylpropane-1,3-diamine, hydrogen peroxide, glutaraldehyde, sodium hypochlorite, boric acid, lactic acid	Sprayable 30% Boramon and 8% Rocima 101 effectively protected the wood against bacterial growth for 12 months; and molds for 3 months	Spraying, fogging	[17]
Consolidant	Methacrylic-siloxane resin and microcrystalline cellulose	Increase of the dynamic moduli, systematic decrease of the thermal expansion coefficient, increase of the flexure stiffness, increase hydrophobicity	Surface application (coating)	[38]
Consolidant	Alum	Naturally occurring, can cause acidic depolymerization of cellulose	Immersion	[25]
Antifungal	Inorganic nanoparticles (TiO_2 suspension, containing silver or copper nanoparticles)	The highest antifungal efficiency was observed for suspensions containing highest level of metallic nanoparticles	Brushing	[18]
Disinfection, consolidant	Polyester acrylate and polyester resin dissolved in styrene, by gamma curing	Changes in thermal, chemical and photochemical stability	Immersion, gamma curing	[39]

3. Materials Used for Restoration and Conservation of Masonry Elements

3.1. General Considerations

Alongside wood, masonry materials (such as natural stones, fired or unfired bricks) represent the basis of traditional constructions. The use of natural stones involves the application of several types of rocks, such as the intrusive, volcanic, sedimentary or metamorphic rocks. A classification of the natural rocks used in traditional constructions is presented in Table 2, together with their compositional characteristics.

Man-made materials can be divided into cob (subsoil, water, straws and lime), adobe (mud and organic materials), mud-bricks (sun-dried materials, composing of loam, mud, sand, water and straws), fired bricks (sometimes called "artificial stones") and mortars (different composition workable paste binding the construction materials). In the following paragraphs, the materials used for restoration and conservation of masonry elements are presented, classified into biocidal materials, cleaning agents, consolidants and protective coatings. (Figure 4).

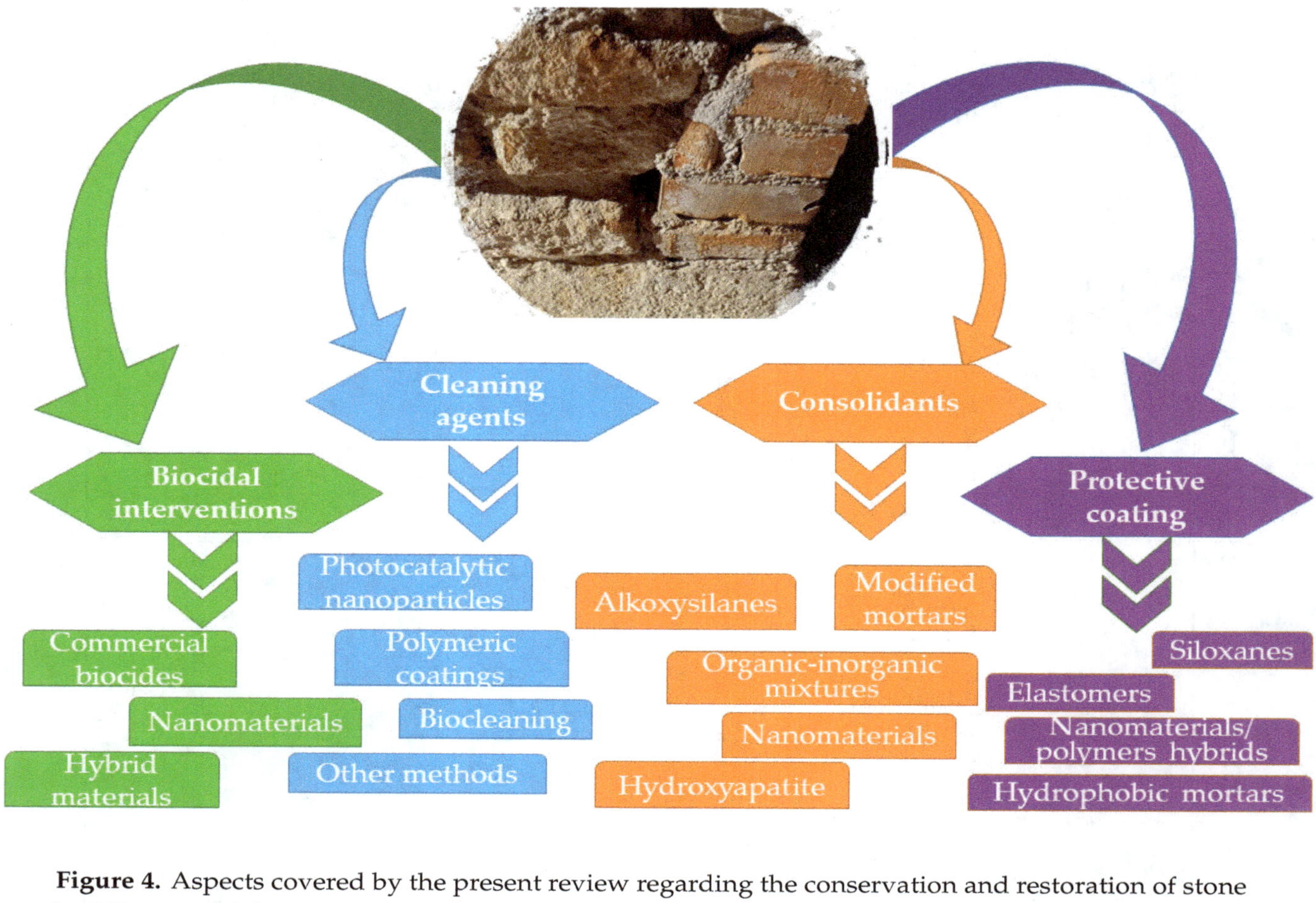

Figure 4. Aspects covered by the present review regarding the conservation and restoration of stone building materials.

Table 2. Characteristics of natural rocks used in vernacular constructions.

Class	Type	Examples	Composition	Characteristics and Uses	Ref.
Intrusive rocks	Felsic	Granite	Quartz (20%–60%), felspars, mica; ratio Plagioclase/(Plagioclase + Alkali feldspar) = 10–65	Granular, phaneritic, massive, hard and tough. Average density 2.65 and 2.75 g/cm^3, compressive strength >200 MPa; used for the construction of pyramids, lumns, door lintels, sills, wall coverings; mostly used as size stone	[41]
	Intermediate	Diorite	Plagioclase feldspar, biotite, hornblende, pyroxene	Phaneritic, occasionally porphyritic, extremely hard, usually used for sculptures, roads, drainage or inscriptions	[42]
	Mafic	Gabbro	Plagioclase and clinopyroxene	Used as ornamental facing or paving stones	[43]
	Ultramafic	Peridotite	Olivine and pyroxene	Coarse-grained, dense, uncommon at the surface, unstable, rarely used	[44]
Extrusive rocks	Felsic	Rhyolite	Quartz (>20%), alkali feldspar (>35%)	Very viscous; used as building, facing or paving stone.	[44]
	Intermediate	Andesite	Plagioclase, pyroxene, hornblende	Porphyritic structure, density 2.11–2.36 g/cm^3, used as filling material, for sculptures or monuments	[45]
	Mafic	Basalt	Pyroxene (augite), plagioclase, olivine	Aphanitic, the most encountered volcanic rock, used as building blocks, cobblestones, for statues	[46]
	Ultramafic	Komatiite	Olivine, pyroxene, anorthite, chromite	Spinifex texture, rare, not usually used for traditional construction	[47]
Sedimentary rocks	Clastic	Sandstone	Quartz or feldspar	Grain size, 0.06–2 mm, variable hardness and color; versatile uses (dependent on the composition)—construction, decoration	[48, 49]
	Biochemical	Limestone	Calcite and aragonite	Variable grain size and texture, hard; used for buildings, decorations or mortars	[50]
	Chemical	Gypsum	Calcium sulfate dihydrate	Variable color and luster, Mohs hardness 2, specific gravity 2.31–2.33; used for plasters, decorations	[51]
Metamorphic rocks		Marble	Calcite or dolomite	Usually white, medium grained, hard, relatively abundant; used for buildings, decorations, sculptures or flooring	[52]

3.2. Biocidal Interventions

3.2.1. Classical Approaches

The biodeterioration of masonry materials can be divided into three main categories, considering the microorganisms involved: deterioration caused by bacteria, by fungi or by lichen [53]. Table 3 presents some examples of biodeterioration induced by different organisms in masonry materials.

Table 3. Biodeterioration of masonry materials (selected examples).

Support Material	Site	Biodeteriogens	Effect	Ref.
Adobe	Capayán ruins (Argentina)	*Centris muralis* Burmeister bee	Massive erosion, high density of cavities	[54]
Limestone and lime stucco	Maya constructions (Mexico)	Fungi, cyanobacteria	Dissolution and recrystallization of calcite, physical breakdown	[55]
Limestone and lime mortar	San Roque church, (Mexico)	Cyanobacteria and Bryophyta	Apparition of dark green to black biofilms after restoration	[56]
Sandstone	La Galea Fortress (Spain)	*Trentepohlia* algae	Reddish biofilm, material disintegration, erosion, discoloration	[57]
Limestone	Chaalis Abbey (France)	Alphaproteobacteria, Actinobacteria, Cyanobacteria, Bacteroidetes, Betaproteobacteria, Deinococcus, Acidobacteria, etc.	Biocorrosion, discoloration, detachment of mineral grains, salt crystallization	[58]
Mortar	Casa Godoy (Porto Alegre, Brazil)	Fungal species: *A. niger*; *T. atroviride*; *T. harzianum*; *Trichoderma* sp.; *C. sphaerospermum*; *Cladosporium* sp.; *Lecanicillium* sp.; *Penicillium oxalicum*; and *Purpureocillium lilacinum*	Chemical alterations in mortar substrates, physical damages due to the growth of filamentous structures	[59]
Tuff and limestone blocks, mortar and plasters, frescoes	Casa della caccia antica (Pompeii, Italy)	Twenty-two lichen species (the most encountered being *Dirina massiliensis*, *Verrucaria macrostoma* and *Lepraria lobificans*)	Physical or chemical interaction with the substrate: hyphal penetration, expansion and contraction of thalli, secretion of metabolites with acidic and chelating functions endolithic growth of other lithobiotic microorganisms	[60]

The literature is much more focused on the identification of the microorganisms affecting the masonry materials than the proposal of new alternative treatments. As there are no substances particularly developed for cultural heritage, the biocidal substances used were typically those developed for other applications (such as commercially available pesticides, based on active ingredients as glyphosate, benzalkonium chloride, *N*-octyl-isothiazolinone, usnic acid, etc.) [61] or natural materials, such as natural extracts and essential oils [62,63] or lipopeptides [64].

Another potential approach is represented by the physical decontamination methods (such as the mechanical cleaning, the use of ionizing or UV-radiation) [65]. These latter methods possess a series of disadvantages (the mechanical cleaning could damage the substrata, UV and gamma radiation can induce color changes). However, with the identification of the potential hazard to human health and to the environment of those methods, the need for dedicated pesticides became evident. Among the first solutions proposed were the natural alternatives (plant extracts, essential oils, etc.), their use being recently reviewed by Fidanza and Caneva [66]; although the approach has some advantages (such as the good efficiency, or a potential eco-friendly character), in our opinion, the lack of supplementary data, regarding the interaction with the material itself, as well as regarding the behavior over longer period of times, makes the solution of natural biocides inapplicable at this time in the field of cultural heritage.

3.2.2. Nanotechnological Approaches

Considering the shortcomings of the classical methods, a viable alternative is represented by the application of nanotechnology. Nowadays, nanotechnology emerged as a research field with successful application in the cultural heritage domain. As our group previously presented [67], the application of nanotechnology for the development of antimicrobial coatings represents an emerging field, which could offer valuable resources for specialists in restoration/conservation. In spite of the wide-spread use of nanomaterials, their application as biocidal agents in cultural heritage stone conservation represents the subject of a surprisingly low number of articles (details presented in Table 4).

Usually, the use of nanomaterials in the field of antimicrobial protection of stone artifacts harvests the known potential of some metallic or metal oxide nanoparticles (such as silver, a wide known antimicrobial, ZnO, TiO_2, CuO, Cu, etc.) [68–72]. Among those examples, the hydrophobic film developed by Ruffolo et al. [68], incorporating a photocatalytic agent (TiO_2) and an antimicrobial agent (Ag), showed very good antifouling properties for application on underwater marble artifacts. The procedure also proved to be reversible, after 20 months no trace of protective film being identified, suggesting a repetition of the treatment every 12–24 months. With a similar approach, Becerra et al. [73] used silver nanoparticles and Ag/TiO_2 nanocomposites as antimicrobial agents on limestone. By using two different synthesis procedures for silver nanoparticles—AgNPs (citrate and sodium borohydride reduction), the authors observed a direct correlation between the antimicrobial effect of the silver nanoparticles/silver nanocomposites and the hydrodynamic diameter. More than that, the AgNPs and Ag/TiO_2 nanocomposites obtained by the citrate method led to superior results regarding the biofouling reduction, even compared with a commercial biocide (>70%, compared with 53% for Biotin T) without significantly altering the aesthetic properties of the limestone (particularly AgNPs). The silver nanoparticles alone were also proven to be an efficient antimicrobial agent against bacteria and fungi developed on different materials (stucco, calcite and basalt), inhibiting the colonization of the stones (0.16%–3.6% colonization, compared with the untreated samples—8%–28% colonization, depending on the material) [74]. The authors observed a dependency on the AgNPs dose, and less on the NPs size (as is the case for the in vitro antimicrobial properties of the nanoparticles) [75,76].

Another potential application of the nanotechnology is represented by the use of nanocapsules or nanoparticles for the controlled release of the biocidal agents. For example, Dresler et al. [77] proposed the use of pristine and functionalized silica (with amino or carboxylic groups) for the controlled release of the widely used biocides New Des 50 and Biotin T. The authors tested the efficiency of the proposed material both in vitro (against Gram-positive and Gram-negative bacteria—*Kokuria rhizophila*, *Staphylococcus aureus* and *Escherichia coli*). The proposed materials were tested by immersing in the solution a stone fragment (not defined) from a fountain located in Diamantina, Minas-Gerais, Brazil. The long-term effect was evaluated by determining the total viable bacterial counts after 1, 6 and 12 months (registering a 98.4% inhibition in bacterial counts after 12 months, compared with an untreated sample). This approach seems to be the most successful for future research works, as it harvests the nanotechnological advantages, together with the proven biocidal effect of commercial products. More than that, the method could be applied for the incorporation and controlled release of other types of biocides (such as, for example, the essential oils), which, without such delivery vehicles, hardly finds application in cultural heritage conservation. Another important aspect related to the potential use of nanoparticles is related to their potential harmful effects. The toxicity of nanoparticles represents to this date a subject of research. However, considering the relatively low levels necessary for the achievement of biocidal effect, the nanomaterials can be considered relatively safe for use [78]. In our opinion, future studies presenting the application of nanomaterials in the cultural heritage area should be accompanied by thorough toxicological studies.

Table 4. Nanomaterials applied as biocidal agents in cultural heritage conservation (references presented in chronological order).

Nanomaterial	Nanomaterial Characteristics	Application	Treated Material	Ref.
Functionalized carbon nanofibers and nanotubes	80–150 nm/1.2–1.4 nm diameter, commercially available	Surface treatment (removal of black and gray patina), finishing cleaning method	Marble	[79]
$Ca(OH)_2$ mixed with ZnO/TiO_2	500/10–30/<50 nm	Antimicrobial (against *Penicillium oxalicum* and *Aspergillus niger*)	Limestone	[80]
Ag nanopowder/ silane/siloxane emulsions	<100 nm, PVP coated, commercially available	Surface treatment emulsion for facades (against algae and cyanobacteria biofouling	Mortar	[81]
TiO_2/SiO_2 nanocomposites	Theoretical proposal, no studies performed	Preventing biodeterioration	Mortar	[56]
Ag nanoparticles	Phytosynthesized, 39 to >100 nm	Surface treatment (against *Pectobacterium carotovorum* and *Alternaria alternate*)	Stucco (pozzolanic material), calcite and basalt	[74]
TiO_2	4 nm, commercially available	Surface treatment (antifouling, against *Chlorella* cf. *mirabilis* Andreeva and *Chroococcidiopsis fissurarum*)	Fired bricks	[82]
TiO_2, TiO_2/Ag, TiO_2/Cu nanoparticles	TiO_2– 40–50 nm	Surface treatment, spraying in three layers (against different algal species)	Travertine	[83]
TiO_2, ZnO and Ag nanoparticles dispersed in melted siloxane wax	Particles mean diameter 100 nm, commercially available	Antifouling agent for underwater stone materials (against epilitic and endolithic micro-organisms)	Marble	[68]
Pristine and functionalized silica (MCM41)	Commercially available	Controlled release of commercially available biocides New Des 50 and Biotin T.	Stone	[77]
ZnO and Ag nanoparticles	30/25 nm, commercially available	Dip-coated mortar disks (against *B. cereus* and *E. coli*)	Mortar	[71]
Si nanocapsules	148 nm	Controlled release of an eco-friendly biocidal agent for antifouling coatings	Proposed for cultural heritage applications	[84]
Si nanocapsules and nanoparticles	128/39 nm, with entrapped active ingredient	Controlled release of a commercial biocidal agent	Proposed for cultural heritage applications	[85]
Ag nanoparticles and Ag/TiO_2 nanocomposites	36/72/94 nm hydrodynamic diameter, depending on the synthesis pathway	Surface treatment (against multiple Chlorophyta and cyanobacteria)	Limestone	[72]

3.3. Restoration/Conservation of Masonry Materials against Abiotic Factors

3.3.1. Deterioration of Masonry Materials by Abiotic Factors—General Considerations

The alteration of stone (either natural or man-made) is mainly related to two types of abiotic factors [86,87]:

- Intrinsic characteristics of the materials:
    - Chemical and mineralogical composition (species solubility and their variation, presence of oxidable species, surface and ionic phenomena);
    - Structure and texture (mainly the pore distribution, resulting in gelifraction and salt crystallization resistance, water absorption and drying rates).
- Extrinsic factors:
    - Water presence;
    - Presence of foreign substances altering the pH or the composition;
    - Pressure and wind;
    - Thermal variations;
    - Anthropogenic abiotic factors (mainly related to pollution products).

The water (in all its forms—solid, liquid or vapors) represents one of the most important factors affecting the integrity of stones: it can interact with the substrate (dissolution, hydrolysis, oxidation-reduction), it can transport other substances (for example sulfates or other pollutants), it represents a medium for chemical reaction (and for the development of microorganisms) or (in case of temperature variation) can produce microcracks in the materials (in its solid form). Closely related to the presence of water, the soluble salts can damage the materials through two pathways (crystallization and hydration).

The main types of damages induced in stone materials are (as classified by the International Council on Monuments and Sites—ICOMOS [87]):

- Cracks (including fracture, star cracks, hair cracks, craquele and splitting) and deformation;
- Detachment (blistering, bursting and delamination);
- Material loss (alveolization, erosion, mechanical damage);
- Discoloration and deposits (crusts, coloration, bleaching, staining, efflorescence and encrustation).

The processes necessary for preserving damaged stone materials usually involves three steps [86]:

- Cleaning, often performed mechanically or with dedicated gentle solutions;
- Consolidation (in order to increase the resistance of the material);
- Protection (generally focused on the use of water repellant solutions, as water represents one of the main factors involved in the degradation, as previously presented).

Some classical and modern materials used for these applications are presented in Table 5, while some representative examples are detailed in the following paragraphs.

3.3.2. Cleaning Agents

Cleaning of stone artifacts is usually performed by mechanical (brushing, projection) [88] or chemical methods (using different solvents, such as methylene, acetone, acids, alkali or even commercial mixtures) [89,90]. More recently, the use of lasers was proposed for a wide variety of stone types (marble, sandstone, etc.) [91]. Removal of biofilms is usually performed using the same techniques, followed by a biocidal treatment (presented in Section 3.1). Regardless of the chosen technique, great care should be paid to not provoke damages to the original material; the processes should be gradual, selective and economically feasible [92].

Among the new materials used for cleaning purposes, the most encountered are the TiO_2 nanoparticles, which, due to their photocatalytic properties, can be used in self-cleaning protective layers. For example, La Russa et al. [93] proposed the use of this property of anatase-form nanoTiO_2 by incorporation in three types of commercial coatings (acrylic polymer in organic solvent—Paraloid B72, fluorinated polymer in an alcoholic solvent—Akeogard P, compared with a commercial aqueous suspension of TiO_2 and an acrylic polymer—Fosbuild). The poorest results were obtained for the material using the nanoparticles/Paraloid composition (intense alteration of the surface, poor photo-degradation effect), while the best results were obtained for the commercial formulation (Fosbuild). The composition based on fluorinated polymers also led to good results, with significant photo-degradation and water repellency. A similar approach was adopted by Quagliarini et al. [94], combining the hydrophilic and photocatalytic photo-induced properties of the nanoparticles into a coating that decreased the water adsorption by 50%. Other examples regarding the use of nano-TiO_2 as self-cleaning materials are provided in Table 5.

A particular and very important area of cleaning procedures is dedicated to the graffiti removing methods. This can be achieved by chemical or mechanical methods, laser cleaning [92] or biocleaning (using, for example, sulfate-reducing bacteria or enzymes) [95]. A preventive action (protective) is represented by the anti-graffiti coatings, based on organic or inorganic agents (either temporary—waxes, silicones, or permanent—polyurethanes, fluorocarbon, alkyl alkoxy silanes) [96].

3.3.3. Consolidation and Protection Agents

Consolidation represents the treatment applied in order to restore its mechanical properties, affected by weathering [86]. At the border between cleaning and consolidation, desalinization is often necessary; the removal of soluble salts is usually made by applying wet poultices consisting of clays or cellulosic mixtures [97,98], the alternative treatments proposed being represented by the application of ohmic technologies [99] and electrochemical methods [98], limewater (calcium hydroxide solution [100], crystallization modifiers (ferrocyanides, borax, etc. [101,102], or diammonium hydrogenphosphate [103].

As it emerges from the literature review, most of the authors treat the desalinization process as a secondary process (accompanying the main consolidation procedure), thus usually the traditional, well-established procedure being followed. This, in turn, leaves serious room for improving, by the development of new materials.

At this time, on the market are available a wide range of both organic and inorganic consolidants [104,105] with proven efficiency in consolidating different types of masonry materials. The stone consolidants can be divided in three main classes: (i) organic products (alkoxysilanes, recently reviewed by Xu et al. [106]); (ii) organic-inorganic mixtures and (iii) purely inorganic products (usually apatitic materials), relevant examples being provided in Table 5.

A good example regarding the application of organic products for the consolidation of stone materials is represented by the study of Liu and Liu [107]. The authors proposed the use of a composite polymeric material (tetraethoxysilane with additives as hydroxyl-terminated polydimethylsiloxane and cetyl trimethyl ammonium bromide) for the consolidation of sandstone. After treatment, the sandstone showed crack-free surface homogeneity, acid resistance, as well as salt crystallization resistance, the polymeric layer also exhibiting very good hydrophobic properties (demonstrated by a water contact angle of 110°). Generally speaking, the organic products (as well as many of the organic/inorganic composites) used as consolidants also possess very good hydrophobicity, which also makes them very good protective compounds. A particularly interesting study was recently published by

Kapridaki et al. [108]. In a three-layer treatment, the authors offered a complex conservation solution for stone monuments, incorporating a self-cleaning (based on TiO_2 nanoparticles), consolidation (based on calcium oxalate nanoparticles) and a protective hydrophobic layer. The approach was successfully applied on different types of masonry materials (limestones of different porosity, ceramic materials and mortars).

The use of in situ formed hydroxyapatite represents a very good alternative to other methods. It is achieved by the reaction of a phosphate salt and the substrate (marble, limestone, sandstone, sulfated stone, gypsum stucco, concrete, etc.). The topic was recently reviewed by Sassoni [109], so a very thorough presentation of the method would be redundant. We have chosen to discuss only a modification of the method, proposed by Pesce et al. [110]. The in situ formation of a hydroxyapatite consolidating layer was achieved using aqueous solution of diammonium phosphate and two types of calcium nanosuspensions, calcium hydroxide and calcite (calcium carbonate), on limestone and sandstone. The authors observed an improvement of the compactness of the stones, accompanied by a risk of incompatibility (proven by the substantial reduction of water sorption, especially for sandstone). The authors also proposed the effective order of reactants deposition on the stone (the phosphate followed by the calcium source) and drew attention to a possible drawback of the method, respectively an elevated risk of yellowing on the surface. A particular approach was developed by our group, regarding the application of synthesized hydroxyapatite and apatitic materials, especially for the protection of limestone, but also for antimicrobial purposes [111–114]. However, due to the significant color changes recorded, the optimization of these recipes still represents a subject for future research.

The incorporation of basalt fibers (chopped fibers, continuous filaments or milled fibers) into mortars (prepared using natural hydraulic lime, dry premix and inert aggregates, respectively dry premix and inert aggregates with crushed bricks and tiles—*cocciopesto*) was proposed by Santarelli et al. [115]. The authors observed that the fibers addition increased the compressive properties of the mortars, while the chopped fibers imparted post-peak stress of the hydraulic lime mortar. For all the proposed materials, was also observed a decrease in the capillary water adsorption coefficient. Similar approaches were published by Moropoulou et al. [116] (using several pozzolanic additives), Andrejkovičová et al. [117] (using palygorskite and metakaolin), Ventolà et al. [118] (using commercial phase changing materials), Rosato et al. [119] (using commercial cellulose nano-fibrils), Gour et al. [120] (using natural polymers) or Pavlík et al. [121] (using crushed lava granulates). However, regarding the reconstruction/consolidation mortars, the best approach, in our opinion, is the one presented by Stefanidou et al. [122], respectively designing the mortars according to the nature of the natural stone (in the cases presented by the authors marl limestone, nummulitic limestone, marble, biogenic calcitic sandstones), and as, often as possible, reclaiming local materials.

A particular case is represented by the earth (unfired) masonry materials (such as earth-blocks—cob, adobe, mud-bricks or mud mortars). In those cases, considering the limited viability of the materials, the best approach is their replacement/grouting with materials as similar as possible, or, at least, compatible [123,124].

As already stated, many of the polymeric composites previously presented possess hydrophobic properties, thus having a protective role for the masonry materials. The usually applied conservation materials are primarily represented by siloxanes [125] and elastomers [126], with proven efficiency over a large variety of masonry materials. More interesting, the use of hybrid nanomaterial/polymeric coatings was proposed by some authors. Corcione et al. [127] proposed a mixture containing commercially available micrometric hydroxyapatite and multiple polymeric components (trimethylolpropane trimethacrylate, trimethoxypropyl silane methacrylate monomer, vinyl terminated polydimethylsiloxane and 3-mercaptopropyltriethoxysilane). The mixture was applied by brushing on two calcarenitic stones, with different porosity and photo-cured by UV-lamp. The coatings led to a surface hydrophobic character of the treated material (demonstrated by the contact angle—129°–136°, depending on hydroxyapatite amount, compared with the untreated sample—96°). Considering the hydrophobicity results and the observed color changes, the authors proposed an upper limit of 5% hydroxyapatite filler for future applications. Cappelletti et al. [128] incorporated nano-titania in a commercial silane resin (Alpha®® SI30). The hybrid mixture was airbrushed on the surface of different types of stones (marble, dolomite), obtaining superior values of hydrophobicity, compared with the pure commercial product. In the case of marble, was even achieved a superhydrophobicity character (contact angle > 150°). The same approach was used by Aslanidou et al. [129]. Using commercially available silica nanoparticles dispersed in commercial alkoxy silanes and organic fluoropolymer emulsion obtained superhydrophobic and superoleophobic coatings, successfully applied for the protection of masonry (marble, sandstone) and other materials (concrete, paper, silk), proving a good versatility of the method. The use of hydrophobic components (such as calcium stearate or silane/stearate) was also proposed Falchi et al. [130] as a viable alternative in order to develop restoration mortars with hydrophobic properties (thus, protection action). The authors introduced the water-repellent admixtures into pozzolana lime-mortar (0.5%–1.5%), obtaining materials with reduced water vapor permeability, without significantly affecting the mechanical properties of the mortar.

When developing functional coatings for masonry materials, particular attention should be payed to the possible formation of a coating surface which could seal in potential degradation agents and interfere with normal wet/dry cycle [70,131].

Table 5. Materials for the treatment of stones used in the vernacular constructions (references presented in chronological order).

Procedure (Effect)	Material	Characteristics and Application	Stone Type	Ref.
Cleaning	TiO_2 nanoparticles	18 nm, aqueous solution, applied by spraying	Travertine	[132]
	TiO_2 nanoparticles	10–20 nm obtained by sol-gel and hydrothermal method, applied by spray coating	Travertine	[94]
	TiO_2/poly(carbonate urethane) nanocomposite	31 nm, commercially available, aqueous dispersion; with associated protective role	-	[133]
	TiO_2 nanoparticles	Commercially available nanoparticles (25 nm), incorporated in different commercial coatings	Marble, calcarenite	[93]
	TiO_2 nanoparticles	5–6 nm, dispersed in water and ethylene glycol, applied by brush	Marble, calcarenite	[134]
	TiO_2 nanoparticles	Commercially available nanoparticles (30 nm), incorporated in different mortars	Lime-, cement- and lime/cement-based mortars	[135]
	TiO_2—tetraethoxysilane-polydimethylsiloxane	25 nm, commercially available, applied by brush	Modica Stone (limestone)	[136]
	TiO_2 nanoparticles	10–40 nm, commercially available, aqueous solution, applied by spraying	Sandstone, concrete slabs	[137]
	TiO_2 nanoparticles	External layer in a multi-purpose solution, with oxalic acid in the interface with environment; surface application, by brush	biomicritic limestone, travertine, calcitic sandstone, ceramic materials, mortars	[108]
Consolidation	Different inorganic additives	Earth of Milos, brick powder, crushed brick, used as pozzolanic additives for lime mortars	Mortar	[116]
	Inorganic additives	Palygorskite and metakaolin, used as pozzolanic additives for lime mortars	Mortar	[117]
	Organic additive	Commercial product (consisting of n-heptadecane core and polymethyl-methacrylate shell, PCM DS 5001 Micronal®®), incorporated in lime mortar	Mortar	[118]
	Basalt fibers	Incorporation in different composition for development of restoration mortars	Mortar	[115]

Table 5. *Cont.*

Procedure (Effect)	Material	Characteristics and Application	Stone Type	Ref.
Consolidation	Silicic acid esters (tetraethyl orthosilicate, dioctyltin dilaurate)	Polymeric coating, commercially available (Tegovakon®® V100), surface application (by brush/drop-by-drop)	Bioclastic calcarenite, chert	[104]
	Ethyl silicate hybrid binder (hydrolyzate)	Polymeric coating, commercially available (Wacker®® Tes 40 WN), surface application (by brush/drop-by-drop)	Bioclastic calcarenite, chert	[105]
	Nano SiO_2	Nanoparticles suspension (<20 nm), commercially available (NanoEstel), surface application (by brush/drop-by-drop)	Bioclastic calcarenite, chert	[104,105]
	Nano $Ca(OH)_2$	Nanoparticles suspension in isopropyl alcohol (Nanorestore®®), surface application (by brush/drop-by-drop)	Bioclastic calcarenite, chert	[104]
	Polymeric composite	Tetraethoxysilane having as additives hydroxyl-terminated polydimethylsiloxane and cetyl trimethyl ammonium bromide; application by immersion	Sandstone	[107]
	SiO_2/polymer	13.5–24 nm, SiO_2 pristine/hydrophobized (methylated or octylated), incorporated in ethoxysilanes mixture; surface application, by pipetting	Sandstone	[138]
	Hydroxyapatite (formed in situ)	Using diammonium phosphate; application by brushing and immersion	Limestone	[100]
	$Ca(OH)_2$	Water solution, limewater poultice for desalinization	Limestone	[100]
	Acrylic resin	Polymeric coating, commercially available (Paraloid®® B72), surface application (drop-by-drop)	Chert	[139]
	Cellulose fibers	Nano-fibrils, commercially available, incorporated in lime mortar	Mortar	[119]
	$Ca(OH)_2$/polymer	7–15 nm, amorphous calcium hydroxide monohydrate nanoparticles incorporated in tetraethoxysilane; first layer in in a multi-purpose solution; surface application, by brush	biomicritic limestone, travertine, calcitic sandstone, ceramic materials, mortars	[108]
	Natural polymer	Areca nut (natural polymer) incorporated in lime mortar	Mortar	[120]
	Diammonium hydrogenphosphate	Water solution, with cellulose pulp, poultice for desalinization	Limestone	[103]
	$Ca(OH)_2$	Commercially available (CaLoSiL®® E25), known as nanolime; surface application (by syringe) to saturation of the sample	Clunch, ooidal limestones (Bath, Barnack, Portland), coarse-grained shelly limestone (Ham), magnesian limestone	[140]

Table 5. *Cont.*

Procedure (Effect)	Material	Characteristics and Application	Stone Type	Ref.
Consolidation	TiO_2 and alkoxysilane	Tetraethyl-orthosilicate and alkyl-trialkoxysilane doped with synthesized (5–40 nm method not disclosed); application by capillary suction	Limestone, sandstone	[141]
	Hydroxyapatite (formed *in situ*)	Using two types of nanomaterials—$Ca(OH)_2$ and $CaCO_3$, and diammonium phosphate; application by capillary suction	Limestone, sandstone	[110]
	Crushed lava granulates	Used as sand replacement in hydrated lime, natural hydraulic lime, or cement-lime binder	Mortar	[111]
	Ternary composition	Commercial SiO_2 nanoparticles (25 nm), $Ca(OH)_2$ nanoparticles (200 nm), hydroxypropyl cellulose, in hydroalcoholic desertion; immersion treatment	Adobe	[124]
Protection	Organosilicons	Long chain polymerized siloxane, short chain polymerized siloxane, alkyl potassium silicate; surface application	Sandstone, dolomite, marble, granite	[125]
	Hydroxyapatite/siloxane-methacrylic formulations	Commercial micrometric hydroxyapatite added in siloxane-modified mixture; application by brush followed by photo-curing	Calcarenitic stones	[127]
	Stearate/silane	Incorporation of calcium stearate and silane/stearate (Silres A®®) in pozzolana-lime binders	Mortar	[130]
	Nano TiO_2/silane resin	TiO_2 nanoparticles solution mixed in a commercial silane resin (Alpha®® SI30); application by airbrush	Marble, dolomite	[128]
	Polymer hybrid coating	Trimethylpropane trimethylacrylate, trimethoxypropyl silane methacrylate, poly(dimethylsiloxane)-terminated vinyl, alkoxy-silane, bis(2,4,6-trimethylbenzoyl)-phenylphosphineoxide, 2-hydroxy-2-methyl-1-phenyl-1-propanone; applied by brushing	Calcarenitic stone	[142]
	Neat and nanomodified coatings	Protective coatings (linseed oil, silane/siloxane, alkosiloxane) neat or with silica nanoparticles (14 nm); applied by full immersion on fired bricks	Fired bricks	[143]
	Boehmite/polymers	Incorporation of organic-modified boehmite mineral in a series of commercially available protective coatings; application by brushing	Calcarenitic stones	[144]

Table 5. *Cont.*

Procedure (Effect)	Material	Characteristics and Application	Stone Type	Ref.
Protection	Oligoamides	Partially fluorinated oligoadipamide, ethylenediamide and hexamethylenediamide, solutions in propanol; adsorption into the stone materials.	Limestone, marble	[126]
	Silanes	Tetraethoxysilane/polydimethylsiloxane composite, intermediate layer in in a multi-purpose solution; surface application, by brush	biomicritic limestone, travertine, calcitic sandstone, ceramic materials, mortars	[108]
	Poly(hydroxyalkanoate)s	Poly(3-hydroxybutyrate) and poly(3-hydroxybutyrate-co-3-hydroxyvalerate co-4-hydroxyvalerate), compared with silane and siloxane commercial formulations—Idrosil®® and Antipluviol®®; applied by dip-coating, poultice, spray.	Sandstone, limestone, marble	[145]
	Siloxanes	Oligomeric ethoxysilane, hydroxyl-terminated polydimethylsiloxane in aqueous n-octylamine solution (original solution proposed by the authors), compared with the commercial product BS290 (Wacker); application by spraying	Limestone	[146]
	SiO_2/polymer	Commercially available SiO_2 nanoparticles (7 nm) dispersed in commercially available emulsion of alkoxy silanes and organic fluoropolymer (Silres BS29A); applied by spraying	Marble, sandstone, concrete	[129]

4. Current Limitations and Future Perspectives

The main goal regarding cultural heritage objects is their preservation (conservation). Restoration appears often necessary, but should always be the subject of professional restorer's decision. As any restorer knows, all the recipes used for the restoration/conservation of cultural heritage objects should meet several critical conditions, among which two are of particular interest: the reversible character of the treatment and authenticity preservation [147]. The specialists working in this area find significant differences between clearly defined terms used (such as *protection*, *preservation*, *conservation*, *consolidation*, *restoration*, *reconstruction* or *anastylosis*) [147]. These terms are not, at this moment, thoroughly integrated in the vocabulary of the researchers working in the field of materials science (from which comes the large majority of the solutions presented in the current review). This represents a sign of a major (and probably the most important) drawback of the current approach, worldwide. Most researchers perform laboratory (or even in situ) experiments without the assistance (or collaboration) from a conservative/restorer. Although with scientific value for sure, this approach could lead to possibly insignificant results for the restorers. The main goal of materials science researchers should be, in our opinion, to provide the necessary tools and the scientific support for the specialists that have the authority and the knowledge to practically perform the interventions on the cultural heritage objects. Thus, closer collaboration should be established between the two types of scientists, as well as the development of a common criteria preferences regarding the materials/treatments, as some studies pointed out the major differences regarding the selection of materials for the protection of heritage objects [148].

Additionally, as a general remark, the wide spread of nanomaterials in their pure state (without incorporation in polymeric matrixes) can, in some instances, violate the reversibility criteria. This could, on the other hand, be explained by its relative wide criticism from conservationists, that consider it a "dubious principle", "chimera", "myth" or "Utopian idea" [149].

From the point of view of materials science, it appears surprising the lack of original materials, implemented at least at laboratory level. As researchers could obtain nanoparticles or nanomaterials (such as, for example, TiO_2, SiO_2, hydroxyapatite, etc.) with the desired morphological characteristics, they should use such synthesized materials instead of commercially available ones. There are a large number of metallic or metal oxide nanoparticles that could exhibit photocatalytic activity (for self-cleaning purposes) or antimicrobial properties. Additionally, the exploration of antimicrobial properties of phytosynthesized nanoparticles could offer new recipes with potential application in this area [78]. The same observations can be made regarding the use of hydroxyapatite/other apatitic materials (which could be tuned for multiple applications). These materials could be incorporated in polymeric films, in order to ensure the previously-stated reversible character.

Especially when talking about materials for cultural heritage buildings, the treatments should also meet some supplementary criteria (easy and deep penetration, resistant to attack, prevention of humidity penetration, allowing exiting of water, no modification alteration, uniform contraction and expansion with the substrate, inexpensive, non-corrosive, non-reactive, lasting properties, easily applicable, resistance to acid and alkaline attack, etc.) [150], which should be all presented by the authors reporting the evaluation of new materials.

Another important aspect for the restoration/conservation of traditional buildings is represented by the engineering assessment of the construction, including the differentiation between the structural and the decorative materials, which opens different treatment routes. For example, in the case of structural materials, the recovery of the material's functionality is mandatory, while for decorative materials, is only necessary the maintaining of their integrity [3,113,151]. If this is not possible, the natural stones can be replaced with natural stones with the same characteristics, or with cast stones, designed to replicate the natural ones [152].

Finally, the current developments in the materials science field in general allows us to envisage a continuous increase of the quality and properties of the materials offered to the specialists working as conservatives of the cultural heritage objects.

5. Conclusions

The present review paper summarizes different aspects related to the recent progress in the field of restoration/conservation of traditional building materials (wooden and masonry elements), including materials used for biocidal interventions, restoration/conservation of materials against abiotic factors, cleaning agents and consolidation and protection agents. This critical review can be considered a starting point for further transdisciplinary studies and experiments, with application in the cultural heritage domain. In this respect, new perspectives have emerged and development of new materials and methods based on classical restoration/conservation approaches will lead to preserving cultural heritage for future generations.

Author Contributions: All authors contributed to data collection and analysis, manuscript design, preparation and revision of the manuscript. All authors have read and agreed to the published version of the manuscript.

References

1. Spišáková, M.; Mačková, D. The use potential of traditional building materials for the realization of structures by modern methods of construction. *SSP J. Civ. Eng.* **2015**, *10*, 127–138. [CrossRef]
2. Doğangün, A.; Tuluk, O.I.; Livaoğlu, R.; Acar, R. Traditional wooden buildings and their damages during earthquakes in Turkey. *Eng. Fail. Anal.* **2006**, *13*, 981–996. [CrossRef]
3. Dutu, A.; Niste, M.; Spatarelu, I.; Dima, D.I.; Kishiki, S. Seismic evaluation of Romanian traditional buildings with timber frame and mud masonry infills by in-plane static cyclic tests. *Eng. Struct.* **2018**, *167*, 655–670. [CrossRef]
4. Mısırlısoy, D.; Günçe, K. A critical look to the adaptive reuse of traditional urban houses in the Walled City of Nicosia. *J. Archit. Conserv.* **2016**, *22*, 149–166. [CrossRef]
5. Nilsson, T.; Rowell, R. Historical wood – structure and properties. *J. Cult. Herit.* **2012**, *13*, S5–S9. [CrossRef]
6. Kim, Y.S.; Singh, A.P. Wood as cultural heritage material and its deterioration by biotic and abiotic agents. In *Secondary Xylem Biology. Origins, Functions, and Applications*; Kim, Y.S., Funada, R., Singh, A.P., Eds.; Academic Press: London, UK, 2016; pp. 233–257.
7. Eken, E.; Taşcı, B.; Gustafsson, C. An evaluation of decision-making process on maintenance of built cultural heritage: The case of Visby, Sweden. *Cities* **2019**, *94*, 24–32. [CrossRef]
8. Morkunaite, Z.; Podvezko, V.; Zavadskas, E.K.; Bausys, R. Contractor selection for renovation of cultural heritage buildings by PROMETHEE method. *Arch. Civ. Mech. Eng.* **2019**, *19*, 1056–1071. [CrossRef]
9. Cronk, Q.C.B.; Forest, F. The Evolution of Angiosperm Trees: From Palaeobotany to Genomics. In *Comparative and Evolutionary Genomics of Angiosperm Trees. Plant Genetics and Genomics: Crops and Models*; Groover, A., Cronk, Q., Eds.; Springer: Cham, Switzerland, 2017; Volume 21, pp. 1–17.
10. Escobin, R.P.; Conda, J.M.; Ramos, M.; Rizare, M.D.; Cortez, R.E., Jr. Scientific restoration of national shrines and landmarks in the Philippines by the Forest Products Research and Development Institute wood identification technique. *Int. J. Conserv. Sci.* **2019**, *10*, 25–38.
11. Hunt, D. Properties of wood in the conservation of historical wooden artifacts. *J. Cult. Herit.* **2012**, *13*, S10–S15. [CrossRef]
12. Unger, A. Decontamination and "deconsolidation" of historical wood preservatives and wood consolidants in cultural heritage. *J. Cult. Herit.* **2012**, *13*, S196–S202. [CrossRef]
13. *PD CEN/TR 15003, Durability of Wood and Wood-Based Products. Criteria for Hot Air Processes for Curative uses against Wood Destroying Organisms*; The British Standards Institution: London, UK, 2012; Available online: https://standards.globalspec.com/std/9693/bs-pd-cen-ts-15003 (accessed on 3 December 2019).
14. Clausi, M.; Crisci, G.M.; La Russa, M.F.; Malagodi, M.; Palermo, A.; Ruffolo, S.A. Protective action against fungal growth of two consolidating products applied to wood. *J. Cult. Herit.* **2011**, *12*, 28–33. [CrossRef]
15. Stejskal, V.; Douda, O.; Zouhar, M.; Manasova, M.; Dlouhy, M.; Simbera, J.; Aulicky, R. Wood penetration ability of hydrogen cyanide and its efficacy for fumigation of *Anoplophora glabripennis*, *Hylotrupes bajulus* (Coleoptera), and *Bursaphelenchus xylophilus* (Nematoda). *Int. Biodeter. Biodegr.* **2014**, *86*, 189–195. [CrossRef]
16. Querner, P.; Simon, S.; Morelli, M.; Fürenkranz, S. Insect pest management programmes and results from

their application in two large museum collections in Berlin and Vienna. *Int. Biodeter. Biodegr.* **2013**, *84*, 275–280. [CrossRef]

17. Koziróg, A.; Rajkowska, K.; Otlewska, A.; Piotrowska, M.; Kunicka-Styczyńska, A.; Brycki, B.; Nowicka-Krawczyk, P.; Kościelniak, M.; Gutarowska, B. Protection of historical wood against microbial degradation-selection and application of microbiocides. *Int. J. Mol. Sci.* **2016**, *22*, 1364. [CrossRef] [PubMed]
18. Goffredo, G.B.; Citterio, B.; Biavasco, F.; Stazi, F.; Barcelli, S.; Munafò, P. Nanotechnology on wood: The effect of photocatalytic nanocoatings against *Aspergillus niger*. *J. Cult. Herit.* **2017**, *27*, 125–136. [CrossRef]
19. Walsh-Korb, Z.; Avérous, L. Recent developments in the conservation of materials properties of historical wood. *Progr. Mater. Sci.* **2019**, *102*, 167–221. [CrossRef]
20. Schönemann, A.; Edwards, H.G.M. Raman and FTIR microspectroscopic study of the alteration of Chinese Tung oil and related drying oils during ageing. *Anal. Bioanal. Chem.* **2011**, *400*, 1173–1180. [CrossRef]
21. Humar, M.; Lesar, B. Efficacy of linseed- and tung-oil-treated wood against wood-decay fungi and water uptake. *Int. Biodeter. Biodegr.* **2013**, *85*, 223–227. [CrossRef]
22. Freitag, C.; Kamke, F.A.; Morrell, J.J. Resistance of resin-impregnated VTC processed hybrid-poplar to fungal attack. *Int. Biodeter. Biodegr.* **2015**, *99*, 174–176. [CrossRef]
23. Hoffmann, P. The Sucrose Method. In *Conservation of Archaeological Ships and Boats. Personal Experiences*; Archetype Publications: London, UK, 2013; pp. 81–90.
24. Hoffmann, P. The Lactitol Method. In *Conservation of Archaeological Ships and Boats. Personal Experiences*; Archetype Publications: London, UK, 2013; pp. 93–96.
25. Andriulo, F.; Braovac, S.; Kutzke, H.; Giorgi, R.; Baglioni, P. Nanotechnologies for the restoration of alum-treated archaeological wood. *Appl. Phys. A* **2016**, *122*, 322. [CrossRef]
26. De Filpo, G.; Maria Palermo, A.; Rachiele, F.; Pasquale Nicoletta, F. Preventing fungal growth in wood by titanium dioxide nanoparticles. *Int. Biodeter. Biodegrad.* **2013**, *85*, 217–222. [CrossRef]
27. Poggi, G.; Toccafondi, N.; Chelazzi, D.; Canton, P.; Giorgi, R.; Baglioni, P. Calcium hydroxide nanoparticles from solvothermal reaction for the deacidification of degraded waterlogged wood. *J. Colloid. Interface Sci.* **2016**, *473*, 1–8. [CrossRef] [PubMed]
28. Schofield, E.J.; Sarangi, R.; Mehta, A.; Jones, A.M.; Smith, A.; Mosselmans, J.F.W.; Chadwick, A.V. Strontium carbonate nanoparticles for the surface treatment of problematic sulfur and iron in waterlogged archaeological wood. *J. Cult. Herit.* **2016**, *18*, 306–312. [CrossRef]
29. Tuduce Traistaru, A.A.; Timar, M.C.; Câmpean, M. Studies upon penetration of Paraloid B72 into poplar wood by cold immersion treatments. *Bull. Transilv. Univ. Brasov* **2011**, *4*, 1–8.
30. Genco, G.; Pelosi, C.; Santamaria, U.; Lo Monaco, A.; Picchio, R. Study of colour change due to accelerated sunlight exposure in consolidated wood samples. *Wood Res.* **2011**, *56*, 511–524.
31. Hoffmann, P. Methods of application of polyethylene glycol. In *Conservation of Archaeological Ships and Boats. Personal Experiences*; Archetype Publications: London, UK, 2013; pp. 43–80.
32. Hoffmann, P.; Wittkoepper, M. The Kauramin Method. In *Conservation of Archaeological Ships and Boats. Personal Experiences*; Archetype Publications: London, UK, 2013; pp. 97–104.
33. Bonifazi, G.; Serranti, S.; Capobianco, G.; Agresti, G.; Calienno, L.; Picchio, R.; Lo Monaco, A.; Santamaria, U.; Pelosi, C. Study of consolidating materials applied on wood by hyperspectral imaging. Proceeding of Advanced Environmental, Chemical, and Biological Sensing Technologies XIII, Baltimore, MD, USA, 17–21 April 2016; Volume 9862. [CrossRef]
34. Alfieri, P.V.; Lofeudo, R.; Canosa, G. Impregnant formulation to the preservation, protection and consolidation of wood heritage assets. *Int. J. Conservat. Sci.* **2018**, *9*, 629–640.
35. Fejfer, M.; Pietrzak, I.; Zborowska, M. Dimensional stabilization of oak and pine waterlogged wood with keratin aqueous solutions. Proceeding of the CONDITION 2015 Conservation and Digitalization, Gdansk, Poland, 19–22 May 2015; Print Group: Gdansk, Poland, 2015; pp. 95–98.
36. Christensen, M.; Larnøy, E.; Kutzke, H.; Hansen, F.K. Treatment of waterlogged archaeological wood using chitosan- and modified chitosan solutions. Part 1: Chemical compatibility and microstructure. *J. Am. Inst. Conserv.* **2015**, *54*, 3–13. [CrossRef]
37. Fulcher, K. An investigation of the use of cellulose-based materials to gap-fill wooden objects. *Stud. Conserv.* **2017**, *62*, 210–222. [CrossRef]
38. Cataldi, A.; Corcione, C.E.; Frigione, M.; Pegorettia, A. Photocurable resin/microcrystalline cellulose composites for wood protection: Physical-mechanical characterization. *Prog. Org. Coat.* **2016**, *99*, 230–239. [CrossRef]

39. Moise, V.; Stanculescu, I.; Vasilca, S.; Cutrubinis, M.; Pincu, E.; Oancea, P.; Raducan, A.; Meltzer, V. Consolidation of very degraded cultural heritage wood artefacts using radiation curing of polyester resins. *Rad. Phys. Chem.* **2019**, *156*, 314–319. [CrossRef]
40. Ndiaye, D.; Tidjani, A. Physical changes associated with gamma doses on wood/ polypropylene composites. *IOP Conf. Ser.: Mater. Sci. Eng.* **2014**, *62*, 012025. [CrossRef]
41. Vlasov, D.Y.; Panova, E.G.; Zelenskaya, M.S.; Vlasov, A.D.; Sazanova, K.V.; Rodina, O.A.; Pavlova, O.A. Biofilms on granite Rapakivi in natural outcrops and urban environment: Biodiversity, metabolism and interaction with substrate. In *Processes and Phenomena on the Boundary between Biogenic and Abiogenic Nature. Lecture Notes in Earth System Sciences*; Frank-Kamenetskaya, O., Vlasov, D., Panova, E., Lessovaia, S., Eds.; Springer: Cham, Switzerland, 2019; pp. 535–559.
42. Rapp, G.R. Building, Monumental, and Statuary Materials. In *Archaeomineralogy. Natural Science in Archaeology*; Rapp, G.R., Ed.; Springer: Berlin/Heidelberg, Germany, 2002; pp. 243–273.
43. Siedel, H. The city of Dresden in the mirror of its building stones: Utilization of natural stone at façades in the course of time. In *Materials, Technologies and Practice in Historic Heritage Structures*; Dan, M.B., Přikryl, R., Török, Á., Eds.; Springer: Dordrecht, Germany, 2010; pp. 137–156.
44. Chen, A.; Ng, Y.; Zhang, E.; Tian, M. *Dictionary of Geotourism*; Springer Nature: Singapore, 2020; pp. 468–516.
45. Koralay, T.; Çelik, S.B. Minero-petrographical, physical, and mechanical properties of moderately welded ignimbrite as a traditional building stone from Uşak Region (SW Turkey). *Arab. J. Geosci.* **2019**, *12*, 732. [CrossRef]
46. Alves, R.; Faria, P.; Simão, J. Experimental characterization of a Madeira Island basalt traditionally applied in a regional decorative mortar. *J. Build. Eng.* **2017**, *13*, 326–335. [CrossRef]
47. Tang, L.; Santosh, M. Neoarchean granite-greenstone belts and related ore mineralization in the North China Craton: An overview. *Geosci. Front.* **2018**, *9*, 751–768. [CrossRef]
48. Dipasquale, L.; Rovero, L.; Fratini, F. Ancient stone masonry constructions. In *Nonconventional and Vernacular Construction Materials. Characterisation, Properties and Applications*; Harries, K.A., Sharma, B., Eds.; Woodhead Publishing: Duxford, UK, 2016; pp. 301–332.
49. Hyslop, E.K.; Albornoz-Parra, L. Developing a future repairs strategy for a sandstone city: A petrographic investigation of building stone in Glasgow, Scotland. *Mater. Character.* **2009**, *60*, 636–643. [CrossRef]
50. Stéphan, E.; Cantin, R.; Caucheteux, A.; Tasca-Guernouti, S.; Michel, P. Experimental assessment of thermal inertia in insulated and non-insulated old limestone buildings. *Build. Environ.* **2014**, *80*, 241–248. [CrossRef]
51. Freire, M.T.; Silva, A.S.; do Rosário Veiga, M.; de Brito, J. Studies in ancient gypsum based plasters towards their repair: Mineralogy and microstructure. *Constr. Build. Mater.* **2019**, *196*, 512–529. [CrossRef]
52. Taelman, D.; Delpino, C.; Antonelli, F. Marble decoration of the Roman theatre of Urvinum Mataurense (Urbino, Marche region, Italy): An archaeological and archaeometric multi-method provenance study. *J. Cult. Herit.* **2019**, *39*, 238–250. [CrossRef]
53. Kakakhel, M.A.; Wu, F.; Gu, J.D.; Feng, H.; Shah, K.; Wang, W. Controlling biodeterioration of cultural heritage objects with biocides: A review. *Int. Biodeter. Biodegr.* **2019**, *143*, 104271. [CrossRef]
54. Rolón, G.; Cilla, G. Adobe wall biodeterioration by the *Centris muralis* Burmeister bee (Insecta: Hymenoptera: Apidae) in a valuable colonial site, the Capayán ruins (La Rioja, Argentina). *Int. Biodeter. Biodegr.* **2012**, *66*, 33–38. [CrossRef]
55. Straulino, L.; Sedov, S.; Michelet, D.; Balanzario, S. Weathering of carbonate materials in ancient Maya constructions (Río Bec and Dzibanché): Limestone and stucco deterioration patterns. *Quatern. Int.* **2013**, *315*, 87–100. [CrossRef]
56. Jurado, V.; Miller, A.Z.; Cuezva, S.; Fernandez-Cortes, A.; Benavente, D.; Rogerio-Candelera, M.A.; Reyes, J.; Cañaveras, J.C.; Sanchez-Moral, S.; Saiz-Jimenez, C. Recolonization of mortars by endolithic organisms on the walls of San Roque church in Campeche (Mexico): A case of tertiary bioreceptivity. *Constr. Build. Mater.* **2014**, *53*, 348–359. [CrossRef]
57. Morillas, H.; Maguregui, M.; Marcaida, I.; Trebolazabala, J.; Salcedo, I.; Madariaga, J.M. Characterization of the main colonizer and biogenic pigments present in the red biofilm from La Galea Fortress sandstone by means of microscopic observations and Raman imaging. *Microchem. J.* **2015**, *121*, 48–55. [CrossRef]
58. Mihajlovski, A.; Gabarre, A.; Seyer, D.; Bousta, F.; Di Martino, P. Bacterial diversity on rock surface of the ruined part of a French historic monument: The Chaalis abbey. *Int. Biodeter. Biodegr.* **2017**, *120*, 161–169. [CrossRef]

59. Guerra, F.L.; Lopes, W.; Cazarolli, J.C.; Lobato, M.; Masuero, A.B.; Dal Molin, D.C.C.; Bento, F.M.; Schrank, A.; Vainstein, M.H. Biodeterioration of mortar coating in historical buildings: Microclimatic characterization, material, and fungal community. *Build. Environ.* **2019**, *155*, 195–209. [CrossRef]
60. Tonon, C.; Favero-Longo, S.E.; Matteucci, E.; Piervittori, R.; Croveri, P.; Appolonia, L.; Meirano, V.; Serino, M.; Elia, D. Microenvironmental features drive the distribution of lichens in the House of the Ancient Hunt, Pompeii, Italy. *Int. Biodeter. Biodegr.* **2019**, *136*, 71–81. [CrossRef]
61. Favero-Longo, S.E.; Benesperi, R.; Bertuzzi, S.; Bianchi, E.; Buffa, G.; Giordani, P.; Loppi, S.; Malaspina, P.; Matteucci, E.; Paoli, L.; et al. Species- and site-specific efficacy of commercial biocides and application solvents against lichens. *Int. Biodeter. Biodegr.* **2017**, *123*, 127–137. [CrossRef]
62. Fierascu, I.; Ion, R.M.; Radu, M.; Dima, S.O.; Bunghez, I.R.; Avramescu, S.M.; Fierascu, R.C. Comparative study of antifungal effect of natural extracts and essential oils of *Ocimum basilicum* on selected artefacts. *Rev. Roum. Chim.* **2014**, *59*, 207–211.
63. Rotolo, V.; Barresi, G.; Di Carlo, E.; Giordano, A.; Lombardo, G.; Crimi, E.; Costa, E.; Bruno, M.; Palla, F. Plant extracts as green potential strategies to control the biodeterioration of cultural heritage. *Int. J. Conserv. Sci.* **2016**, *7*, 839–846.
64. Rosado, T.; Santos, R.; Silva, M.; Galvão, A.; Mirao, J.; Candeias, A.; Caldeira, A.T. Mitigation approach to avoid fungal colonisation of porous limestone. *Int. J. Conserv. Sci.* **2019**, *10*, 3–14.
65. Careddu, N.; Marras, G. The effects of solar UV radiation on the gloss values of polished stone surfaces. *Constr. Build. Mater.* **2013**, *49*, 828–834. [CrossRef]
66. Fidanza, M.R.; Caneva, G. Natural biocides for the conservation of stone cultural heritage: A review. *J. Cult. Herit.* **2019**, *38*, 271–286. [CrossRef]
67. Fierascu, I.; Fierascu, I.C.; Dinu-Pirvu, C.E.; Fierascu, R.C.; Anuta, V.; Velescu, B.S.; Jinga, M.; Jinga, V. A short overview of recent developments on antimicrobial coatings based on phytosynthesized metal nanoparticles. *Coatings* **2019**, *9*, 787. [CrossRef]
68. Ruffolo, S.A.; Ricca, M.; Macchia, A.; La Russa, M.F. Antifouling coatings for underwater archaeological stone materials. *Prog. Org. Coat.* **2017**, *104*, 64–71. [CrossRef]
69. Ruffolo, S.A.; De Leo, F.; Ricca, M.; Arcudi, A.; Silvestri, C.; Bruno, L.; Urzì, C.; La Russa, M.F. Medium-term in situ experiment by using organic biocides and titanium dioxide for the mitigation of microbial colonization on stone surfaces. *Int. Biodeter. Biodegr.* **2017**, *123*, 17–26. [CrossRef]
70. Noeiaghaei, T.; Mukherjee, A.; Dhami, N.; Chae, S.R. Biogenic deterioration of concrete and its mitigation technologies. *Constr. Build. Mater.* **2017**, *149*, 575–586. [CrossRef]
71. Noeiaghaei, T.; Dhami, N.; Mukherjee, A. Nanoparticles surface treatment on cemented materials for inhibition of bacterial growth. *Constr. Build. Mater.* **2017**, *150*, 880–891. [CrossRef]
72. Goffredo, G.B.; Accoroni, S.; Totti, C. Nanotreatments to inhibit microalgal fouling on building stone surfaces. In *Nanotechnology in Eco-efficient Construction*, 2nd ed.; Pacheco-Torgal, F., Diamanti, M.V., Nazari, A., Granqvist, C.G., Pruna, A., Amirkhanian, S., Eds.; Woodhead Publishing: Duxford, UK, 2019; pp. 619–647.
73. Becerra, J.; Mateo, M.; Ortiz, P.; Nicolás, G.; Zaderenko, A.P. Evaluation of the applicability of nano-biocide treatments on limestones used in cultural heritage. *J. Cult. Herit.* **2019**, *38*, 126–135. [CrossRef]
74. Carrillo-González, R.; Martínez-Gómez, M.A.; González-Chávez, M.D.C.A.; Mendoza Hernández, J.C. Inhibition of microorganisms involved in deterioration of an archaeological site by silver nanoparticles produced by a green synthesis method. *Sci. Total Environ.* **2016**, *565*, 872–881. [CrossRef]
75. Fierascu, I.; Georgiev, M.I.; Ortan, A.; Fierascu, R.C.; Avramescu, S.M.; Ionescu, D.; Sutan, A.; Brinzan, A.; Ditu, L.M. Phyto-mediated metallic nanoarchitectures via *Melissa officinalis* L.: Synthesis, characterization and biological properties. *Sci. Rep.* **2017**, *7*, 12428. [CrossRef] [PubMed]
76. Sutan, N.A.; Manolescu, D.S.; Fierascu, I.; Neblea, A.M.; Sutan, C.; Ducu, C.; Soare, L.C.; Negrea, D.; Avramescu, S.M.; Fierascu, R.C. Phytosynthesis of gold and silver nanoparticles enhance in vitro antioxidant and mitostimulatory activity of *Aconitum toxicum* Reichenb Rhizomes alcoholic extracts. *Mater. Sci. Eng. C* **2018**, *93*, 746–758. [CrossRef] [PubMed]
77. Dresler, C.; Saladino, M.L.; Demirbag, C.; Caponetti, E.; Chillura Martino, D.F.; Alduina, R. Development of controlled release systems of biocides for the conservation of cultural heritage. *Int. Biodeter. Biodegr.* **2017**, *125*, 150–156. [CrossRef]
78. Fierascu, I.; Fierascu, I.C.; Brazdis, R.I.; Baroi, A.M.; Fistos, T.; Fierascu, R.C. Phytosynthesized metallic nanoparticles—Between nanomedicine and toxicology. A brief review of 2019's findings. *Materials* **2020**, *13*,

574. [CrossRef]

79. Valentini, F.; Diamanti, A.; Carbone, M.; Bauer, E.M.; Palleschi, G. New cleaning strategies based on carbon nanomaterials applied to the deteriorated marble surfaces: A comparative study with enzyme-based treatments. *Appl. Surf. Sci.* **2012**, *258*, 5965–5980. [CrossRef]
80. Gómez-Ortíz, N.; De la Rosa-García, S.; González-Gómez, W.; Soria-Castro, M.; Quintana, P.; Oskam, G.; Ortega-Morales, B. Antifungal coatings based on $Ca(OH)_2$ mixed with ZnO/TiO_2 nanomaterials for protection of limestone monuments. *ACS Appl Mater Interfaces* **2013**, *5*, 1556–1565. [CrossRef]
81. MacMullen, J.; Zhang, Z.; Dhakal, H.N.; Radulovic, J.; Karabela, A.; Tozzi, G.; Hannant, S.; Alshehri, M.A.; Buhé, V.; Herodotou, C.; et al. Silver nanoparticulate enhanced aqueous silane/siloxane exterior facade emulsions and their efficacy against algae and cyanobacteria biofouling. *Int. Biodeter. Biodegr.* **2014**, *93*, 54–62. [CrossRef]
82. Graziani, L.; Quagliarinin, E.; D'Orazio, M. TiO_2-treated different fired brick surfaces for biofouling prevention: Experimental and modelling results. *Ceramic. Int.* **2016**, *42*, 4002–4010. [CrossRef]
83. Goffredo, G.B.; Accoroni, S.; Totti, C.; Romagnoli, T.; Valentini, L.; Munafò, P. Titanium dioxide based nanotreatments to inhibit microalgal fouling on building stone surfaces. *Build. Environ.* **2017**, *112*, 209–222. [CrossRef]
84. Ruggiero, L.; Crociani, L.; Zendri, E.; El Habra, N.; Guerriero, P. Incorporation of the zosteric sodium salt in silica nanocapsules: Synthesis and characterization of new fillers for antifouling coatings. *Appl. Surf. Sci.* **2018**, *439*, 705–711. [CrossRef]
85. Ruggiero, L.; Di Bartolomeo, E.; Gasperi, G.; Luisetto, I.; Talone, A.; Zurlo, F.; Peddis, D.; Ricci, M.A.; Sodo, A. Silica nanosystems for active antifouling protection: Nanocapsules and mesoporous nanoparticles in controlled release applications. *J. Alloy. Comp.* **2019**, *798*, 144–148. [CrossRef]
86. Rives, V.; García-Talegón, J. Decay and conservation of building stones on cultural heritage monuments. *Mater. Sci. Forum* **2006**, *514–516*, 1689–1694. [CrossRef]
87. Verges-Belmin, V. *Illustrated Glossary on Stone Deterioration Patterns, Glossaire Illustré sur les formes d'altération de la pierre*, English-French ed.; ICOMOS International Scientific Committee for Stone: Paris, France, 2008.
88. Doehne, E.; Price, C.A. *Stone Conservation: An Overview of Current Research*, 2nd ed.; Getty Conservation Institute: Los Angeles, CA, USA, 2010.
89. Grimmer, A.E. *Keeping It Clean: Removing Exterior Dirt, Paint, Stains and Graffiti from Historic Masonry Buildings*; Forgotten Books: London, UK, 2018; Available online: https://www.forgottenbooks.com/en/books/KeepingItClean_10896659 (accessed on 4 December 2019).
90. Rampazzi, L.; Andreotti, A.; Bressan, M.; Colombini, M.P.; Corti, C.; Cuzman, O.; d'Alessandro, N.; Liberatore, L.; Palombi, L.; Raimondi, V.; et al. An interdisciplinary approach to a knowledge-based restoration: The dark alteration on Matera Cathedral (Italy). *Appl. Surf. Sci.* **2018**, *458*, 529–539. [CrossRef]
91. Mazzinghi, P.; Margheri, F. A short pulse, free running, Nd: YAG laser for the cleaning of stone cultural heritage. *Opt. Laser. Eng.* **2003**, *39*, 191–202. [CrossRef]
92. Pozo-Antonio, J.S.; Rivas, T.; López, A.J.; Fiorucci, M.P.; Ramil, A. Effectiveness of granite cleaning procedures in cultural heritage: A review. *Sci. Total Environ.* **2016**, *571*, 1017–1028. [CrossRef] [PubMed]
93. La Russa, M.F.; Rovella, N.; de Buergo, M.A.; Belfiore, C.M.; Pezzino, A.; Crisci, G.M.; Ruffolo, S.A. Nano-TiO_2 coatings for cultural heritage protection: The role of the binder on hydrophobic and self-cleaning efficacy. *Progr. Org. Coat.* **2016**, *91*, 1–8. [CrossRef]
94. Quagliarini, E.; Bondioli, F.; Goffredo, G.B.; Licciulli, A.; Munafò, P. Self-cleaning materials on Architectural Heritage: Compatibility of photo-induced hydrophilicity of TiO_2 coatings on stone surfaces. *J. Cult. Herit.* **2013**, *14*, 1–7. [CrossRef]
95. Giacomucci, L.; Toja, F.; Sanmartín, P.; Toniolo, L.; Prieto, B.; Villa, F.; Cappitelli, F. Degradation of nitrocellulose-based paint by Desulfovibrio desulfuricans ATCC 13541. *Biodegradation* **2012**, *23*, 705–716. [CrossRef]
96. Gomes, V.; Dionísio, A.; Pozo-Antonio, J.S. Conservation strategies against graffiti vandalism on Cultural Heritage stones: Protective coatings and cleaning methods. *Progr. Org. Coat.* **2017**, *113*, 90–109. [CrossRef]
97. Ottosen, L.M.; Christensen, I.V. Electrokinetic desalination of sandstones for NaCl removal—Test of different clay poultices at the electrodes. *Electrochim. Acta* **2012**, *86*, 192–202. [CrossRef]
98. Feijoo, J.; Ottosen, L.M.; Pozo-Antonio, J.S. Influence of the properties of granite and sandstone in the desalination process by electrokinetic technique. *Electrochim. Acta* **2015** *181*, 280–287. [CrossRef]

99. Matyščák, O.; Ottosen, L.M.; Rörig-Dalgaard, I. Desalination of salt damaged Obernkirchen sandstone by an applied DC field. *Constr. Build. Mater.* **2014**, *71*, 561–569. [CrossRef]
100. Graziani, G.; Sassoni, E.; Scherer, G.W.; Franzoni, E. Penetration depth and redistribution of an aqueous ammonium phosphate solution used for porous limestone consolidation by brushing and immersion. *Constr. Build. Mater.* **2017**, *148*, 571–578. [CrossRef]
101. Rivas, T.; Alvarez, E.; Mosquera, M.J.; Alejano, L.; Taboada, J. Crystallization modifiers applied in granite desalination: The role of the stone pore structure. *Constr. Build. Mater.* **2010**, *24*, 766–776. [CrossRef]
102. Granneman, S.J.C.; Lubelli, B.; van Hees, R.P.J. Mitigating salt damage in building materials by the use of crystallization modifiers—A review and outlook. *J. Cult. Herit.* **2019**, *40*, 183–194. [CrossRef]
103. Possenti, E.; Colombo, C.; Conti, C.; Marinoni, N.; Merlini, M.; Negrotti, R.; Realini, M.; Gatta, G.D. Consolidation of building materials with a phosphate-based treatment: Effects on the microstructure and on the 3D pore network. *Mater. Charact.* **2019**, *154*, 315–324. [CrossRef]
104. Zornoza-Indart, A.; Lopez-Arce, P.; Leal, N.; Simão, J.; Zoghlami, K. Consolidation of a Tunisian bioclastic calcarenite: From conventional ethyl silicate products to nanostructured and nanoparticle based consolidants. *Constr. Build. Mater.* **2016**, *150*, 188–202. [CrossRef]
105. Zornoza-Indart, A.; Lopez-Arce, P. Silica nanoparticles (SiO_2): Influence of relative humidity in stone consolidation. *J. Cult. Herit.* **2016**, *18*, 258–270. [CrossRef]
106. Xu, F.; Zeng, W.; Li, D. Recent advance in alkoxysilane-based consolidants for stone. *Progr. Org. Coat.* **2019**, *127*, 45–54. [CrossRef]
107. Liu, Y.; Liu, J. Synthesis of TEOS/PDMS-OH/CTAB composite coating material as a new stone consolidant formulation. *Constr. Build. Mater.* **2016**, *122*, 90–94. [CrossRef]
108. Kapridaki, C.; Verganelaki, A.; Dimitriadou, P.; Maravelaki-Kalaitzaki, P. Conservation of monuments by a three-layered compatible treatment of TEOS-nano-calcium oxalate consolidant and TEOS-PDMS-TiO2 hydrophobic/photoactive hybrid nanomaterials. *Materials* **2018**, *11*, 684. [CrossRef] [PubMed]
109. Sassoni, E. Hydroxyapatite and other calcium phosphates for the conservation of cultural heritage: A review. *Materials* **2018**, *11*, 557. [CrossRef] [PubMed]
110. Pesce, C.; Moretto, L.M.; Orsega, E.F.; Pesce, G.L.; Corradi, M.; Weber, J. Effectiveness and compatibility of a novel sustainable method for stone consolidation based on di-ammonium phosphate and calcium-based nanomaterials. *Materials* **2019**, *12*, 3025. [CrossRef] [PubMed]
111. Fierascu, I.; Fierascu, R.C.; Ion, R.M.; Radovici, C. Synthesized apatitic materials for artefacts protection against biodeterioration. *Rom. J. Mater.* **2014**, *44*, 292–297.
112. Ion, R.M.; Turcanu-Caruţiu, D.; Fierascu, R.C.; Fierascu, I.; Bunghez, I.R.; Ion, M.L.; Teodorescu, S.; Vasilievici, G.; Raditoiu, V. Caoxite-hydroxyapatite composition as consolidating material for the chalk stone from Basarabi-Murfatlar churches ensemble. *Appl. Surf. Sci.* **2015**, *358*, 612–618. [CrossRef]
113. Ion, R.M.; Ion, M.L.; Radu, A.; Şuică-Bunghez, R.I.; Fierascu, R.C.; Fierascu, I.; Teodorescu, S. Nanomaterials-based mortars for building façades preservation. *Rom. J. Mater.* **2016**, *46*, 412–418.
114. Fierascu, I.; Fierascu, R.C.; Somoghi, R.; Ion, R.M.; Moanta, A.; Avramescu, S.M.; Damian, C.M.; Ditu, L.M. Tuned apatitic materials: Synthesis, characterization and potential antimicrobial applications. *Appl. Surf. Sci.* **2018**, *438*, 127–135. [CrossRef]
115. Santarelli, M.L.; Sbardella, F.; Zuena, M.; Tirillò, J.; Sarasini, F. Basalt fiber reinforced natural hydraulic lime mortars: A potential bio-based material for restoration. *Mater. Des.* **2014**, *63*, 398–406. [CrossRef]
116. Moropoulou, A.; Bakolas, A.; Moundoulas, P.; Aggelakopoulou, E.; Anagnostopoulou, S. Optimization of compatible restoration mortars for the earthquake protection of Hagia Sophia. *J. Cult. Herit.* **2013**, *14S*, 147–152. [CrossRef]
117. Andrejkovičová, S.; Velosa, A.; Gameiro, A.; Ferraz, E.; Rocha, F. Palygorskite as an admixture to air lime–metakaolin mortars for restoration purposes. *Appl. Clay Sci.* **2013**, *83–84*, 368–374.
118. Ventolà, L.; Vendrell, M.; Giraldez, P. Newly-designed traditional lime mortar with a phase change material as an additive. *Constr. Build. Mater.* **2013**, *47*, 1210–1216. [CrossRef]
119. Rosato, L.; Stefanidou, M.; Milazzoc, G.; Fernandez, F.; Livreri, P.; Muratore, N.; Terranova, L.M. Study and evaluation of nano-structured cellulose fibers as additive for restoration of historical mortars and plasters. *Mater. Today Proc.* **2017**, *4*, 6954–6965. [CrossRef]
120. Gour, K.A.; Ramadoss, R.; Selvaraj, T. Revamping the traditional air lime mortar using the natural polymer—Areca nut for restoration application. *Constr. Build. Mater.* **2018**, *164*, 255–264. [CrossRef]

121. Pavlík, Z.; Pokorný, J.; Pavlíková, M.; Zemanová, L.; Záleská, M.; Vyšvaril, M.; Žižlavský, T. Mortars with crushed lava granulate for repair of damp historical buildings. *Materials* **2019**, *12*, 3557. [CrossRef] [PubMed]
122. Stefanidou, M.; Pachta, V.; Papayianni, I. Design and testing of artificial stone for the restoration of stone elements in monuments and historic buildings. *Constr. Build. Mater.* **2015**, *93*, 957–965. [CrossRef]
123. Papayianni, I.; Pachta, V. Earth block houses of historic centers. A sustainable upgrading with compatible repair materials. *Proc. Environ. Sci.* **2017**, *38*, 274–282. [CrossRef]
124. Camerini, R.; Chelazzi, D.; Giorgi, R.; Baglioni, P. Hybrid nano-composites for the consolidation of earthen masonry. *J. Colloid Interf. Sci.* **2019**, *539*, 504–515. [CrossRef]
125. Zhang, H.; Liu, Q.; Liu, T.; Zhang, B. The preservation damage of hydrophobic polymer coating materials in conservation of stone relics. *Progr. Org. Coat.* **2013**, *76*, 1127–1134. [CrossRef]
126. Cao, Y.; Salvini, A.; Camaiti, M. Oligoamide grafted with perfluoropolyether blocks: A potential protective coating for stone materials. *Progr. Org. Coat.* **2017**, *111*, 164–174. [CrossRef]
127. Corcione, C.E.; Striani, R.; Frigione, M. UV-cured siloxane-modified methacrylic system containing hydroxyapatite as potential protective coating for carbonate stones. *Progr. Org. Coat.* **2013**, *76*, 1236–1242. [CrossRef]
128. Cappelletti, G.; Fermo, P.; Camiloni, M. Smart hybrid coatings for natural stones conservation. *Progr. Org. Coat.* **2015**, *78*, 511–516. [CrossRef]
129. Aslanidou, D.; Karapanagiotis, I.; Lampakis, D. Waterborne superhydrophobic and superoleophobic coatings for the protection of marble and sandstone. *Materials* **2018**, *11*, 585. [CrossRef] [PubMed]
130. Falchi, L.; Müller, U.; Fontana, P.; Izzo, F.C.; Zendri, E. Influence and effectiveness of water-repellent admixtures on pozzolana–lime mortars for restoration application. *Constr. Build. Mater.* **2013**, *49*, 272–280. [CrossRef]
131. Camuffo, D. *Microclimate for Cultural Heritage. Measurement, Risk Assessment, Conservation, Restoration, and Maintenance of Indoor and Outdoor Monuments*, 3rd ed.; Elsevier: Amsterdam, The Netherlands, 2019; pp. 125–152.
132. Quagliarini, E.; Bondioli, F.; Goffredo, G.B.; Cordoni, C.; Munafò, P. Self-cleaning and de-polluting stone surfaces: TiO_2 nanoparticles for limestone. *Constr. Build. Mater.* **2012**, *37*, 51–57. [CrossRef]
133. D'Orazio, L.; Grippo, A. A water dispersed Titanium dioxide/poly(carbonate urethane) nanocomposite for protecting cultural heritage: Preparation and properties. *Progr. Org. Coat.* **2015**, *79*, 1–7. [CrossRef]
134. Gherardi, F.; Colombo, A.; D'Arienzo, M.; Di Credico, B.; Goidanich, S.; Morazzoni, F.; Simonutti, R.; Toniolo, L. Efficient self-cleaning treatments for built heritage based on highly photo-active and well-dispersible TiO_2 nanocrystals. *Microchem. J.* **2016**, *126*, 54–62. [CrossRef]
135. Pozo-Antonio, J.S.; Dionísio, A. Physical-mechanical properties of mortars with addition of TiO_2 nanoparticles. *Constr. Build. Mater.* **2017**, *148*, 261–272. [CrossRef]
136. Crupi, V.; Fazio, B.; Gessini, A.; Kis, Z.; La Russa, M.F.; Majolino, D.; Masciovecchio, C.; Ricca, M.; Rossi, B.; Ruffolo, S.A.; et al. TiO_2–SiO_2–PDMS nanocomposite coating with self-cleaning effect for stone material: Finding the optimal amount of TiO_2. *Constr. Build. Mater.* **2018**, *166*, 464–471. [CrossRef]
137. Carmona-Quiroga, P.M.; Martínez-Ramírez, S.; Viles, H.A. Efficiency and durability of a self-cleaning coating on concrete and stones under both natural and artificial ageing trials. *Appl. Surf. Sci.* **2018**, *433*, 312–320. [CrossRef]
138. Remzova, M.; Sasek, P.; Frankeova, D.; Slizkova, Z.; Rathousky, J. Effect of modified ethylsilicate consolidants on the mechanical properties of sandstone. *Constr. Build. Mater.* **2016**, *112*, 674–681. [CrossRef]
139. Zornoza-Indart, A.; Lopez-Arce, P.; López-Polín, L. Durability of traditional and new nanoparticle based consolidating products for the treatment of archaeological stone tools: Chert artifacts from Atapuerca sites (Burgos, Spain). *J. Cult. Herit.* **2017**, *24*, 9–21. [CrossRef]
140. Tzavellos, S.; Pesce, G.L.; Wu, Y.; Henry, A.; Robson, S.; Ball, R.J. Effectiveness of nanolime as a Stone Consolidant: A 4-Year Study of Six Common UK Limestones. *Materials* **2019**, *12*, 2673. [CrossRef] [PubMed]
141. Ban, M.; Mascha, E.; Weber, J.; Rohatsch, A.; Rodrigues, J.D. Efficiency and compatibility of selected alkoxysilanes on porous carbonate and silicate stones. *Materials* **2019**, *12*, 156. [CrossRef] [PubMed]
142. Striani, R.; Corcione, C.E.; Dell'Anna Muia, G.; Frigione, M. Durability of a sunlight-curable organic–inorganic hybrid protective coating for porous stones in natural and artificial weathering conditions. *Progr. Org. Coat.* **2016**, *101*, 1–14. [CrossRef]
143. Stefanidou, M.; Karozou, A. Testing the effectiveness of protective coatings on traditional bricks. *Constr. Build. Mater.* **2016**, *111*, 482–487. [CrossRef]

144. Corcione, C.E.; De Simone, N.; Santarelli, M.L.; Frigione, M. Protective properties and durability characteristics of experimental and commercial organic coatings for the preservation of porous stone. *Progr. Org. Coat.* **2017**, *103*, 193–203. [CrossRef]
145. Andreotti, S.; Franzoni, E.; Esposti, M.D.; Fabbri, P. Poly(hydroxyalkanoate)s-based hydrophobic coatings for the protection of stone in cultural heritage. *Materials* **2018**, *11*, 165. [CrossRef] [PubMed]
146. Elhaddad, F.; Carrascosa, L.A.M.; Mosquera, M.J. Long-term effectiveness, under a mountain environment, of a novel conservation nanomaterial applied on limestone from a roman archaeological site. *Materials* **2018**, *11*, 694. [CrossRef] [PubMed]
147. Feilden, B.M.; Jokilehto, J. *Management Guidelines for World Cultural Heritage Sites*, 2nd ed.; ICCROM—International Centre for the Study of the Preservation and Restoration of Cultural Property: Rome, Italy, 1998; pp. 59–76.
148. Turk, J.; Pranjić, A.M.; Hursthouse, A.; Turner, R.; Hughes, J.J. Decision support criteria and the development of a decision support tool for the selection of conservation materials for the built cultural heritage. *J. Cult. Herit.* **2019**, *37*, 44–53. [CrossRef]
149. Viñas, S.M. Contemporary theory of conservation. *Stud. Conserv.* **2002**, *47*, 25–34. [CrossRef]
150. Rivas, T.; Pozo, S.; Fiorucci, M.P.; López, A.J.; Ramil, A. Nd:YVO4 laser removal of graffiti from granite. Influence of paint and rock properties on cleaning efficacy. *Appl. Surf. Sci.* **2012**, *263*, 563–572. [CrossRef]
151. Evola, G.; Marletta, L.; Natarajan, S.; Patanè, E.M. Thermal inertia of heavyweight traditional buildings: Experimental measurements and simulated scenarios. *Energy Procedia* **2017**, 42–52. [CrossRef]
152. Bugini, R.; Corti, C.; Folli, L.; Rampazzi, L. The use of mortar to imitate white marble and other stones. *Int. J. Conserv. Sci.* **2018**, *9*, 3–12.

3

Analysis of Cylindrical Masonry Shell in St. Jacob's Church in Dolenja Trebuša, Slovenia

Mojmir Uranjek [1,*], Tadej Lorenci [2] and Matjaž Skrinar [1]

1 Chair of Structural Mechanics, Faculty of Civil Engineering, Transportation Engineering and Arhitecture University of Maribor, Maribor 2000, Slovenia; matjaz.skrinar@um.si
2 Department of Health and Safety, Abrasiv Muta, Muta 2366, Slovenia; tadej.lorenci@gmail.com
* Correspondence: mojmir.uranjek@um.si

Abstract: This paper focuses on identifying key reasons for the damage of the cylindrical masonry shell structure in St. Jacob's church in Dolenja Trebuša, Slovenia. Typical damage patterns which can be formed in shell structures and may affect the load bearing capacity are outlined. Several stress states (membrane, bending and also combined stress state) that can occur in the shell structure are described. Load cases such as the vertical displacement of the support structure, temperature loading, weight of maintenance team and also seismic loading are taken into account in order to identify the actual cause for the registered crack pattern in the shell structure. Analysis of the shell structure is performed using the SAP2000 structural software. Based on the obtained results, which highlighted key reasons for registered damage, the monitoring of cracks is recommended in the first phase, and, in continuation, the most appropriate repair and strengthening measures are proposed.

Keywords: masonry shell; cracks in shells; static analysis; strengthening

1. Introduction

Motivation for this work is the fact that masonry shell structures as a part of historical buildings such as churches, monasteries and castles are present in a relatively large number not just in Slovenia but also worldwide. Over the years, due to various reasons (change in magnitude and distribution of loading, earthquake loading, foundation settlements, etc.) these structures can be damaged. Consequently, inner forces are redistributed and in many cases the static system is changed. Nevertheless, most of these structures maintain their function also afterwards [1].

In the case of reconstruction, the load bearing capacity of the shell structure should be checked in accordance with current regulations and, if necessary, appropriate measures have to be undertaken. In this paper, computational analysis of the actual shell structure is demonstrated. By using the results of this analysis, causes for damage are identified and possible rehabilitation and consolidation measures are proposed.

2. Stresses and Typical Damage Patterns of Masonry Shells

To describe the geometry of the shell, the spatial position of the midsurface and the thickness of the shell at each point have to be known. The analysis of shell structures is usually performed using two theories: Membrane theory, which is usually valid for lager part of the shell and bending theory, which includes the effect of bending [2]. In areas where due to local disturbances (supports, point loads, changes of curvature or thickness) the ideal membrane stress state is no longer valid (Figure 1), the membrane theory should be supplemented by the bending theory.

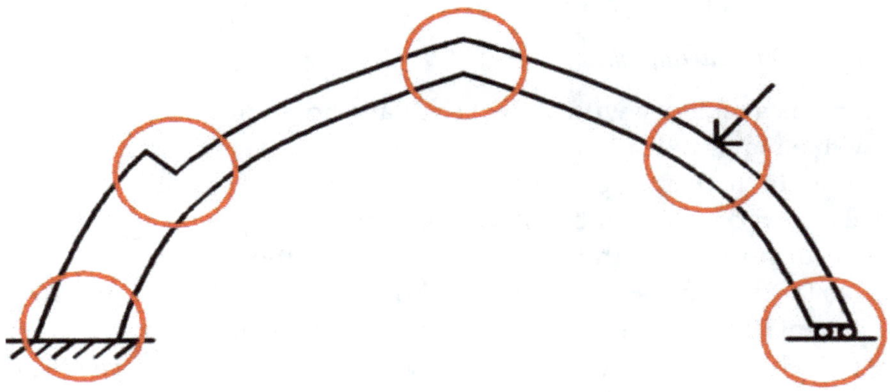

Figure 1. Critical parts of the shell where the bending stress state can be formed (adapted after [3]).

For shell structures in historical buildings in most cases the membrane theory has to be supplemented by the bending theory. As the result of the simultaneous effect of membrane forces N_x, N_y, N_{xy} and bending moments M_x, M_y, M_{xy} the stresses which occur in the shell can be written as:

$$\sigma_x = \frac{N_x}{t} - \frac{12M_x z}{t^3} \tag{1}$$

$$\sigma_y = \frac{N_y}{t} - \frac{12M_y z}{t^3} \tag{2}$$

$$\tau_{xy} = \frac{N_{xy}}{t} - \frac{12M_{xy} z}{t^3} \tag{3}$$

The first terms in the above expressions represent the membrane stress while the second terms, the bending stress. The distribution of stresses σ_x, σ_y and τ_{xy} within the shell thickness is linear. Perpendicular shear stresses τ_{xz} in τ_{yz} as a consequence of shear forces have a parabolic distribution, but their values are generally small in comparison with other stress components and can usually be neglected.

Damage of shells, often in the form of cracks, occurs due to various reasons: Foundation settlement, horizontal displacements of supporting walls, material deficiencies and degradation or local overloading as well as their combinations. Some typical crack configurations due to various reasons in the case of barrel vault are shown in Figure 2.

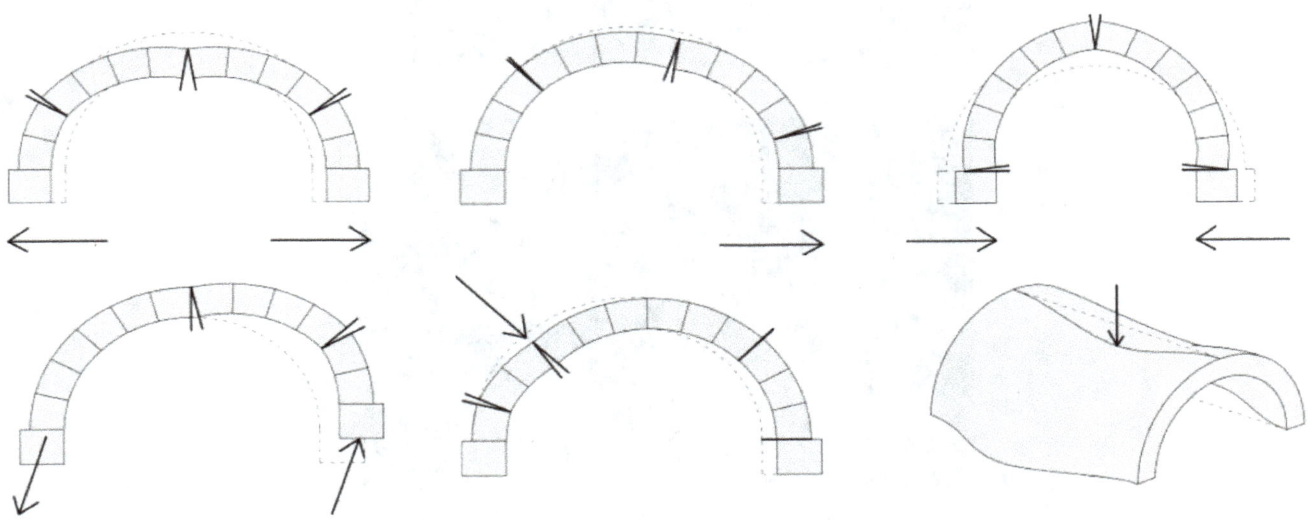

Figure 2. Some typical crack configurations for barrel vault [4].

3. Analysis of Masonry Shell in St. Jacobs Church

3.1. Characteristics of the 2004 Earthquake

Generally, Slovenia is a territory with almost regular (moderate) seismic activities. On an average day, an earthquake with a magnitude higher than 1 is normally detected. Further, Dolenja Trebuša lies in a region with some more than average seismic activities. Nevertheless, although this area is not one of the most seismically exposed parts of our country; it still lies quite close to the region, which was severely damaged by the 1976 earthquake (with the magnitude of 6.5 the strongest earthquake that has been recorded in Slovenia so far). The considered 2004 earthquake occurred on 12 July 2004 with an epicenter to the northwest of Dolenja Trebuša (areal distance was about 35 km). It occurred on the same place of the 1998 earthquake (which had the magnitude of 5.7). Both these earthquakes are listed among the three last severe earthquakes that were recorded in Slovenia. Although most of the damage resulting from the 2004 earthquake occurred in the vicinity of Bovec mainly due to local geological conditions, other areas such as Dolenja Trebuša were also affected. The 2004 earthquake caused material damage primarily on older buildings, which in this region have a very low earthquake resistance. Namely, the building stock consists mainly of old stone masonry houses with one or two storeys built largely from two-leaf masonry with two-leaf stone masonry with weak lime mortar and wooden floors [5].

3.2. Structural Characteristics and Damage of Analyzed Masonry Shell

For the analysis, a barrel vault in St. Jacob Church in Dolenja Trebuša in Slovenia was chosen. The Church was built in 1786 and retrofitted in 2013 in order to simultaneously rehabilitate and strengthen its structure after the earthquake in 2004. In spite of the renovation in 2013, cracks in the considered vault reopened. The stone masonry barrel vault is built out of limestone and leech stone in lime mortar. Geometry and materials of the vault are shown in Figure 3. The length of the vault is 10 m (above the nave) with a span of 7.75 m. The thickness of the vault is changing from 30 cm at the base and to 8 cm at the top.

(a)

(b)

Figure 3. Analysed stone masonry barrel vault from bottom (**a**) and top (**b**).

Before rehabilitation and strengthening of the church in 2013, numerous cracks in the walls and vaults were listed. In the middle area of the analysed vault there was a crack running along the entire length of the nave, which continued through the supporting arch and presbytery vault with length of 13.5 m and thickness of 1.0 mm. Other cracks were also present on the vaults, especially in the area of the transverse vaults although of smaller width and length (Figure 4).

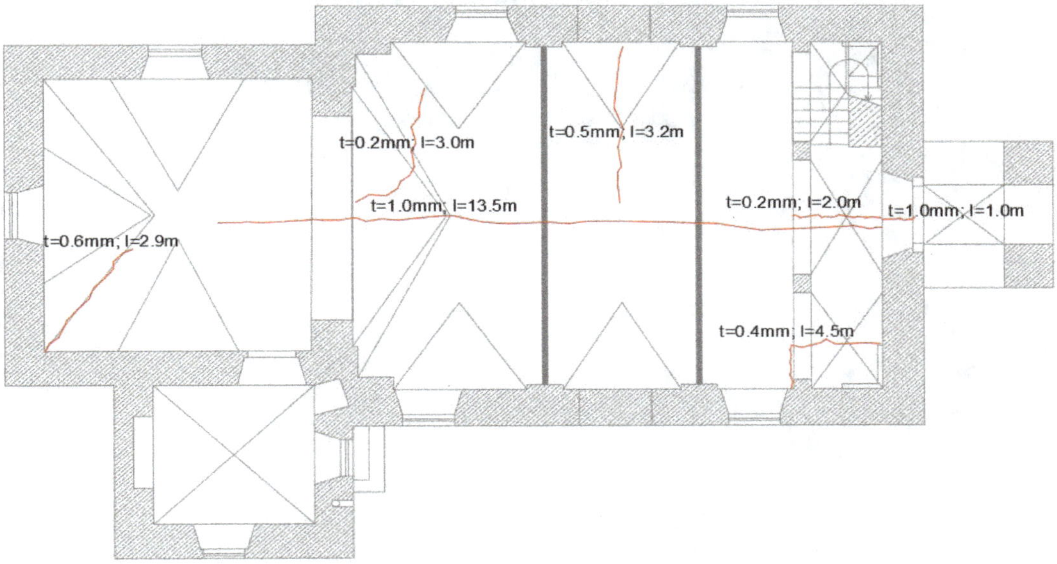

Figure 4. Registered crack pattern of analysed barrel vault before rehabilitation in 2013 [6].

The cracks although to a lesser extent, reopened in the same places after the repair works in 2013 (Figure 5).

(a) (b)

Figure 5. Reopened crack in the analysed shell (**a**) and in wall near transverse vault (**b**).

3.3. *Static Analysis of Masonry Shell*

The stone masonry barrel vault was analysed by using the SAP2000 software [7]. A combination of three and four-node shell finite elements combined with linear finite elements was used in the model. The structure was modelled with varying thickness according to actual geometry. Supports (main arch, closed arches at the side, supporting wall) were modelled using linear finite elements of the corresponding cross section. The barrel vault was modelled with 2211 shell and 436 linear finite elements interconnected in 2499 nodes. Horizontal steel ties were modelled as linear elements with circular cross section of ϕ 32–45 mm (Figure 6).

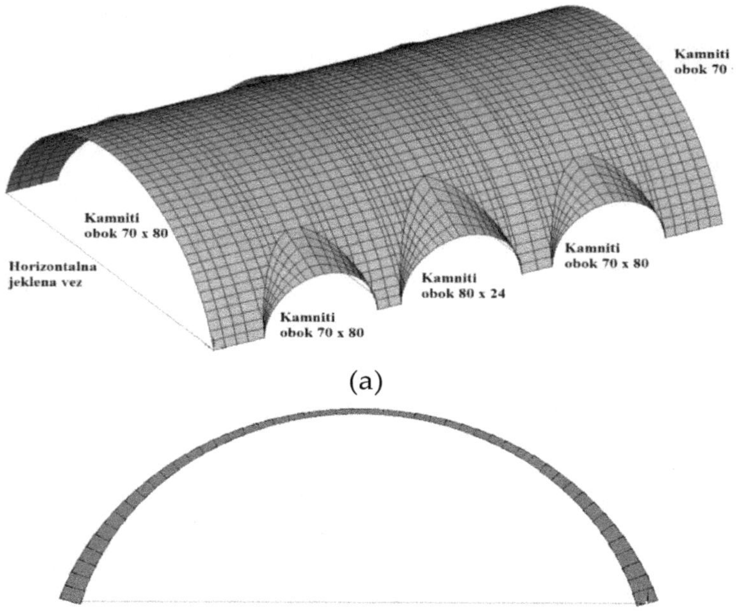

(a)

(b)

Figure 6. 3D FE model (**a**) and cross section (**b**) showing changing thickness of analysed barrel vault.

When the registered crack pattern is compared with possible crack configurations shown in Figure 2, several actions can be identified as the potential cause of such damage. The longitudinal crack at the bottom middle part of the shell may thus be caused either by simultaneous horizontal divergent displacement of supporting walls (Figure 2 top left), differential settlement of foundations (Figure 2 bottom left) or point load in the central part of the shell (Figure 2 bottom right).

Since the increase of the structures weight was eliminated as a possible cause for the registered crack pattern, other possible causes were analysed in the numerical analysis: Self weight of structure and rubble layer, differential settlement of foundations, seismic loading, maintenance loading and temperature difference (Figures 7 and 8).

The implementation of non-linear behaviour would be the most appropriate approach for the analysis of the considered shell. However, we were not involved in any material properties acquisition (all the mechanical parameters were taken from the Technical Report, [6]) for the analysed shell. Furthermore, our analyses were not part of any official study, and were thus not financially supported. Therefore, the decision was met to perform only linear elastic analyses, which nevertheless provided satisfactory results and enabled the identification of possible causes for the damage.

Five load combinations were considered altogether:

- Load case »1«: Self weight and rubble layer:

$$\sum 1.35 * G + \sum 1.35 * G_N$$

- Load case »2«: Self weight, rubble layer, vertical displacement of outer supports on one side by $\Delta_z = -0.5$ cm:

$$\sum 1.35 * G + \sum 1.35 * G_N + \sum 1.5 * \Delta_z$$

- Load case »3«: Self weight, rubble layer, maintenance team:

$$\sum 1.35 * G + \sum 1.35 * G_N + \sum 1.5 * Q_V$$

- Load case »4«: Self weight, rubble layer, horizontal displacement of all supports on one side by Δ_y = 0.5 cm:

$$\sum 1.35 * G + \sum 1.35 * G_N + \sum 1.5 * \Delta_y$$

- Load case »5«: Self weight, rubble layer, temperature difference:

$$\sum 1.35 * G + \sum 1.35 * G_N + \sum 1.5 * \Delta_T$$

The discrete results' values are presented for seven points visible in Figure 9.

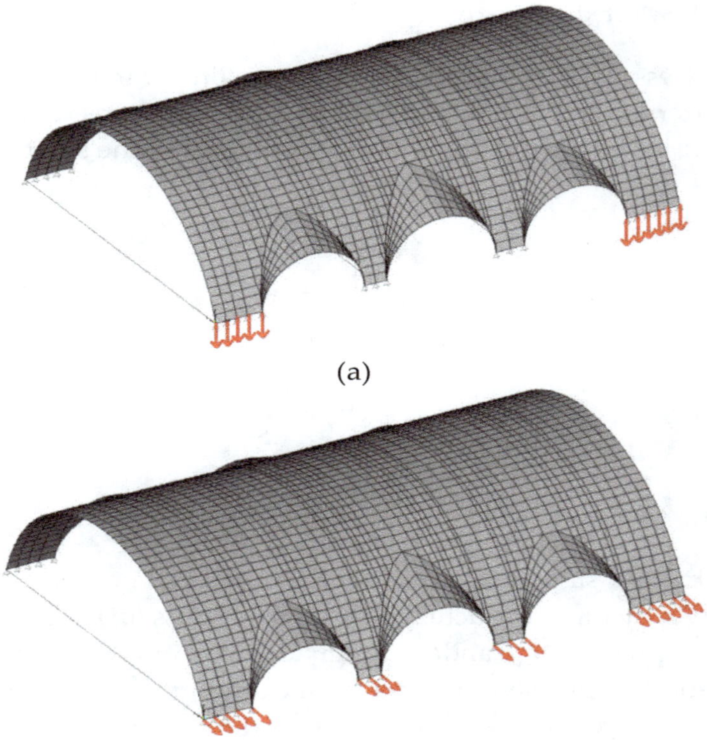

(a) (b)

Figure 7. Load case simulating vertical displacement of outer supports on one side (**a**) and horizontal displacement of all supports on one side (**b**).

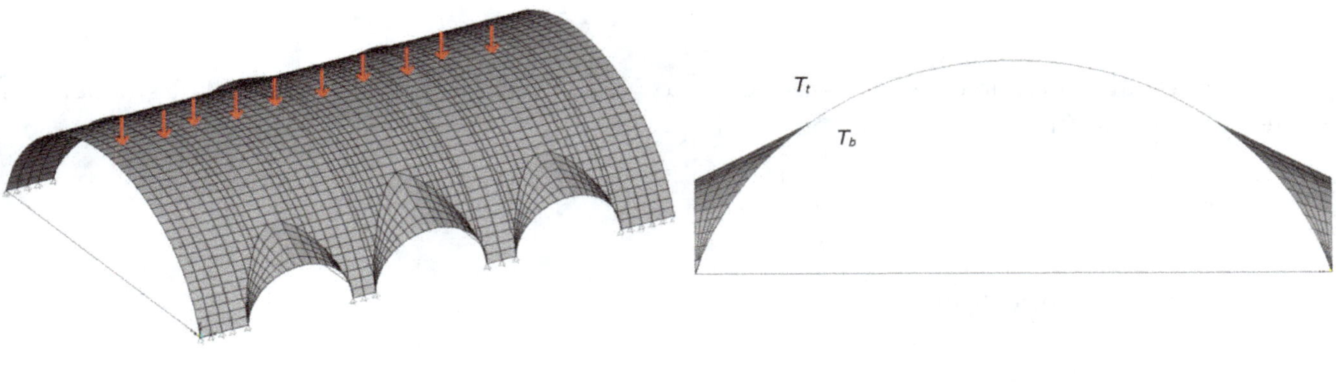

(a) (b)

Figure 8. Load case simulating maintenance team (**a**) and temperature difference (**b**).

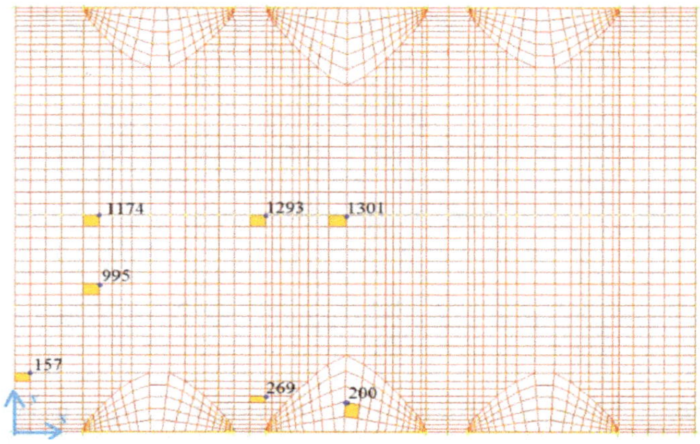

Figure 9. Selected points corresponding to stress values given in Table 1.

The distribution of stresses across the shell thickness is linear. In the evaluation of results, values obtained in the upper (external) and bottom part of the shell were considered. By taking into account the values of z = t/2 and z = −t/2 for the upper and bottom part of the shell, respectively we obtain:

$$\sigma_{11up} = \frac{F_{11}}{t} - \frac{6M_{11}}{t^2} \tag{4}$$

$$\sigma_{11bot} = \frac{F_{11}}{t} + \frac{6M_{11}}{t^2} \tag{5}$$

$$\sigma_{22up} = \frac{F_{22}}{t} - \frac{6M_{22}}{t^2} \tag{6}$$

$$\sigma_{22bot} = \frac{F_{22}}{t} + \frac{6M_{22}}{t^2} \tag{7}$$

where designations »*up*« in »*bot*« stand for stresses on the upper and bottom part of the shell. According to Eurocode 6 for masonry structures, the actual compressive stress (σ_c) and tensile stress (σ_t) should be smaller than the design compressive (f_{cd}) and/or design tensile strength (f_{td}) of the material, respectively. Values of the characteristic compressive (f_c) and tensile strength (f_t) were taken from the Technical Report, reference No. 6 and were obtained from investigations of similarly built stone masonry walls and vaults studied both in situ and in the laboratory of ZRMK. Design values are obtained by dividing the characteristic compressive (f_c) and tensile strength (f_t) of the material with the material safety factor γ_M, as given by the code:

$$\sigma_c < \frac{f_c}{\gamma_M}, \ \sigma_t < \frac{f_t}{\gamma_M} \tag{8}$$

The material safety factor γ_M is assessed as $\gamma_M = 2.0$. Calculated stresses should be smaller than permissible values. In the case of tension we obtain:

$$\sigma_t < f_{td} = \frac{f_t}{\gamma_M} = \frac{0.08}{2} = 0.04 \text{ MPa} = 40 \text{ kN/m}^2 \tag{9}$$

And in the case of compression:

$$\sigma_c < f_{cd} = \frac{f_c}{\gamma_M} = \frac{0.5}{2} = 0.25 \text{ MPa} = 250 \text{ kN/m}^2.$$

The results obtained in selected points are summarised in Table 1.

Table 1. Evaluated discrete stress values in seven points for load cases considered.

	Point	**157**	**200**	**269**	**955**	**1174**	**1293**	**1301**
	σ_{11top} [kN/m^2]	−41.28	−36.73	52.70	−33.07	−33.53	−71.39	−68.44
LC1	σ_{11bot} [kN/m^2]	44.92	7.10	−97.54	1.48	−2.49	−26.30	−30.91
	σ_{22top} [kN/m^2]	−119.26	124.74	−12.98	−147.14	−156.93	−177.86	−179.27
	σ_{22bot} [kN/m^2]	−181.40	-28.88	-259.61	−64.44	−86.14	−61.21	−61.43
	σ_{11top} [kN/m^2]	−120.11	−265.17	567.59	120.82	13.75	−180.22	−177.48
LC2	σ_{11bot} [kN/m^2]	50.19	513.14	−447.47	29.13	86.94	85.03	83.61
	σ_{22top} [kN/m^2]	−240.61	344.94	321.98	−158.93	−334.60	−531.38	−560.17
	σ_{22bot} [kN/m^2]	−282.48	113.76	−886.58	−257.32	−47.17	164.13	199.86
	σ_{11top} [kN/m^2]	−75.67	−66.08	91.63	−44.53	−136.14	−354.65	−158.88
LC3	σ_{11bot} [kN/m^2]	82.37	12.28	−161.91	−17.07	50.18	148.84	−42.61
	σ_{22top} [kN/m^2]	−204.97	196.11	−32.57	−209.95	−363.76	−608.23	−436.52
	σ_{22bot} [kN/m^2]	−305.60	−48.38	−431.64	−167.37	−86.61	174.44	17.66
	σ_{11top} [kN/m^2]	−63.50	−66.64	82.45	−60.25	−57.92	−112.90	−107.81
LC4	σ_{11bot} [kN/m^2]	71.79	5.53	−153.84	10.56	1.21	-38.65	−46.36
	σ_{22top} [kN/m^2]	−212.72	186.28	−39.18	−269.19	−277.32	−301.05	−302.35
	σ_{22bot} [kN/m^2]	−272.82	−49.41	−421.87	−88.60	−131.64	−103.18	−104.44
	σ_{11top} [kN/m^2]	−84.97	−103.62	42.07	−72.93	−67.18	−123.88	−118.42
LC5	σ_{11bot} [kN/m^2]	94.06	54.80	−120.62	20.64	8.32	−30.73	−38.42
	σ_{22top} [kN/m^2]	−228.95	167.47	−67.97	−265.28	−269.97	−296.05	−297.56
	σ_{22bot} [kN/m^2]	−268.14	−3.97	−404.63	−91.91	−138.79	−107.81	−109.23

In addition to the discrete stress values' presentation in Table 1, the distribution of stresses for load cases "2" and "3" is shown in Figures 10–13.

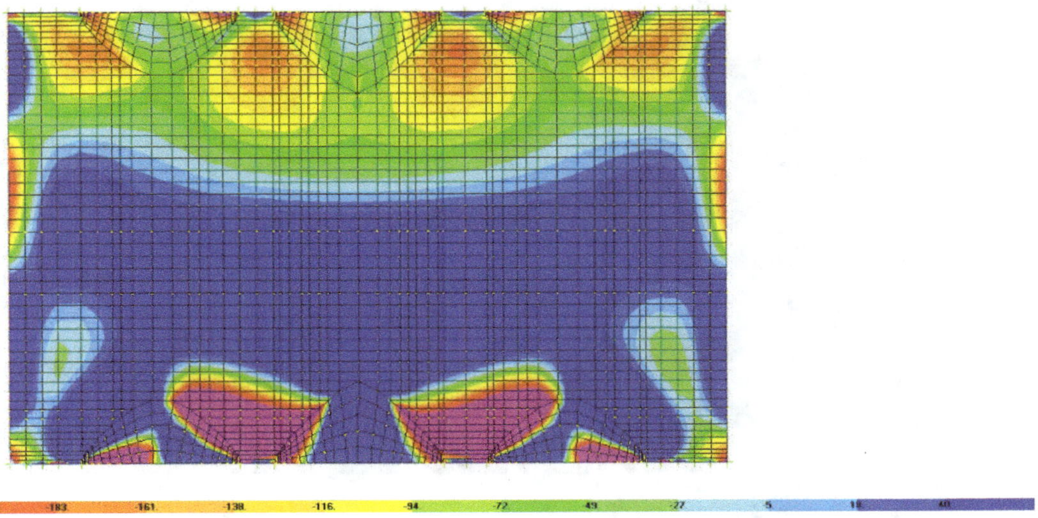

Figure 10. Stresses σ_{11bot} [kN/m^2]—load case »2«.

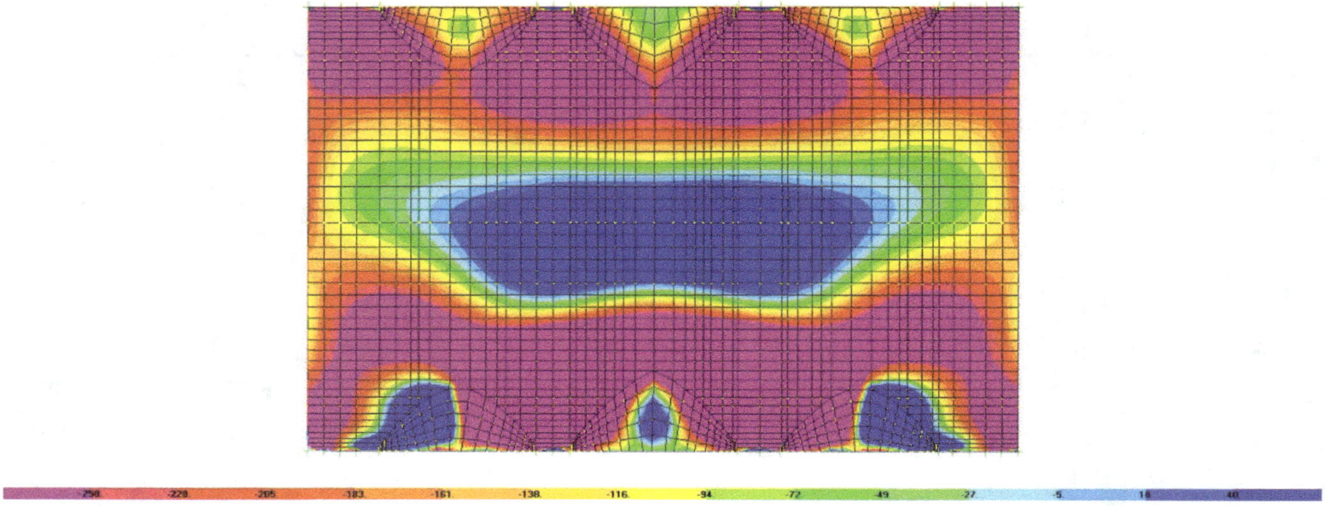

Figure 11. Stresses σ_{22bot} [kN/m^2]—load case »2«.

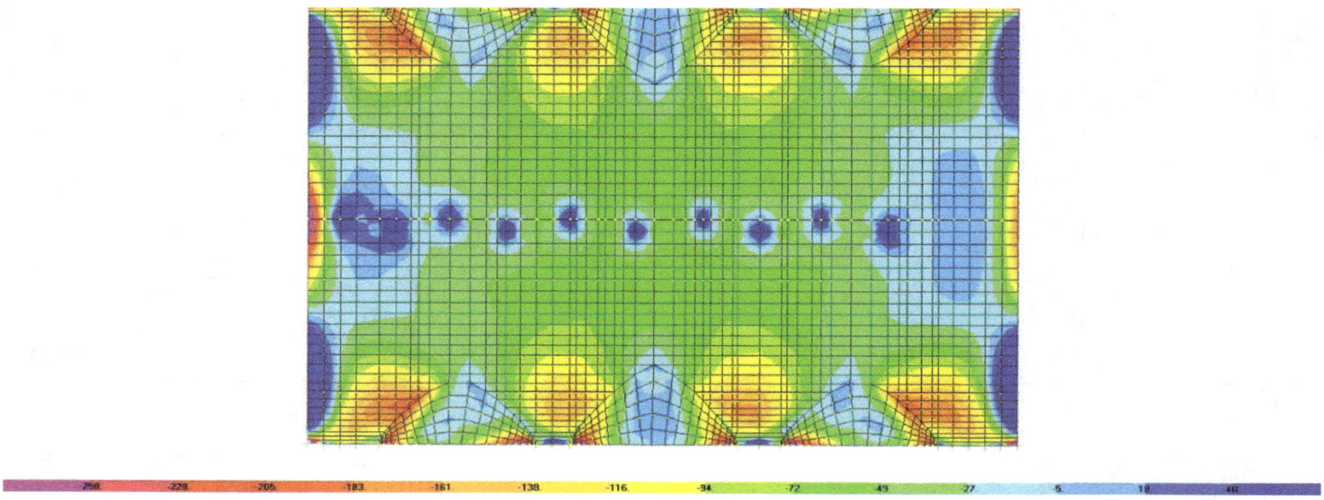

Figure 12. Stresses σ_{11bot} [kN/m^2]—load case »3«.

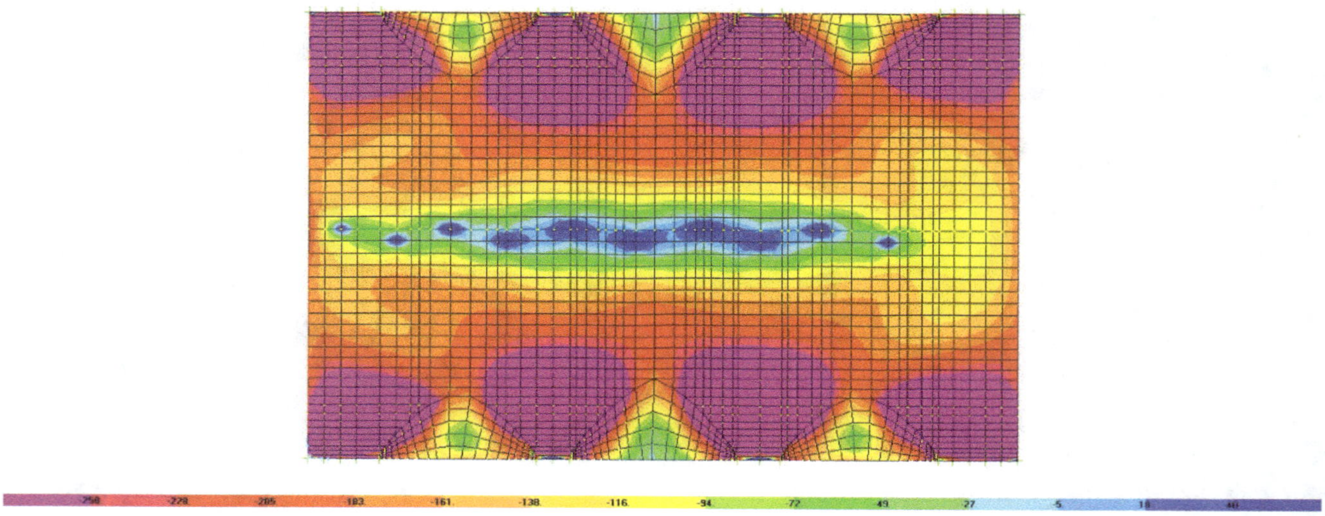

Figure 13. Stresses σ_{22bot} [kN/m^2]—load case »3«.

4. Discussion of the Results

As already stated, five load combinations were considered in order to find the most probable cause for the formation of the registered crack pattern.

In load case »1« self weight and rubble layer was considered. Numerical results show that stresses are locally exceeded in areas where no actual damage was registered on the shell. On the other hand, calculated stresses in areas with cracks are within the permissible limits. However, it should be noted, that the cracks on the upper side of the shell are poorly visible due to the rubble layer and rough surface.

In load case »2« vertical displacement ($\Delta_Z = -0.5$ cm) of outer supports on one side was considered which simulates differential settlement of foundations. In this case, high tensile stresses occur in the top area of the shell (Figure 11), which may cause the recorded longitudinal crack. Differential settlement of foundations can thus be one of possible causes for the registered crack pattern. However, the results of this load case otherwise show high stress levels also in other parts where no damage of the actual shell has been recorded.

Load case "3" took into account the self weight of the structure, weight of the rubble layer and weight of the maintenance team. In this case, compressive stresses in the upper middle part (Figure 13) as well as tensile stresses in the lower middle part (Figure 12) of the analysed shell are exceeded, which coincides with the registered crack pattern. This loading combination can therefore be considered as the most possible cause for the formation of a longitudinal crack [8].

With load case "4" the impact of seismic loading on the analysed structure was examined. Results show that stresses in this case are not exceeded, and, consequently, the influence of the seismic load as a possible cause for the formation of longitudinal crack can be excluded.

Results of load case "5" which took into account self weight of the structure, rubble layer and temperature change ($\Delta_T = 15$ °C) show that such loading combination is critical for collateral vaults but cannot be responsible for the registered longitudinal crack.

The executed numerical analyses thus show that load cases "2" and "3" are the most likely ones to cause the registered crack pattern. The longitudinal crack in the middle area of the analysed shell most likely formed due to separated or combined effect of maintenance team loading and differential settlement of foundations.

5. Possible Repair and Strengthening Measures

Based on the results of the analysis of the considered shell, it can be concluded that the cracks formed mainly due to direct loading of the middle upper part of the shell (simulation of maintenance works in load case "3"). Considering the fact that cracks reappeared in the same areas after repair works in 2013, it is recommended that monitoring of the shell be carried out with an emphasis on the longitudinal crack in the middle part of the shell. In the case that the widening of that crack continues, strengthening of the shell will be required. Strengthening could be done by applying reinforced concrete layer either using FRP (fiber reinforced polymer) or FRM (fiber reinforced mortar) on the critical area. However, if the monitoring does not confirm any propagation (widening and/or length extension) of cracks, strengthening considering unchanged loading condition is not necessary. Nevertheless, the problem in the case of loading of the shell during maintenance works should be solved for any potential future works as the numerical simulations show that the shell is locally overloaded even by weight of a single maintenance worker. A possible simple solution for this problem is a working platform not only for access of the maintenance team but generally also for any secure access to the area above the shell [8]. The weight of the platform could be partially carried by the existing wooden roof structure and mainly by the stronger stone-masonry walls of the church structure.

6. Conclusions

The paper demonstrates the simulation modelling as a powerful tool in addressing problems of structures' reconstructions. Although the implementation of non-linear behaviour would be the most appropriate or rather more common approach for the analysis of the considered shell, the analyses carried out in this work showed that it is possible to identify the causes for damage, even with the simple elastic material model, without performing more complex non-linear analyses. As found, the longitudinal crack in the middle area of the analysed shell most likely formed due to the separate or combined effect of maintenance team loading and differential settlement of foundations. However, none of these two most probable causes for the registered crack pattern were considered in the numerical analysis within the official rehabilitation project. Consequently, the main longitudinal crack re-opened after more or less cosmetic patching as a part of rehabilitation works. This shows that repair and/or strengthening measures of existing structures should be planned after a thorough analysis of all possible causes for the damage, since only by using such approach any further damage and propagation of cracks can be prevented. Registered damage of existing structures should thus not be addressed partially by merely cosmetic corrections such as patching of visible cracks, but firstly by performing numerical analysis aiming to identify the possible causes for damage, by monitoring of the structure if necessary and eventually by strengthening of critical parts or the structure as a whole.

Author Contributions: Conceptualization, M.U.; Methodology, M.U. and M.S.; Software, T.L.; Validation, M.U. and M.S.; Formal Analysis, T.L.; Investigation, T.L.; Resources, M.U. and T.L.; Data Curation, M.U. and T.L.; Writing—Original Draft Preparation, M.U. and M.S.; Writing—Review and Editing, M.U. and M.S.; Visualization, M.U. and M.S.; Supervision, M.U. and M.S.; Project Administration, M.U. and M.S.; Writing—original draft, M.U. and M.S.; Writing—review & editing, M.U. and M.S.

Acknowledgments: The authors wish to thank the firm GI ZRMK for providing insight into the technical report on reconstruction of the Church of St. Jakob in Dolenja Trebuša.

References

1. Holzer, S. *Statische Beurteilung historischer Tragverke. Band 1-Mauerwerkskonstruktionen*; Ernst & Sohn: Berlin, Germany, 2013.
2. Ugural, A.C. *Stresses in plates and shells*; McGraw-Hill: Boston, MA, UAS, 1999.
3. Peerdeman, B. Analysis of thin Concrete Shells Revisited: Opportunities due to Innovations in Materials and Analysis Methods. Master's Thesis, Delft University of Technology, Delft, The Netherlands, June 2008.
4. De Vent, I.A.E. Prototype of a diagnostic decision support tool for structural damage in masonry. Ph.D. Thesis, Delft University of Technology, Delft, The Netherlands, June 2011.
5. Gostič, S.; Dolinšek, B. Lessons learned after 1998 and 2004 earthquake in Posočje region. In Proceedings of the 4th International i-Rec Conference Building Resilience, Christchurch, New Zealand, 30 April–2 May 2008; pp. 11–25.
6. Štampfl, A. *Reconstruction of the Church of St. Jakob, Dolenja Trebuša*; Technical Report; GI ZRMK: Ljubljana, Slovenia, December 2011.
7. Computers & Structures. CSI Analysis Reference Manual for SAP2000, ETABS, SAFE and CSiBridge, Berkeley. Available online: http://docs.csiamerica.com/manuals/misc/CSI%20Analysis%20Reference%20Manual%202011-12.pdf (accessed on 10 December 2015).
8. Lorenci, T. Masonry shell structures in historical buildings. Maribor, 2016; p. 104. Available online: https://dk.um.si/IzpisGradiva.php?id=61543 (accessed on 12 December 2018).

4

Mechanical and Typological Characterization of Traditional Stone Masonry Walls in Old Urban Centres

José Carlos Domingues [1], Tiago Miguel Ferreira [2,*], Romeu Vicente [3] and João Negrão [4]

1 Department of Civil Engineering, University of Coimbra, 3030-788 Coimbra, Portugal; jcarl.domingues@gmail.com
2 ISISE, Institute of Science and Innovation for Bio-Sustainability (IB-S), Department of Civil Engineering, University of Minho, 4800-058 Guimarães, Portugal
3 RISCO, Department of Civil Engineering, University of Aveiro, 3810-193 Aveiro, Portugal; romvic@ua.pt
4 Department of Civil Engineering, University of Coimbra, 3030-788 Coimbra, Portugal; jhnegrao@dec.uc.pt
* Correspondence: tmferreira@civil.uminho.pt

Abstract: Essential for any intervention in existing buildings, a thorough knowledge of both structural and material characteristics is even more important in the case of traditional stone masonry buildings, due both to the variability of this technology's properties and the degradation buildings might have sustained. In Portugal, a number of in situ and laboratory experimental campaigns has allowed us in recent years to expand the knowledge on the mechanical properties of stone masonry walls. Nevertheless, the existence of different wall typologies built with the same material necessitates that this characterization takes into account the various regional constructive cultures. This paper presents the results obtained through an in-situ characterization campaign carried out in the old urban center of Viseu, for which there is no information available in the literature. Granite stone masonry walls of two different buildings were analyzed and characterized considering their geometrical and material features, contributing to the identification of stone masonry typologies present in the city's old urban center. Flat-jack testing yielded resistance and deformability parameters to be used both in safety evaluation and intervention design. The properties obtained can be said to be consistent with those deriving from other experimental campaigns, conducted in granite walls of different typologies, throughout the country. Simultaneously, relevant conclusions about the use of flat-jacks to characterize this type of stone masonry were drawn.

Keywords: masonry characterization; mechanical properties; in situ test campaign; granite masonry; flat-jack testing; old urban center; regional constructive cultures

---

1. Introduction

When intervening in existing buildings, and with a view to preserving historical and material authenticity, no action should be conducted without being absolutely necessary. To ensure this, a deep knowledge of structural and material characteristics, constructive technologies and techniques, changes made over time, and current state of the building is fundamental [1]. This necessity is all the more pressing when dealing with stone masonry buildings, as their structural behavior is significantly different from recent buildings (generally consisting of reticulated reinforced concrete structures), besides also presenting high typological and mechanical variability, which are quite dependent of the local material availability and constructive tradition.

The characterization of the mechanical behavior of granite stone masonry, a typology common in the North and Center-North of Portugal, has already been the subject of some studies conducted both in situ [2,3] and in laboratory [4–7]. Nevertheless, the existence of various regional constructive cultures, resulting in different wall typologies built with the same material, leads to the need for such characterization to take into account the specific locations where these walls are representative. Regarding the specific case of the city of Viseu, it is worth mentioning that information concerning the mechanical properties of granite stone masonry walls in its old urban center is absent. Therefore, and aiming at providing data that can help to retain the original structural system of these buildings, while ensuring structural safety, this work intends to offer a first contribution to the characterization of these elements. It is intended, namely, and apart from geometrical and material characterization which can contribute to the identification of stone masonry typologies frequent in the old urban center of Viseu, to quantify resistance and deformability parameters to be used both in safety evaluation and intervention design. Thus, this paper presents and discusses the results obtained from an in situ test campaign conducted in the old urban center of Viseu. Granite stone masonry walls of two different buildings were analyzed and characterized considering their geometrical and material characteristics. Flat-jack testing technique was used to obtain resistance and deformability parameters to be used both in safety evaluation and intervention design. Relevant conclusions about the use of flat-jacks to characterize this type of stone masonry were also drawn.

2. The City of Viseu: Brief Urbanistic and Constructive Characterization

The city of Viseu plays an important role in the history of Portugal, as it is markedly linked to the figure of Viriathus, a tribal leader who resisted the Roman invasion. In fact, since Roman times, Viseu has assumed a prominence in the nearby settlements. Following the ancient Roman urbanistic tradition, the city developed around two main roads, the *cardus*, oriented North-South, and the *decumanus*, oriented East-West. This influence can still be seen today, as both roads evolved to be prominent axis of the city's old urban center (Figure 1).

Figure 1. View of the old urban center of Viseu.

It was during the XVI century that Viseu started expanding outwards towards its ancient walls. It was also by this time that the area nowadays known as Rossio started gaining importance; by the XIX century, it would have become the new civic center of the city, with the construction of the town hall building. Nowadays, with 73,518 inhabitants in its urban perimeter [8], Viseu is the second biggest city in the Região Centro of Portugal, as well as the largest urban center outside both metropolitan areas and the littoral of the country.

As to the constructive system of the buildings from the old urban center of the city of Viseu, namely in what pertains to the vertical resisting elements, the abundance of granite stone makes this the most common material with which to build masonry walls, as is common in the Northern Region of Portugal. The granite used in the construction of the traditional masonry buildings of the old urban center of Viseu came from different quarries between the Douro and Mondego rivers. The larger and well-cut granite stone blocks were used in the construction of churches, palaces (today museums), cathedrals since they were not covered or rendered. The fragmented granite stone was commonly used for traditional building construction from one to four stories high inside the city walls. Nevertheless, and despite its abundance, no information related to the mechanical properties of granite stone masonry walls from the old urban center of Viseu exists in the literature. Besides stone, different materials and constructive processes can be found, namely timber "tabique" walls for the inner (often non-resisting) walls, or "taipa", an earth-based material, for the walls in the upper floors.

Aiming at safeguarding and potentiating the city's cultural heritage, both tangible and intangible, the Viseu City Council has recently promoted the project *Viseu Património*. It is in this scope that the campaign described in this paper was conducted.

3. Typological, Geometrical and Material Characterization of Stone Masonry Walls

Due to its architectural and historical significance, the buildings studied in this campaign (named *Orfeão de Viseu* and *Águas de Viseu*) have been selected by the City Council to be part of the first phase of the project *Viseu Património*, within the framework of which they will be object of intervention resorting to the "best practices" in heritage conservation. The main objective is that these two buildings/interventions can become models of interventions for future conservation and rehabilitation works in the old urban center of Viseu.

Figure 2 shows the location of these buildings in the old urban center of the city. In the following subsections, each building is briefly described, and the geometrical and material characteristics of the walls studied in each one of them are presented.

Figure 2. Buildings studied in this campaign: Orfeão de Viseu (green) and Águas de Viseu (dark blue). The light blue line marks the old urban center's limit (adapted from map kindly made available by the Viseu Urban Rehabilitation Society Viseu Novo).

3.1. *Orfeão de Viseu*

The building located in the number 149 of Rua Direita (the ancient commercial artery in the city) housed between 1955 and 2006 the headquarters of *Orfeão de Viseu*, a cultural society dedicated to arts and music. Structurally, the building follows the traditional constructive technology, with load-bearing stone masonry walls and timber floors and roofs. Of unusually large dimensions for the area in which it is located, it is known that it was partially rebuilt after a fire in 1926 [9]. Figure 3 presents the main façade of the building, as well as the plans of the ground floor and the first floor.

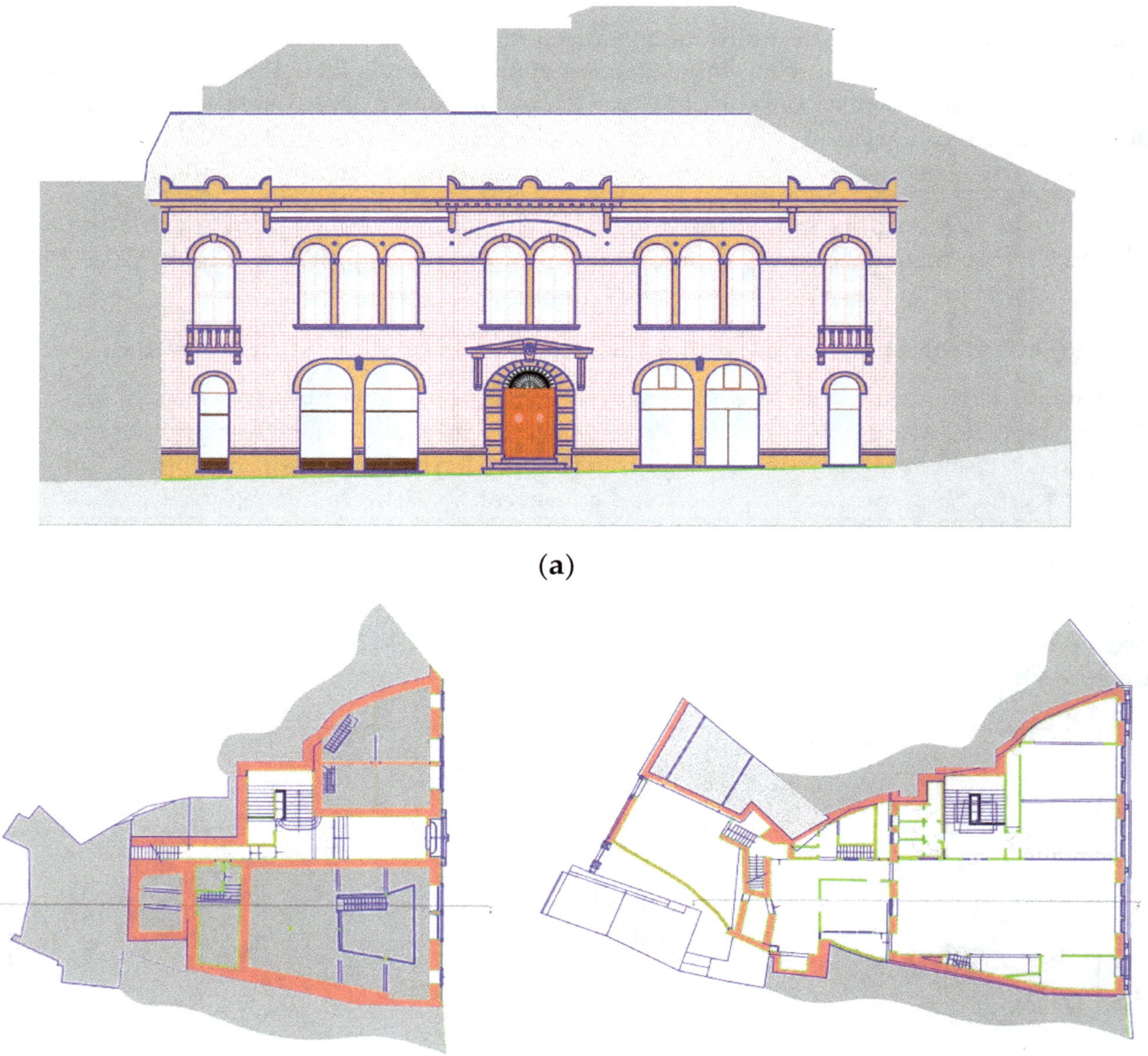

(a) (b) (c)

Figure 3. *Orfeão de Viseu*: (**a**) main façade of the building, (**b**) plan of the ground floor and (**c**) the first floor.

In its interior, it is worth mentioning the stairs, with a balustrade in wrought iron, and the stairwell covered with flower motive tiles, dating from 1912 (Figure 4). In the main hall, located in the first floor, it is possible to observe railroad tracks that were used as part of the walls' structural system. As for the partition walls, and apart from stone masonry, "tabique" walls are present. The presence of a structurally elaborate timber roof is also worth mentioning.

(a)

(b)

Figure 4. View to (**a**) the staircase with the tiles; and (**b**) the "tabique" walls.

A 50 cm thick interior load-bearing wall, located in the main hall and parallel to the main façade, was studied in this building, Figure 5a. As can be seen in Figure 5b, stone masonry was visible without the need to remove any plaster. The masonry consists of gray granite stones, with both rectangular and square shape, and cream color mortar. In general, a length of around 70 cm, for the biggest rectangular stones, and around 20 cm, for the square ones, can be observed. Rectangular stones present a length/height ratio of approximately 3. The presence of wedges was not significant.

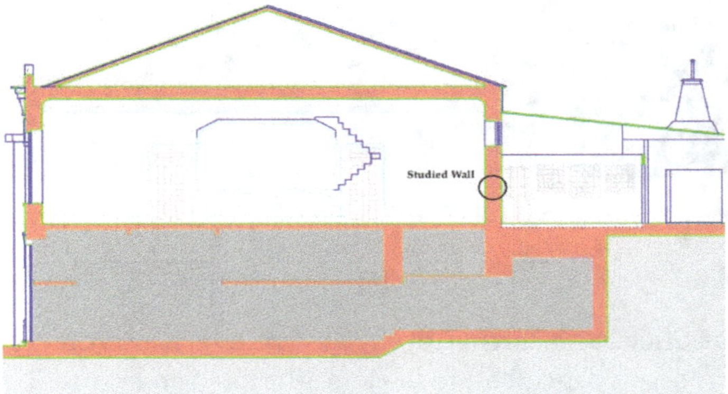

(a)

(b)

Figure 5. Identification of the studied wall. (**a**) location in the building; (**b**) testing area.

Due to access limitations, it was not possible to observe the inner side of the wall, and thus it was not possible to infer about the existence of multiple leaves and/or through stones. It must be pointed that, due to repointing, the apparent dimension of the stones may not correspond to their actual dimension, as a result of mortar overlapping along their contour. As for the cross-section, the wall was found to be composed of two leaves, with a poor to reasonable degree of connection. The existence of an inner core, of weaker quality, was in any event ruled out. This last conclusion was further reinforced by the observation, resorting to videoscopy, of another wall's interior in the same room.

3.2. Águas de Viseu

The building dates back to the 1920s, and has three fronts, facing Rua Dr. Luís Ferreira (the ancient Rua do Comércio), Rua D. Duarte and Travessa de São Domingos (Figure 6). Besides the ground floor, it counts three floors and an underground basement.

(a) (b)

Figure 6. *Águas de Viseu*: (**a**) view to Rua do Comércio; and (**b**) view to Rua D. Duarte.

Despite it being classified as a structure with "relevant architectural and heritage value", the building is presently in an advanced state of degradation, see Figure 7.

Figure 7. *Cont.*

Figure 7. Severely decayed structural elements.

As in the previous case study, an internal load-bearing wall was studied, this time located at the ground floor and perpendicular to the façade facing Rua D. Duarte, Figure 8a. As can be seen in Figure 8b, also in this case, stone masonry was visible without the need to remove any plaster. Due to the presence of different minerals, such as quartz, mica and feldspar, stones present various colors, ranging from gray to yellow and pink. The mortar, of a gray color, appears to evidence recent repointing. Stones were rather irregular, both in shape and dimensions. The assemblage was markedly irregular, at times with stones laid with their larger dimension alongside the vertical. The presence of wedges was not significant.

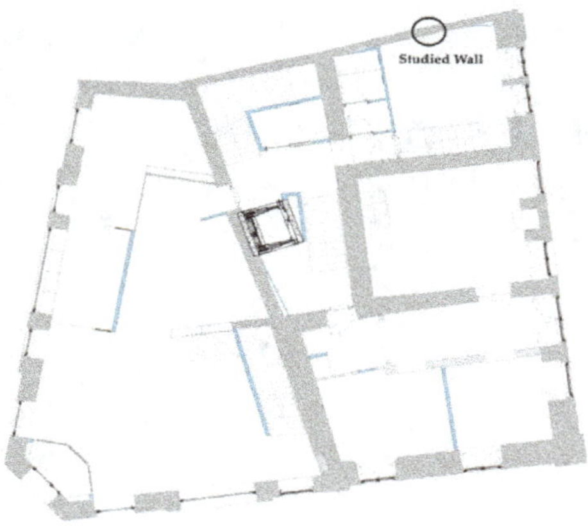

(a)

(b)

Figure 8. *Águas de Viseu:* (**a**) building plan with the identification of the studied wall; (**b**) view to the wall.

The wall was found to be composed of two leaves, with an interior non-cohesive core (*sacco* masonry). Both the interior leaf and the inner core are 25 cm thick. It was not possible to measure the outer leaf thickness, but a 25 cm thickness is equally plausible. A total thickness of 75 cm, at ground level, would then be expected. Regarding the inner core, it was possible to observe that the wall was filled with roughly-shaped granite stones of about 10–15 cm long, laid in a random manner (Figure 9a), linked by lime mortar and earth (Figure 9b). Some small wooden elements, such as wood splinters, were also found in the inner core of the masonry wall, see Figure 9c.

(**a**) (**b**) (**c**)

Figure 9. Wall's inner core. (**a**) roughly-shaped granite stones; (**b**) lime mortar and earth; (**c**) wood splinters.

3.3. Mechanical and Visual Characterisation of the Stone

The need to locally disassemble the masonry required collecting some granite stones as well, which could be then used to perform mechanical characterization tests in laboratory, namely to estimate their compressive strength. For this purpose, one sample of granite stone was extracted from the wall and prepared in laboratory to perform a uniaxial compression test, which took place in the Laboratory of the Civil Engineering Department of the University of Aveiro, Portugal, following the Portuguese standard NP EN 1926 [10].

Figure 10 presents the test sample, the test procedure and the final appearance of the sample after tested. The results obtained in the compressive test are presented in Table 1, where L, D, m and γ stand respectively for the length, diameter, mass and specific weight of the test specimen and fcb is the maximum compressive strength recorded in the compressive test.

(a) (b)

(c)

Figure 10. Compression test. (**a**) stone specimen; (**b**) experimental apparatus; (**c**) appearance of the specimen after the test.

Table 1. Results of compression tests.

L (mm)	D (mm)	m (kg)	γ (kN/m^3)	f_{cb} (N/mm^2)
126	84	1.77	24.85	32.7

Despite being low, the value of compressive strength obtained can be considered to be consistent with the range of values expected for ancient granites [4,11]. It is however important to stress that since only one sample could be tested (due to several constraints, namely related with the extraction of stone material in quantity and with enough quality/integrity to be tested in lab) the representativeness of the result is not fully guaranteed. Even so, this result was important to identify current state of meteorization of this granite and to better understand its mechanical characteristics, which is considered a relevant output since no other similar tests could be found in the literature for the granite present in the old urban center of Viseu.

Visual analysis of the granite samples showed the simultaneous occurrence of various types of granite. It was possible to note the existence both of coarse-grained porfiroid granite, composed of biotite, muscovite and feldspar fenocrystals, as well as of medium-grained gneissic granite, also composed of biotite and muscovite. In both types, weathering was observed, namely in the yellow coloration around the biotite crystals, signaling the oxidation of ferrous oxides. This could help to explain the low compressive strength value obtained in the uniaxial test, as granites experience a reduction of strength when subject to this phenomenon [12].

Moreover, it is worth noting that compressive and shear strength of masonry are dependent on the masonry fabric and arrangement, mechanical properties of the stone, mortar and their percentage, as well as the presence of voids. Besides, it is important to underline the fact that the dimension of the stone blocks and their arrangement are responsible for the development of preferential and alternative load. Masonry walls with the lowest percentage of mortar and voids lead to the highest values of maximum stress and Elastic Modulus.

4. Mechanical Characterization of Stone Masonry Walls Through Flat-Jack Technique

Flat-jack testing is recognized as one of the most versatile techniques for the in situ characterization of the mechanical behavior of masonry walls under compression [13,14]. Through flat-jack testing it is possible, namely, to obtain information regarding the in situ stress level, as well as resistance and deformability characteristics. Because it involves only the execution of one or two cuts of small dimension and thickness, which can be filled after the test is finished, this technique is usually considered as a semi-destructive procedure.

Flat-jack technique has been widely and successfully applied in several past studies, namely in regular brick and stone masonry walls [15]. However, the use of this technique in irregular stone masonry has been the subject of relatively few studies [13,16–19]. In these walls, not complying with all standards' requirements is a contingency that must be taken into account. Among these, instructions regarding the minimum length of the jack, which is defined on the basis of the single masonry units' length, are often difficult to follow. In this regard, it is worth remembering that standards directing this procedure date back to the beginning of the 1990s, when this technique, already widely used in brick masonry, was still rarely applied in stone masonries, and especially in those with irregular morphology (as still today). Moreover, geometrical variability of stone masonry contributes for the low reliability of results obtained in a localized area in the wall [18]. Difficulty in finding regular mortar joints represents a further complexity when applying flat-jack in irregular masonries, as the cut(s) must then be made through the stones [20]. Geometrical characteristics of stone masonry also influence the choice of the points where measuring devices are to be located. Likewise, causing irregular distributions of stone and mortar, and, thus, areas with different behavior, they can cause localized anomalies through rotations of the devices, which influence the results. Besides these, the experience acquired through this campaign has allowed us to identify further limitations of the application of flat-jack testing to irregular masonries. It is acknowledged that a larger number of flat-jack tests is necessary to obtain more reliable results to better match with morphological and constitution features of masonry fabric. Complementarily, other non-destructive testing techniques, such as sonic tests, tube-jack testing as well as destructive testing techniques such as: shear tests, in-plane and out-of-plane testing, chemical tests, etc., should be considered and combined namely for structural evaluation and damage assessment of historical constructions.

In Portugal, flat-jack testing has been used in several studies, namely those conducted by Roque [21], Pagaimo [22], Vicente et al. [13], Ferreira [23], Miranda [3] and Simões et al. [19]. Nevertheless, taking into account the aim of the present paper, it must be noted that flat-jack testing campaigns aiming at characterizing granite stone masonry are still scarce.

4.1. In Situ Experimental Campaign

As already referred, flat-jack testing allows obtaining information regarding both the in situ stress level of the wall section as well as resistance and deformability characteristics. For this, different testing procedures are conducted, namely, the single and the double flat-jack test. These procedures are described in the American standards ASTM C1196-91 [24] and ASTM C1197-91 [25] and in the European recommendations RILEM MDT. D. 4 [26] and RILEM MDT. D. 5 [27].

4.1.1. Single and Double Flat-Jack Testing

Single flat-jack testing is based on determining the stress needed to restore the initial geometry of a wall, which has been disturbed by eliminating the wall's initial vertical stress by means of a horizontal cut. After this cut, and the corresponding lowering of the distance between reference points located, above and below, in the same vertical alignment transversal to the cut, a flat-jack is introduced in the slot and afterwards pressurized by means of a hydraulic system, so that stress is being transmitted to the surrounding masonry and the initial distance between reference points is eventually restored. The average existing stress level (σ_m) can thus be computed as:

$$\sigma_m = k_m \times k_a \times p \tag{1}$$

where k_m is a dimensionless factor taking into account the flat-jack rigidity, k_a is the ratio between the area of the jack effectively in contact with the masonry and the total area of the jack, and p is the pressure measured within the flat-jack.

Due to the anisotropic behavior of masonry, single flat-jack testing is not a suitable procedure for the determination of deformability properties. For this purpose, the double flat-jack testing, consisting of two vertically aligned flat-jacks introduced in parallel slots, is used. The portion of the wall located between the jacks can be thus tested under uniaxial compression, therefore obtaining a relatively reliable estimate of the Young's modulus (E) value from the stress-strain curve. Using a horizontal displacement transducer, it is further possible to compute the Poisson's ratio (ν).

According to ASTM C1197-91 [25], double flat-jack testing overestimates the value of the Young's modulus by 15 to 20%. It must nevertheless be recalled that this value applies for brick masonry; it is to be expected that the deviation will be greater in stone masonry [14].

4.1.2. Equipment and Testing Protocol

The following equipment was used:

- Flat-jacks: BOVIAR MPA—A semicircular flat-jacks, with dimensions 350 × 260 × 4 mm, and maximum service pressure 60 bar, Figure 11a. According to the calibration certificate, a km value of 0.902 was assumed for all the flat-jacks used in this campaign.
- Masonry cutting equipment: HUSQVARNA k970—Ring cutting machine, with a 300 mm eccentric diamond blade disk, allowing a cut depth of 270 mm.
- Pressurizing equipment: manual hydraulic ENERPAC pump, with a 500 bar capacity and a 3 L reservoir, Figure 11b. The pressure applied is controlled using a pressure reading manometer. The hydraulic system and the flat-jacks are connected via high pressure hoses.
- Displacement transducers: GEFRAN displacement transducers, with a 50 mm stroke, provided with self-aligning ball-joints. The displacement transducers were attached to the masonry by means of 5 mm threaded rods, applied with chemical anchor, at a depth of 50 mm, Figure 12.
- Data acquisition and log system, consisting of a load cell, a NATIONAL INSTRUMENTS acquisition unit, a laptop and acquisition software developed in the LabVIEW environment [28].

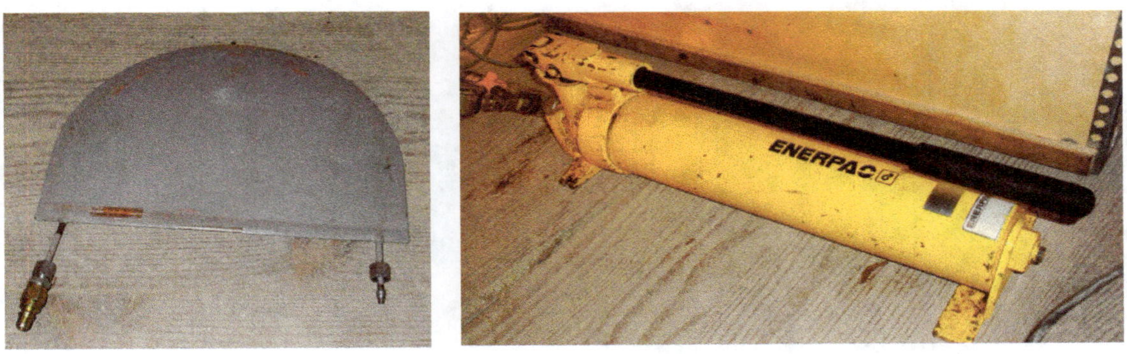

(a) (b)

Figure 11. Equipment used in this campaign: (**a**) semicircular flat-jacks and (**b**) hydraulic pump.

Figure 12. Displacement transducers attached to the wall.

Figure 13 shows the single flat-jack testing. As presented, three vertical reference alignments were considered. In order to estimate the effective flat-jack loaded area, the procedure proposed by Gregorczyk and Lourenço [29] was adopted: a sheet of carbon paper, sandwiched between two sheets of ordinary paper, was placed between the jack's surface and the surrounding masonry (this set was prepared in the shape of the jack). In this way, the areas where the jack comes into contact with the masonry are marked in the paper and can be measured (Figure 14).

(**a**)

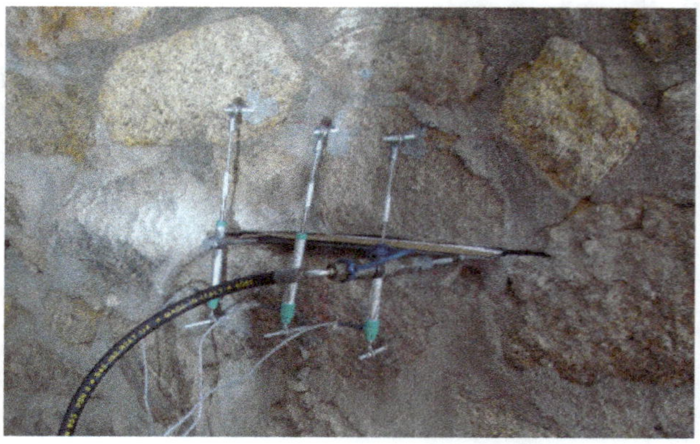

(**b**)

Figure 13. Single flat-jack testing: (**a**) conducted in the Órfeão de Viseu and (**b**) in the Águas de Viseu (**b**).

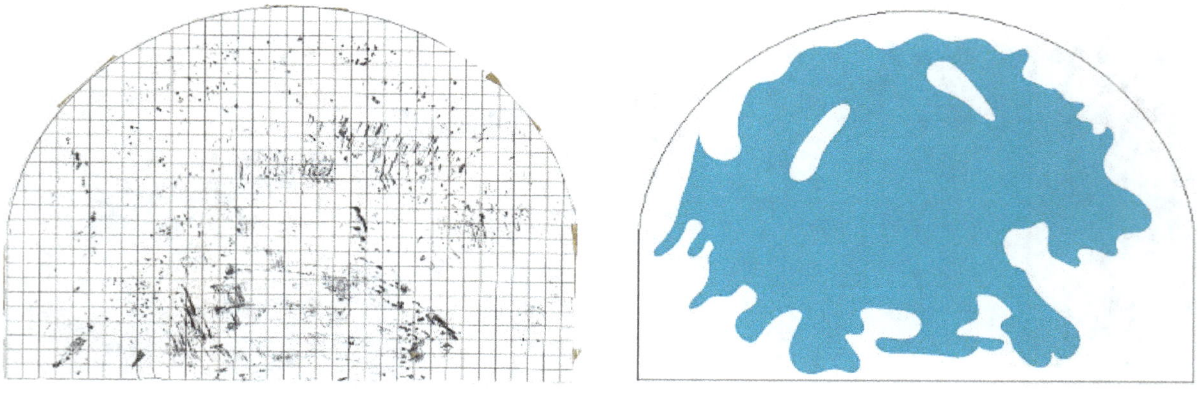

Figure 14. Contact area between the flat-jack and surrounding masonry.

Due to the manifest poor condition of some of the carbon papers after removing of the flat-jacks and considering the deformation visible in the flat-jacks, in those cases, coefficient k_a was estimated based on visual observation.

It is important to mention that, although humidity and temperature of the walls can slightly affect the results [13], such conditions were not possible to control during the tests.

Usually, the initial distance between reference points is not attained simultaneously in the various alignments. This is due, among other factors, to different stone and mortar distributions (areas with a larger proportion of mortar will sustain greater deformation, comparatively to others with a greater quantity of stone). To take this into account, the stress level that restores the initial wall geometry was assumed to be the average of the values that do so in the various alignments. Double flat-jack tests were conducted in the same areas where single flat-jacks had already been conducted, thus taking advantage of the existing cut. Figure 15 shows the double flat-jack testing being conducted.

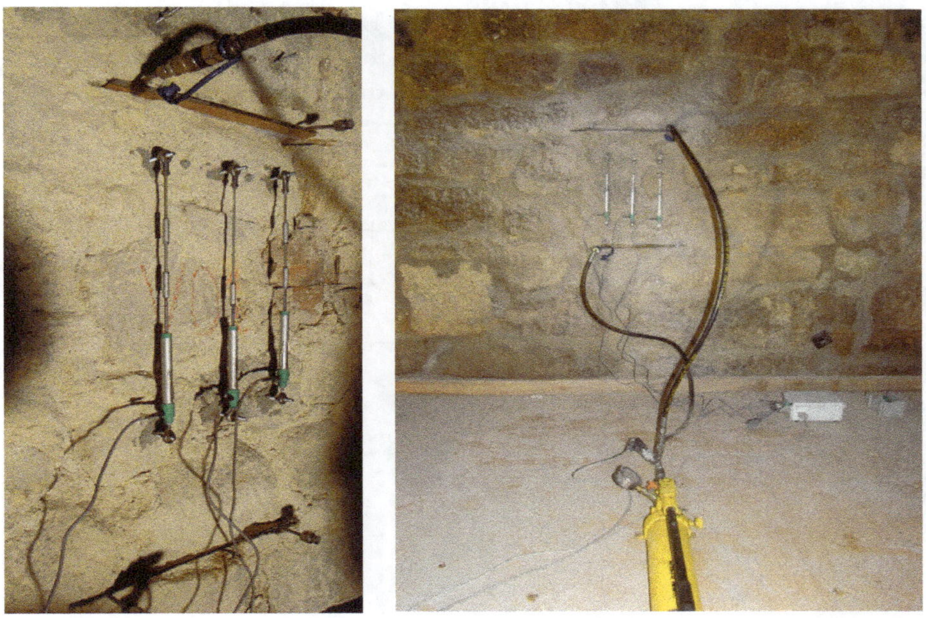

(a) (b)

Figure 15. Double flat jack testing conducted in Orfeão de Viseu (**a**) and in Águas de Viseu (**b**).

5. Results and Data Interpretation

Two single flat-jack tests and one double flat-jack test were conducted in Orfeão de Viseu, whereas one single flat-jack test and two double flat-jack tests were conducted in Águas de Viseu. Tests are named according to the buildings (O—Orfeão de Viseu; A—Águas de Viseu) and type of test (S—single flat-jack test; D—double flat-jack test). Figures 16–18 show the evolution of the relative displacements in the different alignments, depending on the flat-jack pressure and the estimation of the corresponding in situ stress level, for all tests conducted. Pressure increments of 0.5 MPa, corresponding to 5 bar, were applied in the double flat-jack tests.

From the displacement results, the in-plane stress is estimated by the average of the four deformation measurements obtained from the three alignments. In the case of the *Orfeão de Viseu*, the low deviance of the displacement readings is a good indicator of the test evolution. Lower values for the single flat-jack test OS1 and OS2 where obtained since the test was carried out on first floor. As for the *Águas de Viseu*, the test was carried out on the ground floor, which is why the in situ stress value are higher than those obtained in *Orfeão de Viseu*. Moreover, in the case of *Águas de Viseu* the displacement readings were more variable.

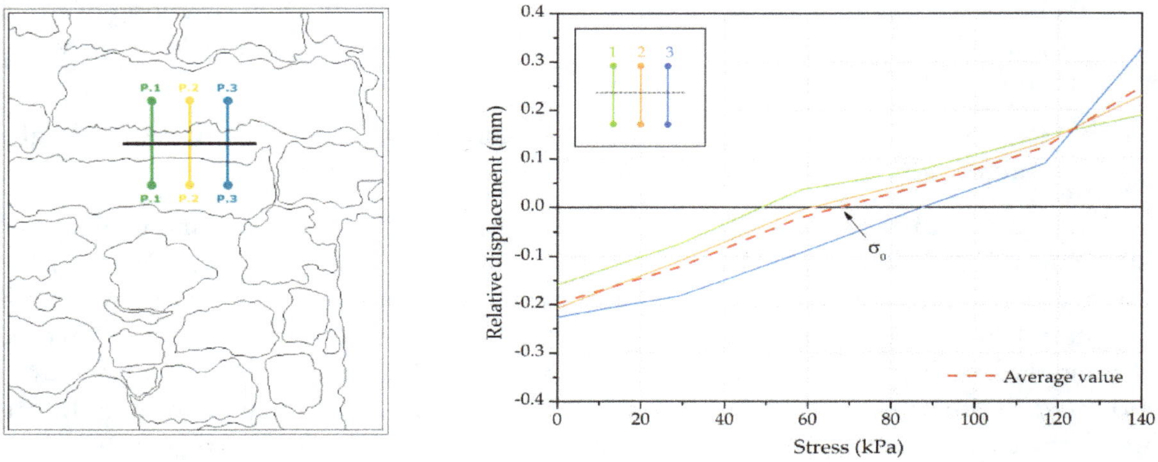

(a) (b)

Figure 16. Single flat-jack test OS1: (**a**) identification of the vertical alignments; (**b**) evolution of the relative displacements for each vertical alignment and estimation of the in-situ stress.

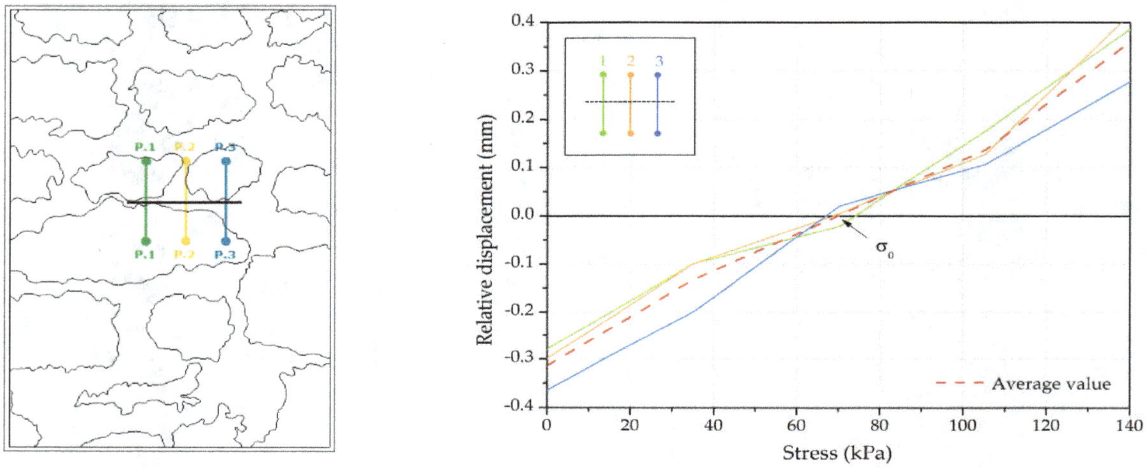

(a) (b)

Figure 17. Single flat-jack test OS2: (**a**) identification of the vertical alignments; (**b**) evolution of the relative displacements for each vertical alignment and estimation of the in-situ stress.

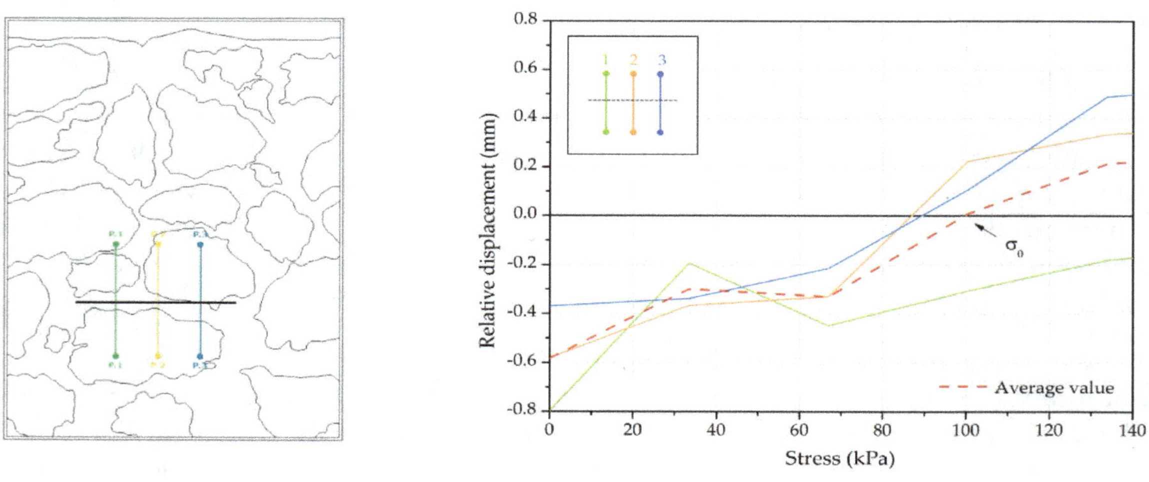

(a) (b)

Figure 18. Single flat-jack test AS1: (**a**) identification of the vertical alignments; (**b**) evolution of the relative displacements for each vertical alignment and estimation of the in-situ stress.

Figures 19–21 present the average stress-strain curves obtained for the double flat-jack tests conducted for the same walls and locations. For both buildings, failure was not reached and the maximum stress values attained range between 0.75 to 1.0 MPa. For the three tests, the masonry presented a high percentage of stone in the tested areas and also good fabric, even though there were generalized small dimensions and irregular stone blocks used. In the case of test AD1, this location revealed lower deformability in comparison to the other tested locations.

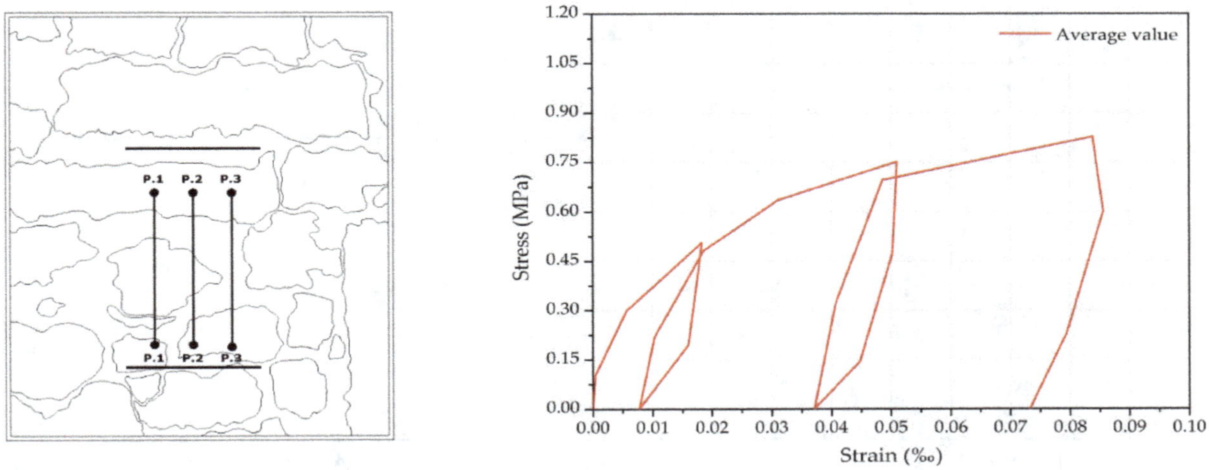

(a) (b)

Figure 19. Double flat-jack test OD1: (**a**) identification of the vertical alignments; (**b**) average stress-strain curve.

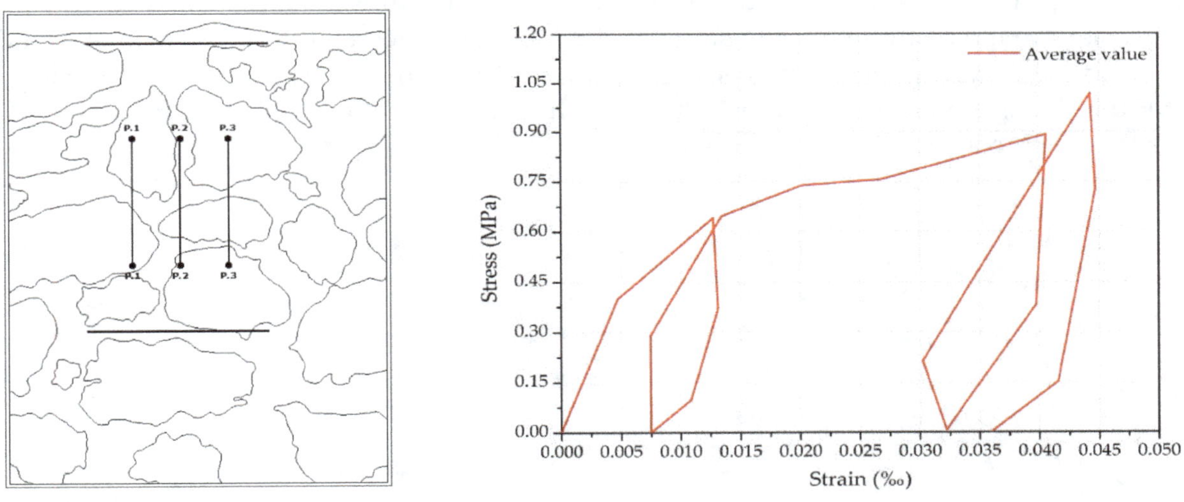

(a) (b)

Figure 20. Double flat-jack test AD1: (**a**) identification of the vertical alignments; (**b**) average stress-strain curve.

Table 2 shows the estimated in situ stress that was obtained in the single flat-jack tests, and the maximum stress applied (which represents a lower limit of the wall's compressive strength). As one can understand from the analysis of the values given in Table 1, in the case of *Orfeão de Viseu*, the value of maximum stress recorded in the double flat-jack test is more than 10 times higher than those determined in the single flat-jack test, which is revealing of the high compressive safety level that these structures generally present.

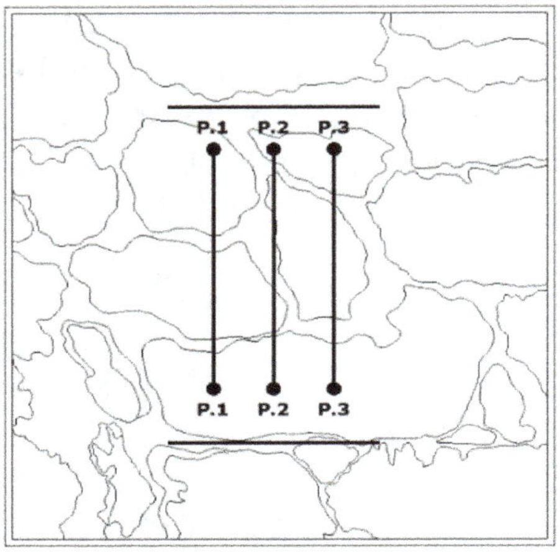

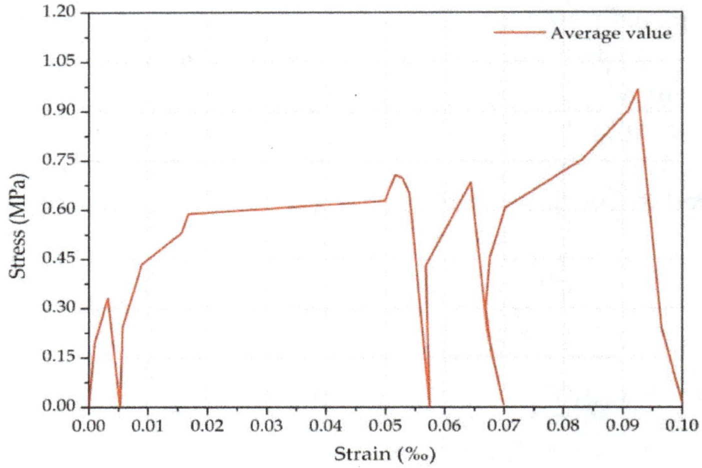

(a) (b)

Figure 21. Double flat-jack test AD2: (**a**) identification of the vertical alignments; (**b**) average stress-strain curve.

Table 2. Flat-jack test results (in situ stress).

	Orfeão de Viseu			*Águas de Viseu*		
	OS1	**OS2**	**OD1**	**AS1**	**AD1**	**AD2**
in situ stress (kPa)	68	70	–	99	–	–
Maximum stress (kPa)	–	–	915	–	1098	1119

The Young's modulus was obtained through graphical analysis, taking into account the slope of the stress-strain curves, namely the slope of the secant stiffness. Different portions of the curve (corresponding to different percentages of the maximum strength) can be analyzed, leading to different values for the Young's modulus. Figure 22 presents the different options for evaluation of the Young's modulus that were considered in this paper.

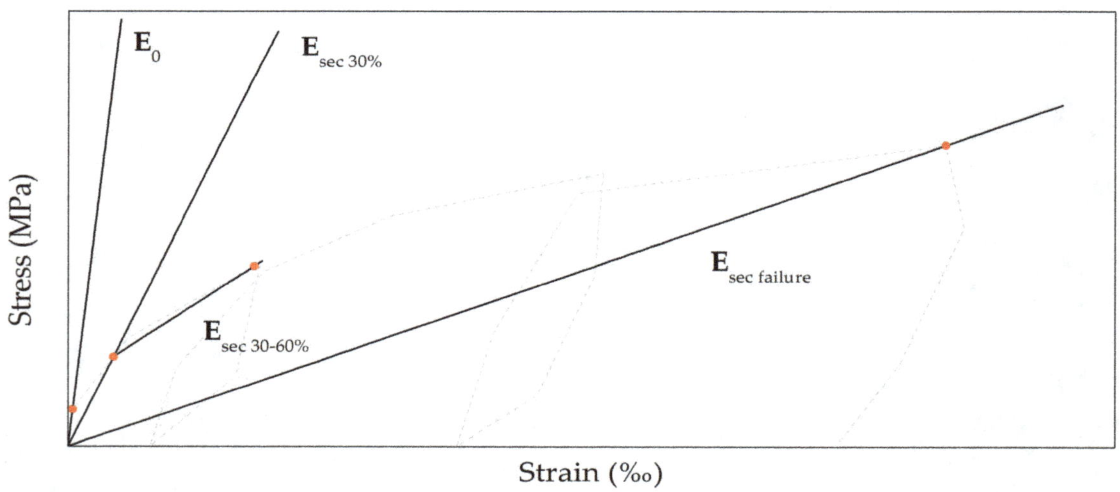

Figure 22. Representation of the different Young's modulus.

Table 3 presents the estimation of the values for the different Young's modulus that were computed, with a 15% reduction coefficient [26].

Regarding these values, it is important to stress here that several authors and code documents suggest that they can be affected by a non-neglectable variation resulting from the lateral confinement conditions of the masonry portion tested. ASTM standard C197-91 [25], RILEM TC 76-LUM [30], Binda [31], Noland et al. [32], all refer to variations for in situ stress up to 24%, overestimation of maximum stress of about 15% and overestimation from 10 to 20% of the elastic modulus.

Table 3. Flat-jack test results (Young's modulus).

Building	*Orfeão de Viseu*			*Águas de Viseu*		
Flat-jack Test	**OS1**	**OS2**	**OD1**	**AS1**	**AD1**	**AD2**
E_0 (MPa)	–	–	1994	–	833	1716
$E_{sec\ 30\%}$ (MPa)	–	–	486	–	833	983
$E_{sec\ 30–60\%}$ (MPa)	–	–	157	–	319	175
$E_{sec\ failure}$ (MPa)	–	–	84	–	195	89

Finally, Table 4 presents the values of maximum stress applied and Young's modulus obtained in other experimental campaigns aiming at characterizing granite stone masonry walls in the North of Portugal, in situ [3] and in laboratory, conducted in wall panels collected from existing buildings [6].

Table 4. Granite stone masonry experimental campaigns results.

Experimental Campaign	Maximum Stress (kPa)	Young's Modulus (MPa)
Miranda, 2011 [3]	49–840	800–3300
Almeida, 2013 [6]	2500–3940	220–320

As can be observed, the values obtained in the experimental campaign presented herein are consistent with those reported by other authors for granite stone masonry walls in Portugal. Nevertheless, it must be recalled that some typological differences exist between the walls studied in this paper and in those campaigns, namely, in what concerns the number of leaves and their connection level. This feature must be taken in account when comparing mechanical properties. In fact, the studies by Miranda [3] and Almeida [6] mentioned above concern walls typical from the old urban center of Porto, representing a peculiar typology, based on granite stones with medium to large dimensions (with a diagonal length up to 1.10 m), and assembled in roughly horizontal alignments. As to the construction technique, stones are first put in place and only then is mortar applied along the stone's border. This represents a constructive process unique to a specific geography, and the results obtained in those walls are not easily extrapolated to stone masonry walls from different areas in Northern Portugal. This peculiarity reinforces the pioneering character of the research described in this paper.

6. Conclusions

The work presented in this paper intends to offer a contribution to the typological and mechanical characterization of granite stone masonry walls frequent in the Portuguese old urban centers. With this intent, an in situ characterization campaign was carried out in the old city center of Viseu, Portugal, under the scope of which two traditional stone masonry buildings were used as case studies. From this experimental campaign, some general conclusions can be drawn:

Flat-jack testing is a relatively simple and economical testing procedure that is extremely useful in in situ characterization of stone masonry walls. In fact, as discussed in this manuscript, the level of rigor achieved with this technique is clearly consistent with that needed for practical rehabilitation or strengthening interventions [23,33]. It must nevertheless be mentioned that a skilled operator is necessary in order to guarantee the desired reliability.

Moreover, when dealing with irregular stone masonry walls, the difficulty in following all the requirements stated in the standards are evident. Among these, one must mention the difficulty in locating a suitable area to open the cuts (preferably a horizontal joint, difficult to locate with a suitable extension in this kind of masonry), the dimension of the stones (which conditions the dimension of the jacks to be used), the possibility that the cutting releases filling fragments (which affects the original load paths) or causes local stone crushing, and the difficulty in positioning the displacement measuring equipment. All of this introduces subjectivity in the results interpretation procedure. A uniformized flat-jack testing procedure suitable for application in irregular stone masonry walls is thus desirable in order to guarantee that not only are the values obtained representative of the wall, but also that values resulting from different campaigns are comparable [14,34,35].

Determining the effective area of the flat-jack in contact with the masonry is determined to be a decisive conditioner of the results' reliability. Some considerations on the adopted procedure are thus desirable. It was found that, besides requiring sensitivity from the analyst and thus being thus open to subjectivity, this procedure presents the disadvantage that the flat-jack removal may damage the carbon paper set, thus rendering its analysis even more difficult. Estimating the effective loaded area and resorting to visual observation of the jack, after its removal, can be a suitable alternative. On the other hand, and even when it is possible to keep the integrity of the carbon papers during flat-jack removal, this operation may cause the marking of areas exclusively derived from this procedure, and not from the jack pressurizing, resulting in underestimated stress values.

Regarding the use of this experimental technique to characterize specific type of masonry (granite stone masonry), it is worth noting that due to the characteristics of these walls, there is often the need to resort to heavy techniques/equipment to remove the flat-jacks or the cutting equipment from the wall, leading to the disassembly of the testing area. In some cases, the damage suffered by the walls as a result of these operations is clearly unacceptable. Furthermore, when the disassembling is required during the slot cutting, conducting the flat-jack test becomes altogether impossible.

The relationship between stress values measured from the single flat-jack test and maximum stress attained for each test site are indicators of the level of safety of the masonry tested in respect to vertical loading. The test results show that maximum stress values are considerably higher than the in situ stress, which leads to high safety factors (ratio between maximum stress value reached and the in situ stress) in respect to vertical actions, characteristic feature of this type of masonry, with a ratio between 11 and 13.

As a final note, the authors would like to underline that this work reports and discusses a first study of this kind conducted in the old urban center of Viseu, so further characterization campaigns are encouraged in order to improve the reliability of the assumptions made and to validate some of the results and conclusions presented.

Author Contributions: All authors contributed to every part of the research described in this paper.

Acknowledgments: This research was carried out in the scope of the project "Viseu Património". The authors wish to thank the support of the Viseu City Council, through the Urban Rehabilitation Society Viseu Novo. Engineer Jorge Fonseca is gratefully acknowledged for his contribution in the experimental component of the campaign. The constructive comments and suggestions provided by the two anonymous reviewers are also acknowledged by the authors.

References

1. ICOMOS. *ICOMOS Chapter—Principles for the Analysis, Conservation and Structural Restoration of Architectural Heritage*; International Council on Monuments and Sites (ICOMOS): Paris, France, 2003.
2. Sousa, L. Caraterização e Parametrização de Paredes Portantes de Alvenarias de Pedra Quanto à Regularidade Geométrica no seu Plano. Master's Thesis, Universidade do Porto, Porto, Portugal, 2010.
3. Miranda, L. Ensaios Acústicos e de Macacos Planos em Alvenarias Resistentes. Ph.D. Thesis, Universidade do Porto, Porto, Portugal, 2011.
4. Vasconcelos, G. Experimental Investigations on the Mechanics of Stone Masonry: Characterization of Granites and Behavior of Ancient Masonry Shear Walls. Ph.D. Thesis, Universidade do Minho, Guimarães, Portugal, 2005.

5. Oliveira, D.; Lourenço, P.B. Experimental Behaviour of Three-Leaf Stone Masonry Walls. In *Proceedings of the "The Construction Aspects of Built Heritage Protection" Conference and Brokerage Event, Dubrovnik, Croatia, 14–17 October 2006*; Radić, J., Rajčić, V., Žarnić, R., Eds.; European Construction Technology Platform: Brussels, Belgium, 2006; pp. 356–362.
6. Almeida, C. Paredes de Alvenaria do Porto: Tipificação e Caraterização Experimental. Ph.D. Thesis, Universidade do Porto, Porto, Portugal, 2013.
7. Ferreira, T.; Costa, A.A.; Arêde, A.; Gomes, A.; Costa, A. Experimental characterization of the out-of-plane performance of regular stone masonry walls, including test setups and axial load influence. *Bull. Earthq. Eng.* **2015**, *13*, 2667–2692. [CrossRef]
8. Instituto Nacional de Estatística. *Censos 2011 Resultados Definitivos—Centro*; Instituto Nacional de Estatística (INE), I.P.: Lisboa, Portugal, 2012.
9. Aragão, M. *Viseu, Instituições Sociais*; Seara Nova: Lisbon, Portugal, 1936.
10. IPQ. *NP EN 1926. Métodos de Ensaio Para Pedra Natural. Determinação da Resistência à Compressão (Natural Stone Test Methods. Determination of Compressive Strength)*; Instituto Português da Qualidade (Portuguese Institute for Quality): Caparica, Portugal, 2000.
11. Casella, G. *Gramáticas de Pedra. Levantamento de Tipologias de Construção Murária*; Centro Regional de Artes Tradicionais: Porto, Portugal, 2003.
12. Bell, F. *Engineering Geolog*; Elsevier: Oxford, UK, 2007.
13. Vicente, R.; Ferreira, T.; Mendes da Silva, J.; Varum, H. In situ flat-jack testing of traditional masonry walls: Case study of the old city center of Coimbra, Portugal. *Int. J. Archit. Herit.* **2015**, *9*, 794–810. [CrossRef]
14. Cescatti, E.; Dalla Benetta, M.; Modena, C.; Casarin, F. Analysis and evaluations of flat jack test on a wide existing masonry buildings sample. In *Proceedings of the 16th International Brick and Block Masonry Conference, Padova, Italy, 26–30 June 2016*; Modena, C., da Porto, F., Valluzzi, M.R., Eds.; CRC Press: Boca Raton, FL, USA, 2016; pp. 1485–1491.
15. Binda, L.; Tiraboschi, C. Flat-jack test as a slightly destructive technique for the diagnosis of brick and stone masonry structures. *Int. J. Restor. Build. Monum.* **1999**, *5*, 449–472.
16. Uranjek, M.; Bosiljkov, V.; Žarnić, R.; Bokan-Bosiljkov, V. In situ tests and seismic assessment of a stone-masonry building. *Mater. Struct.* **2012**, *45*, 861–879. [CrossRef]
17. Lombillo, I.; Thomas, C.; Villegas, L.; Fernández-Álvarez, J.; Norambuena-Contreras, J. Mechanical characterization of rubble stone masonry walls using non and minor destructive tests. *Constr. Build. Mater.* **2013**, *43*, 266–277. [CrossRef]
18. Andreini, M.; de Falco, A.; Giresini, L.; Sassu, M. Mechanical characterization of masonry walls with chaotic texture: Procedures and results of in-situ tests. *Int. J. Archit. Herit.* **2014**, *8*, 376–407. [CrossRef]
19. Simões, A.; Bento, R.; Gago, A.; Lopes, M. Mechanical characterization of masonry walls with flat-jack tests. *Exp. Tech.* **2016**, *40*, 1163–1178. [CrossRef]
20. Gelmi, C.; Modena, C.; Rossi, P.P.; Zaninetti, A. Mechanical characterization of stone masonry structures in old urban nuclei. In *Proceedings of the 6th North American Masonry Conference, Philadelphia, PA, USA, 6–9 June 1993*; Hamid, A.A., Harris, H.G., Eds.; The Masonry Society: Longmont, CO, USA, 1993; pp. 505–516.
21. Roque, J. Reabilitação estrutural de paredes antigas de alvenaria. Master's Thesis, Universidade do Minho, Guimarães, Portugal, 2002.
22. Pagaimo, F. Caracterização morfológica e mecânica de alvenarias antigas: Caso de estudo da vila histórica de Tentúgal. Master's Thesis, Universidade de Coimbra, Coimbra, Portugal, 2004.
23. Ferreira, T. Avaliação da vulnerabilidade sísmica de núcleos urbanos antigos: Aplicação ao núcleo urbano antigo do Seixal. Advanced Studies Thesis in Rehabilitation of Built Heritage, Universidade do Porto, Porto, Portugal, 2010.
24. ASTM. *ASTM C 1196-91: Standard Test ASTM C1196-91, Standard Test Method for in Situ Measurement of Masonry Deformability Properties Using the Flatjack Metho*; ASTM: West Conshohocken, PA, USA, 1991.
25. ASTM. *ASTM C 1197-91: Standard Test ASTM C1197-91, In-Situ Compressive Stress within Solid Unit Masonry Estimated Using Flat-Jack Measurements*; ASTM: West Conshohocken, PA, USA, 1991.
26. RILEM. RILEM Recommendation MDT.D.4: In-situ stress tests based on the flat jack. International union of laboratories and experts in construction materials, systems and structures (RILEM), RILEM TC 177-MDT: Masonry durability and on-site testing. *Mater. Struct.* **2004**, *37*, 491–496. [CrossRef]

27. RILEM. RILEM Recommendation MDT.D.5: In-situ stress-strain behaviour tests based on the flat jack. International union of laboratories and experts in construction materials, systems and structures (RILEM), RILEM TC 177-MDT: Masonry durability and on-site testing. *Mater. Struct.* **2004**, *37*, 497–501. [CrossRef]
28. *LabView SignalExpress*; National Instruments: Austin, TX, USA, 2010.
29. Gregorczyk, P.; Lourenço, P.B. A review on flat-jack testing. *Engenharia Civ.* **2000**, *9*, 39–50.
30. De Vekey, R.C. General recommendations for methods of testing load-bearing unit masonry. *Mater. Struct.* **1988**, *21*, 229–231. [CrossRef]
31. Binda, L. Sonic tomography and flat-jack tests as complementary investigation procedures for the stone pillars of the temple of S. Nicolò l'Arena (Italy). *NDT E Int.* **2003**, *36*, 215–227. [CrossRef]
32. Noland, J.; Atkinson, R.; Schuller, M. A review of the flat-jack method for nondestructive evaluation. In Proceedings of the Nondestructive Evaluation of Civil Structures and Materials, Boulder, CO, USA, 15–17 October 1990.
33. Santhakumar, A.R.; Mathews, M.S.; Thirumurugan, S.; Uma, R. Seismic Retrofitting of Historic Masonry Buildings—Case Study. *Adv. Mater. Res.* **2010**, *133*, 991–996. [CrossRef]
34. Giordano, A.; Cascardi, A.; Micelli, F.; Aiello, M.A. Theoretical study for the strengthening of a series of vaults in a cultural heritage masonry building: A case study in Italy. In Proceedings of the 10th International Masonry Conference (10IMC), Milan, Italy, 9–11 July 2018; pp. 2510–2531.
35. La Mendola, L.; Lo Giudice, E.; Minafò, G. Experimental calibration of flat jacks for in-situ testing of masonry. *Int. J. Archit. Herit.* **2018**, 1–11. [CrossRef]

5

Prediction of the Seismic Response of Multi-Storey Multi-Bay Masonry Infilled Frames using Artificial Neural Networks and a Bilinear Approximation

Tanja Kalman Šipoš * and Kristina Strukar

Faculty of Civil Engineering and Architecture Osijek, Department for Technical Mechanics, 31000 Osijek, Croatia; kstrukar@gfos.hr
* Correspondence: tkalman@gfos.hr

Abstract: In order to test the reliability of neural networks for the prediction of the behaviour of multi-storey multi-bay infilled frames, neural network processing was done on an experimental database of one-storey one-bay reinforced-concrete (RC) frames with masonry infills. From the obtained results it is demonstrated that they are acceptable for the prediction of base shear (BS) and inter-storey drift ratios (IDR) in characteristic points of the primary curve of infilled frame behaviour under seismic loads. The results obtained on one-storey one-bay infilled frames was extended to multi-bay infilled frames by evaluating and comparing numerical finite element modelling(FEM) modelling and neural network results with suggested approximating equations for the definition of bilinear capacity by defined BS and IDRs. The main goal of this paper is to offer an interpretation of the behaviour of multi-storey multi-bay masonry infilled frames according to a bilinear capacity curve, and to present the infilled frame's response according to the contributions of frame and infill. The presented methodology is validated by experimental results from multi-storey multi-bay masonry infilled frames.

Keywords: masonry; infilled frames; capacity curve; bilinear approximation; neural networks; database

---

1. Introduction

The use of masonry infilled frames is very common for most types of building, accordingly state of the art of masonry infilled frame behaviour [1–3] in general is known but there is still no suggestion of regulations on how to model or use it properly in structural analysis.

The use of neural networks in the civil engineering field is already approved [4,5] however the application of neural networks for the prediction of infilled frame behaviour is rare. There are only a few studies [6–8] that have explored this topic. With a lack of available data from experiments of masonry infilled frames and with the uncertainty of numerical modelling, this research area needs to be further investigated. In order to connect most of the previously published data with new valuable conclusions, an experimental database of masonry infilled frames was collected. It was limited to only one-storey, one-bay infilled frames according to the availability and uniformity of the structural type.

Previous assumptions based on the monolithic behaviour of confined masonry [9] will be evaluated and validated for the possible monolithic behaviour of masonry infilled frames in this study. The methodology is based on the processing of the collected database with neural networks, accurate FEM modelling of infilled frames, a bilinear approximation of observed results and the capacity curve range definition with validation by multi-storey multi-bay infilled frame experimental results.

The experimental database of infilled frames is expressed with general geometric and material input and output data based on inter-storey drift ratios (IDR) and base shear (BS) in two stages of behaviour. First, damage indication introduced by the first major crack in the infill, which is presented by a decline of the initial stiffness of infilled frames. The second stage is related to the maximum strength of the infilled frame where the infill and RC frames still act together as monolithic elements. Inputs and outputs are then used for the prediction of behaviour by neural network processing. The number of hidden neurons is defined according to suggested equations by the amount of input and output data. The Levenberg-Marquardt learning algorithm is used for neural network processing to obtain output data.

Numerical FEM models of infilled frames are defined in order to connect predicted results from neural networks based on one-storey one-bay infilled frame with results from calibrated numerical multi-bay models.

A comparison and definition of the relation of IDRs and BSs between one- and multi-bay infilled frames is based on the contribution of infill in infilled frames in regard to bare frames. According to the results, new equations are defined, and their accuracy confirmed for both regular and irregular bay lengths.

The applicability of presented equations for capacity curve definition is verified on multi-storey multi-bay experimental models.

These relations present the connection between behaviour measures expressed by IDRs and BSs from one-storey one-bay to multi-storey multi-bay real buildings. They present the first proposed relations which are limited to RC frames with fully infilled frames and behaviour based on capacity.

2. Experimental Database of Infilled Frames, EDIF

The results of experimental tests conducted on one-story one-bay infilled reinforced-concrete frames with masonry infill are collected, systematized and processed in the Experimental Database of Infilled Frames called EDIF [7]. The tests that were observed had no shear connection, outside adhesion between the frame and infill, and no openings in the infill.

The database was obtained from a previous work by the authors [7] in which a large number of in-plane lateral tests reported in the literature were considered. Although many tests have been performed since then, the database is still representative of the typical lateral behaviour of masonry infilled frames.

The collected experimental database contains 113 published tests (Table 1) based on all available data including material and geometric properties from RC frame and masonry infill, type and size of load, failure mode, and capacity values obtained from the capacity curves. Although the initial goal was to create a database that has identical parameters for a large number of samples, some parameters were omitted as they were incomplete or unavailable (transverse reinforcement of columns and beams, material properties of mortar and masonry units, masonry shear strength, the maximum drift).

Table 1. Experimental database: authors and samples list.

Author	Year	Laboratory	Scale	Load	No of Samples
Combescure [10]	2000.	LNEC, Lisbon	1:1.5	C	1
Colangelo [11]	1999.	L'aquila, Italy	1:2	C	11
Cavaleri [12]	2004.	-	1:2	C	1
Lafuente [13]	1998.	U.C.V. Caracas, Venezuela	1:2	C	10
Kakaletsis [14]	2007.	-	1:3	C	2
Dukuze [15]	2000.	-	1:3	M	23
Žarnić [16,17]	1985.	Institute for Testing and Research in Materials and Structures (ZRMK), Ljubljana	1:2	C	1
	1992.		1:3		3
Al-Charr [18]	1998.	USACERL, Illinois	1:2	M	2
Angel [19]	1994.	University of Illinois, Champaign	1:1	C	7

Table 1. *Cont.*

Author	Year	Laboratory	Scale	Load	No of Samples
Mehrabi [20]	1994.	University of Colorado, Boulder	1:2	C	8
				M	3
Crisafulli [21]	1997.	-	1:1.33	M	2
Fiorato [22]	1970.	University of Illinois, Urbana	1:8	C	3
Yorulmaz [23]	1968.	University of Illinois, Urbana	1:8	M	7
Benjamin [24]	1958.	Stanford University, California	1:2.94	M	5
			1:1.33		2
			1:1		1
			1:4		5
			1:2.38		7
Zovkić [25]	2012.	Faculty of Civil Engineering, Osijek, Croatia	1:2.5	C	9

According to the analysis of the most influenced parameters [7], the input data (Table 2) that were used for this study are:

- a—height to length ratio,
- b—ratio of moments of inertia of beam to column,
- g—ratio of column width to the thickness of masonry infill,
- r_c—reinforcement ratio of column,
- f_y—yield strength of the reinforcing steel,
- λ_h—stiffness ratio (Equation (1)),
- V—axial load on columns.

The parameter λ_h is a measure of the relative stiffness between the frame and masonry infill; a greater λ_h corresponds to a more flexible frame. It can be determined using the following equation:

$$\lambda_h = h\sqrt[4]{\frac{E_i \cdot t \cdot \sin(2\theta)}{4E_c \cdot I \cdot h_w}} \tag{1}$$

where:

- h—height of frame between the beam axis,
- h_w—height of masonry infill,
- E_c—modulus of elasticity of column,
- E_i—modulus of elasticity of masonry infill,
- I—moment of inertia of the column,
- θ—whose tangent is equal to the relation between height and length of infill.

For evaluation of the performance of infilled frame structures the measured resistance envelope curve from experiments is presented by bilinear curve using the equal energy rule. Therefore, two points are obtained on the idealized bilinear curve defined by IDR and BS in the first cracking point (IDR_c and BS_c) and maximum point (IDR_m and BS_m) of the capacity curve. The first cracking point is characterized by a sudden decline in stiffness and the maximum point is associated with the maximum lateral capacity of the system. Post-ultimate behaviour could not be determined from the available data since that region was not observed in most of the tests. In Table 2 are given output data including both IDR_c and BS_c, and IDR_m and BS_m from neural network processing and they are expressed as dimensionless because of the simplicity in neural network modelling.

Table 2. The range of input and output data in the EDIF Database.

Statistical Function	Input Data								Output Data		
	a	b	g	r_c	f_y (MPa)	λ_h	V (kN)	IDR_c	IDR_c	BS_c (kN)	BS_c (kN)
min	0.33	0.60	1	0.01	203.37	1.78	0	0.01	0.03	55.9	76.95
max	2.28	8.00	6.1	0.04	607	8.56	2343.75	0.55	2.91	2278.4	2563.2
average	0.74	2.04	2.01	0.02	406.94	3.68	599.06	0.13	0.72	594.64	878.22

3. Neural Network Modelling

Neural networks have demonstrated their capability in dealing with highly nonlinear relationships between the applied load and the measured displacement values [26]. In this paper, neural networks are designed to examine their capability in estimating the values of output data from EDIF database for the performance of infilled frames.

The employed neural network is a multi-layer neural network with a fully connected configuration. Totally, 113 data sets were obtained from experimental tests. The early stopping method is used for training neural networks. According to this method, data are divided into three groups of training, testing and validation. Herein, 70% of data was randomly allocated to training and the rest are equally divided between validation and testing data. The testing data set presents the unseen data set that was not included in the training of neural networks.

The Levenberg–Marquardt (LM) backpropagation algorithm was employed for training the neural networks by using the Matlab ANN toolbox. Moreover, the gradient descent weight/bias learning function was used with a hyperbolic tangent function as the activation function for neural network processing. To avoid the saturation of neural networks, the input and output data were scaled to [0, 1]. The training of data was stopped when the MSE in the validation set started to increase signifying that the ANN generalization stopped increasing. General neural network parameters of used NN model are presented in Table 3.

Table 3. The values of neural network parameters used in NN model.

Parameters	ANN
Number of input layer units	7
Number of hidden layers	1
Number of hidden layer units	3, 5, 8
Number of output layer units	1
Learning rate	0.01
Performance goal	0
Maximum number of epochs	10,000

It has been shown in many damage identification studies that the application of one hidden layer is adequate for accurate prediction [27,28]. The number of hidden layers was determined by the analysis of empirical criteria (Table 4) according to the suggestions from different authors [29,30] based on the number of input data N_i and number of outputs (in this study No = 1, neural networks processing was done always with only one output). As is visible in Figure 1, the most frequent values are three, five and eight neurons. Accordingly, further analysis was done with those three suggestions in order to obtain the best results with neural network processing.

As it can be seen from Table 5, the best performance was achieved when the hidden layer had five neurons. For the evaluation of accuracy, performance measures used are: mean absolute error (MAE), root mean squared error (RMSE), mean absolute percentage error (MAPE) and coefficient of correlation (R). These performance measures are based on average values from four output data observed.

Table 4. Empirical criteria for the number of hidden neurons N_h.

No	Method and Reference	N_h	Number of N_h
1.	Hecht-Nielsen (1987) [30]	$\leq 2 \cdot N_i$	14
2.	Hush (1989) [30]	$3 \cdot N_i$	21
3.	Popovics (1990) [31]	$(N_i + N_o)/2$	4
4.	Gallant (1993) [31]	$2 \cdot N_i$	11
5.	Wang (1994) [30]	$2 \cdot N_i/3$	5
6.	Masters (1994) [30]	$(N_i + N_o)^{1/2}$	3
7.	Li (1995) [29]	$((1 + 8\,N_i)^{1/2} - 1)/2$	3
8.	Tamura (1997) [29]	$N_i + 1$	8
9.	Lai (1997) [31]	N_i	7
10.	Nagendra (1998) [31]	$N_i + N_o$	8
11.	Zhang (2003) [29]	$2^{Ni}/n + 1$	19
12.	Shibata (2009) [29]	$(N_i \cdot N_o)^{1/2}$	3
13.	Sheela (2013) [29]	$(4\,N_i^2 + 3)/(N_i^2 - 8)$	5

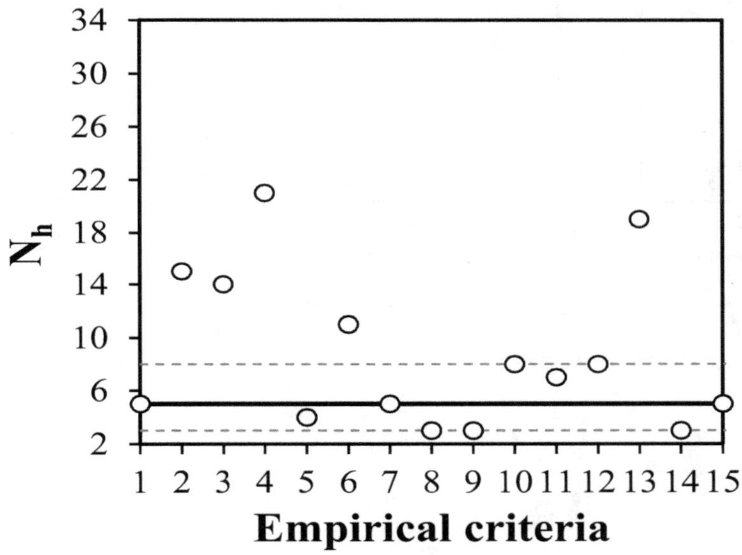

Figure 1. Range of number of hidden neurons according to Ni = 7, No = 1.

The obtained results from the processed neural networks are presented in Figure 2. All output data (BS and IDR) results for training, validation and the testing set are presented. As it can be seen from the correlation coefficients (R), all four output data are successfully predicted. It is evident that the trained neural networks demonstrate capability for the generalization of processed neural networks. It can be concluded that NNs can be used for accurate estimation of capacity curve of masonry infilled frames.

Table 5. Statistical performance of ANN models.

ANN Model						
Neural Network Label	Learning Algorithm	No. of Hidden Nodes	MAE [1]	RMSE [2]	MAPE [3] (%)	R [4]
LM_3	Levenberg-Marquardt	3	13.709	0.926	13.157	0.829
LM_5		5	10.0265	0.481	11.633	0.919
LM_8		8	12.189	0.558	17.844	0.886

[1] Mean absolute error (MAE), [2] root mean squared error (RMSE), [3] mean absolute percentage error (MAPE) and [4] coefficient of correlation (R).

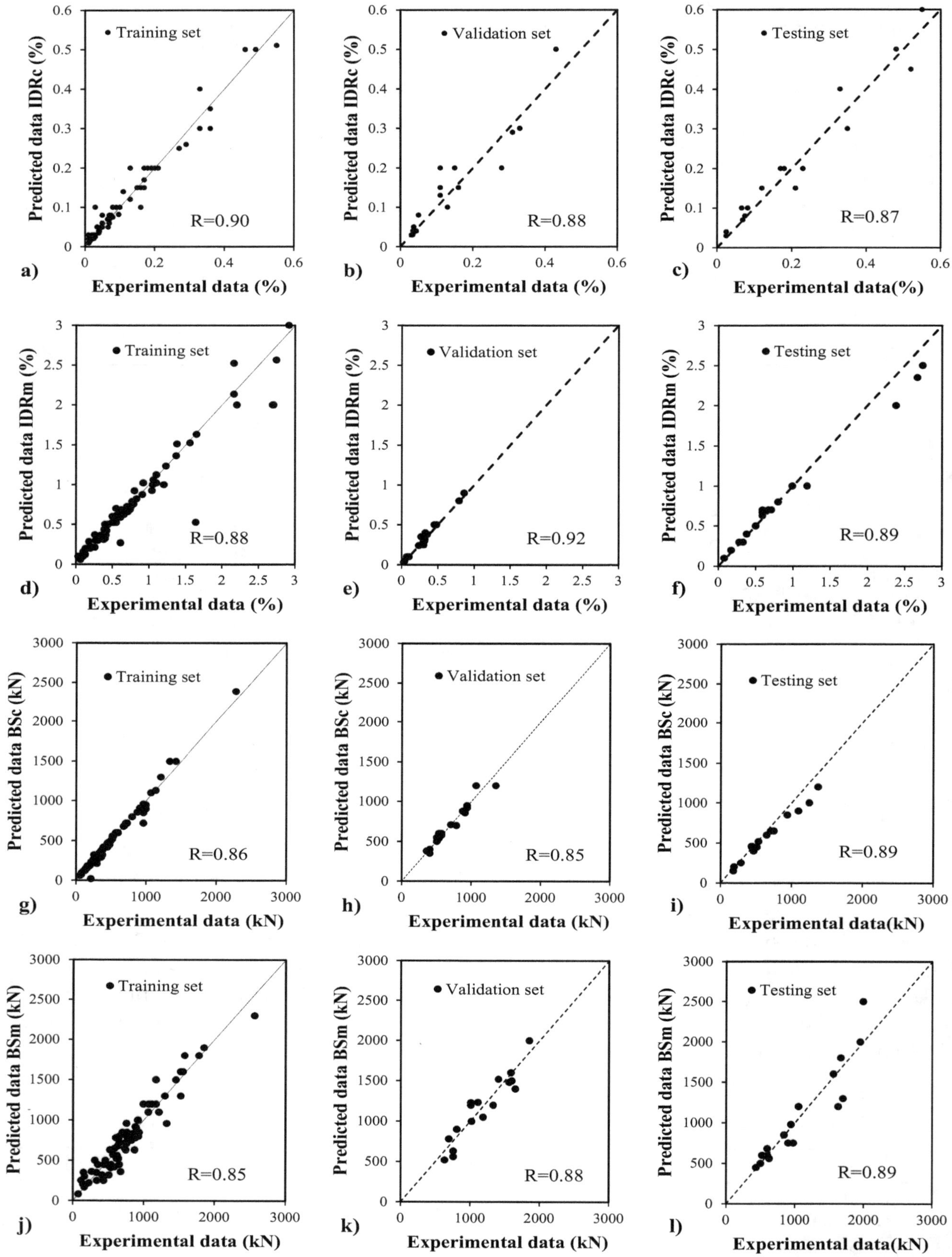

Figure 2. Results of neural network processing for 5 hidden neurons.(**a**) for IDRc training set; (**b**) for IDRc validation set; (**c**) for IDRc testing set; (**d**) for IDRm training set; (**e**) for IDRm validation set; (**f**) for IDRm testing set; (**g**) for BSc training set; (**h**) for BSc validation set; (**i**) for BSc testing set; (**j**) for BSm training set; (**k**) for BSm validation set; (**l**) for BSm testing set.

4. Definition of Models for Numerical Nonlinear FEM Analysis and Neural Network Processing

The accepted numerical FEM model for masonry infill is implemented in Seismostruct 2018 [32] and calibrated according to multi-storey multi-bay infilled frame experimental results [33]. Infilled framed models (Figure 3) applied for analysis with neural networks and nonlinear numerical analysis are based on different length to height ratios of frames with a medium type of masonry infill (according to value of compressive strength). The accuracy and relation between numerical models and prediction was carried out for equivalent and various frames with two and three bays.

Reinforced concrete frames with height of 375 m has different bay length (A frame–3 m, B frame –4 m, C frame–5 m). Cross-section dimensions are kept constant for all three bay lengths, columns with 50 × 50 cm, and beams with 30 × 50 cm. The RC frames are designed according to seismic regulations [34]. Therefore, material properties for concrete members are based on concrete class C30/37 with reinforcement B500, with reinforcement ratio of 2% of the cross sections. Pushover analysis was carried out in order to evaluate the capacity of infilled frames.

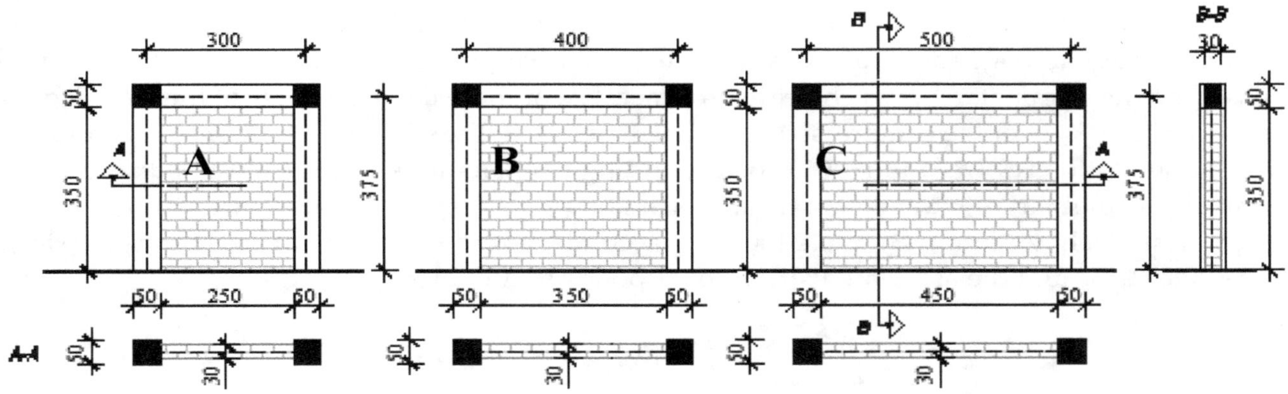

Figure 3. Infilled frame models.

The masonry infill type is defined according to compressive strength values. In order to define the basic material properties of the masonry infill (compressive strength of the masonry and the stress-strain relation in compression), verified recommendations are used for determining the compressive strength of masonry infill based on the masonry and mortar strength according to Hendry and Malek [35] presented in Equation (2):

$$f_k = 0.334 f_b{}^{0.778} f_m{}^{0.234} \tag{2}$$

For the analysis, as is stated earlier and according to Table 6, the medium type of masonry infill was used, which is the most commonly used masonry unit—a clay block with vertical hollow perforations, with dimensions of 29 × 19 × 19 cm and 10 MPa masonry unit compressive strength.

Table 6. Definition of masonry infill type based on the compressive strength.

Masonry Infill Type	Compressive Strength		
	Masonry Unit f_b (MPa)	Mortar f_m (MPa)	Masonry Infill f_k (MPa)–Equation (2)
Weak—ytong block	3	10	1.35
Medium—hollow clay block	**10**	**5**	**2.92**
Strong—solid brick	20	5	5.01

A definition of the behaviour in compression is determined according to the recommendations of Kaushik [36] as is presented in Figure 4.

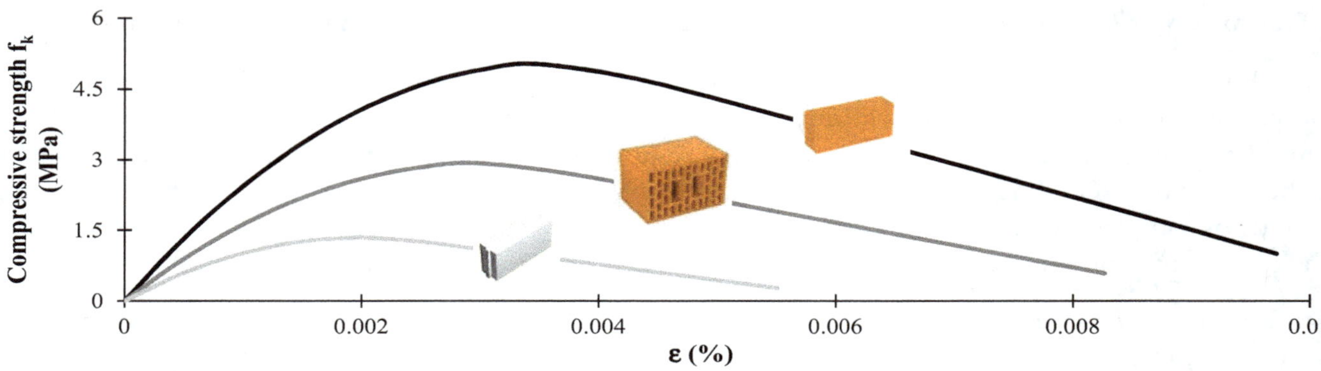

Figure 4. Stress-strain relation for three masonry infill types according to Kaushik [36].

For nonlinear FEM analysis, Mander's model [37] for concrete and Menegotto-Pinto's model [38] for reinforcement are used to define the force-based plastic hinge (FBPH) fiber elements, while masonry infill model is based on inelastic infill panel element (Table 7) defined by strut/shear curve properties [32]. The masonry infill wall was modelled as the infill panel model with calibrated parameters of hysteretic behavior. The initial diagonal width w_1 was determined according to proposal of Stafford et al. [39]. It is based on the parameter λ_h which presents a measure of the relative stiffness of the frame to infill. The reduced area A_{ms2} of the compressed diagonal depends on the stiffness coefficient λ_h, according to the recommendations of Decanini [40]. Corresponding deformations were determined according to the limit states: the start of reduction of the initial area A_{ms1} corresponds to the deformation at the end of linear elastic behaviour ($\varepsilon_m/3$) while the A_{ms2} secondary area is reached at 70% of the maximum compressive stress and the associated strain corresponds to the $1.5 \times \varepsilon_m$ (Figure 5).

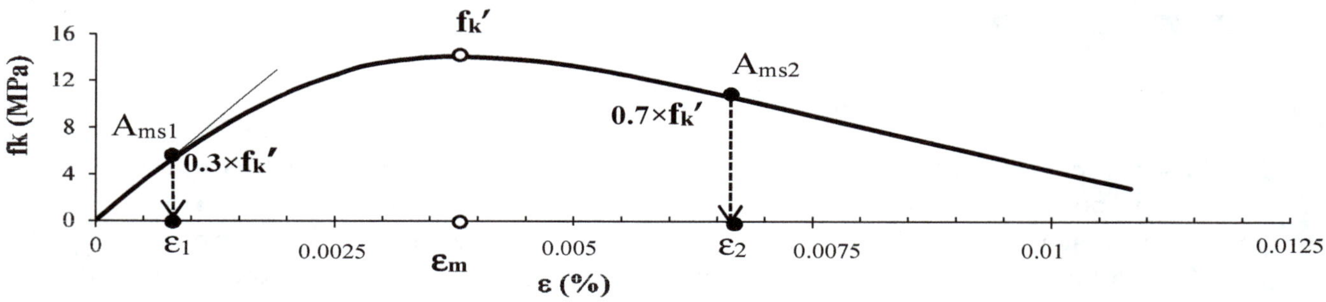

Figure 5. Stress-strain relation for masonry axial compression and the definition of limit states corresponded to the variability of areas and related axial strain.

Table 7. Material and geometric properties for nonlinear FEM model of masonry infill.

Masonry Infill	f_k (MPa)	E_i (MPa)	ε_m	ε_u	ε_1	ε_2	λ_h	$f_{m\theta}$* (MPa)	A_{ms1} (m^2)	A_{ms2} (%A_{ms1})
A							2.47		0.494	76.37
B	2.92	1610	0.0030	0.0083	0.001	0.0045	2.51	0.273	0.563	76.17
C							2.48		0.637	76.28

f_k—compressive strength of masonry; E_i—modulus of elasticity of masonry; ε_m—strain at maximum axial stress; ε_u—ultimate strain; ε_1—strut area reduction strain; ε_2—residual strut area strain; λ_h—relative panel-to-frame stiffness parameter; $f_{m\theta}$*—compressive strength; A_{ms1}—initial area of strut; A_{ms2}—final area of equivalent diagonal strut.

Unfortunately, the damage pattern type cannot be evaluated with this numerical model. This model can only give the information of RC members damage, but if we have only masonry infill damage it cannot be directly defined. Therefore, the structural performance levels for masonry infilled frames defined by values of IDRs [41] can be used for evaluation of possible damage in a composite system but not separately for every component individually.

4.1. Results of Analysis and Prediction of Neural Networks

The results of the predicted characteristic points of the capacity curve and the numerical calculation are verified on the basis of a relative error between values obtained from neural networks and numerical modelling values.

In Figure 6 the force-displacement curves for reinforced concrete frames without infills (bare frame—BF) and with infills are shown in order to detect the contribution of bearing capacity and the stiffness of infill and frame separately. An infilled frame has eight-fold greater stiffness and 1.5-fold larger ultimate capacity than the bare frame. Verification of the use of neural networks for the one-storey, one-bay frame is shown in Table 8, where relative errors are shown for each of the models. By comparing the results of neural networks and modelling, it can be concluded that yield strengths are very close, while, secondary stiffness is somewhat smaller (25%), unlike initial stiffness. The average relative error values are satisfactorily small (<10%), so it can be concluded that numerically obtained results, and prognosis results using neural networks are equivalent.

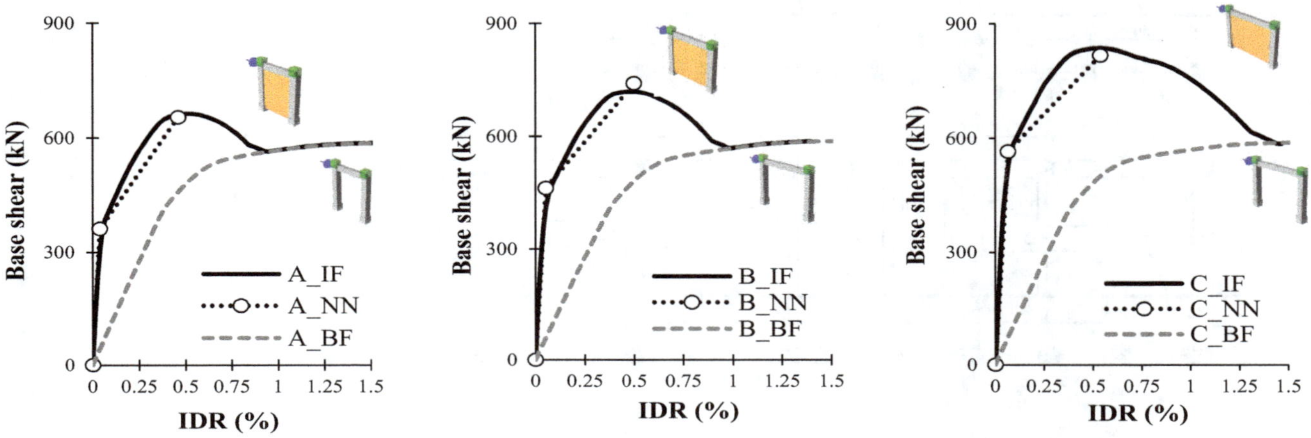

(a) (b) (c)

Figure 6. Capacity curves of RC infilled frames A (**a**), B (**b**) and C (**c**) with and without masonry infill in comparison with NN results.

Table 8. Evaluation of compatibility of neural networks processing for infilled frames A, B and C.

Capacity Curve Data	A_IF	B_IF	C_IF	A_NN	B_NN	C_NN
IDR_c (%)	0.04	0.054	0.067	0.038{5}	0.053{2}	0.066{1}
IDR_m (%)	0.49	0.50	0.51	0.46{8}	0.5{2}	0.54{6}
BS_c (kN)	357.92	447.04	545.96	363{2}	462{3}	565{3}
BS_m (kN)	663.83	719.69	815.47	658{1}	742{3}	818{1}

[1]—values in braces { } presents relative error from data obtained by neural networks in regard to results from numerical analysis.

From the comparison of the bare frame with an infilled frame by observing the two areas (yield and ultimate) can be concluded that:

- in the cracking area the infill wall retains 80% load capacity, while the RC frame assumes 20% load capacity (10% per column);
- in the area of the maximum strength, the infill takes on average 40%, while the frame or each of the columns assume 30% of the load capacity.

According to these values it can be concluded that by adding an additional frame to the initial one, the bearing capacity in the yield area will be increased by 90%, while in the range of the maximum load for each new additional frame bearing capacity will be increased by 70%.

4.2. Application of Neural Networks on Multi-bay Frames with Same Bay Length

Previous considerations and conclusions will be extended to multi-bay frames with the same bay length for two or three bays (Figure 7). Multi-bay frames cannot be directly applied and processed by neural networks in this paper, however they are expressed and included in approximation formulas with factors depending on the number of bays.

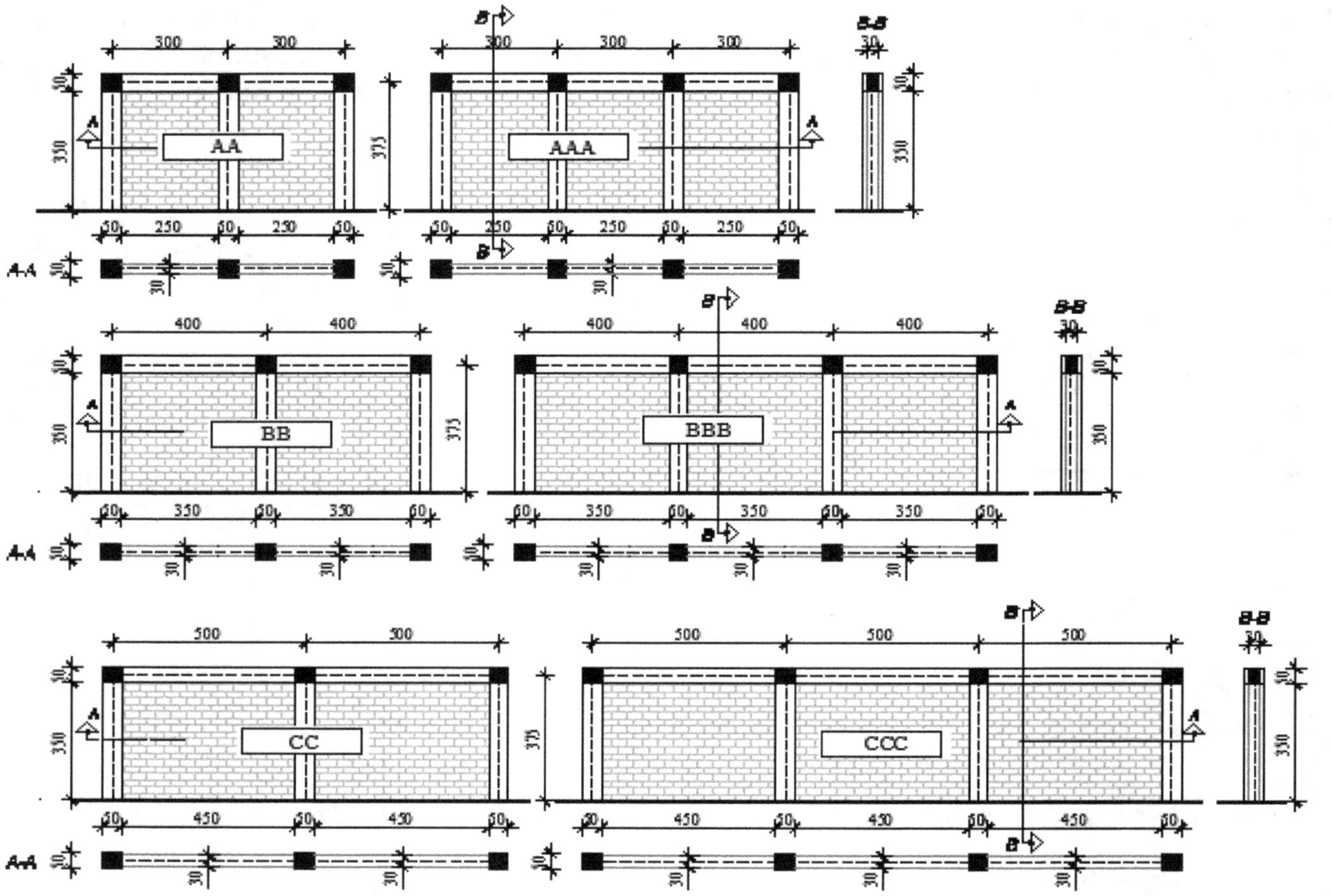

Figure 7. Infilled frame models—multi-bay models with same bay length.

Bilinear approximation (BA) equations:

$$IDR_{c,i} = IDR_c(NN) \cdot (1 + 0.3 \cdot (i - 1)) \tag{3}$$

$$IDR_{m,i} = IDRm(NN) \tag{4}$$

$$BS_{c,i} = BS_c(NN) \cdot (1 + 0.9 \cdot (i - 1)) \tag{5}$$

$$BS_{m,i} = BS_m(NN) \cdot (1 + 0.7 \cdot (i - 1)) \tag{6}$$

where:

- $IDR_{c,i}$—inter-storey drift Ratio at cracking point of multi-bay infilled frame,
- $IDR_{m,i}$—inter-storey drift Ratio at maximum capacity point of multi-bay infilled frame,
- $BS_{c,I}$—base shear at first cracking point of multi-bay infilled frame,
- $BS_{m,i}$—base shear at maximum capacity point of multi-bay infilled frame,

- $IDRc_y$ (NN)—inter-storey drift ratio at cracking point of one story one bay infilled frame obtained by NN
- IDR_m (NN)—inter-storey Drift Ratio at maximum point of one story one bay infilled frame obtained by NN
- BS_c (NN)—base shear at cracking point of one story one bay infilled frame obtained by NN
- BS_m (NN)—base shear at maximum point of one story one bay infilled frame obtained by NN
- i = 2 ..., n—number of bays of multi-bay frame.

Multi-bay frames with equal bays (Figure 6) are obtained by adding one (AA, BB, CC) or two (AAA, BBB, CCC) additional equal ranges to the initial frame (A, B, C). For each of the models, results from nonlinear analysis are obtained by pushover method in Seismostruct 2018 [32] and are shown in the following pictures and tables.

Bilinear approximation equations are then used for the calculation of the same responses by the use of neural network results (Table 8) from one-story one-bay infilled frames.

An acceptability of the Equations (3)–(6) is shown in Figure 8 and Table 9, for multi-bay frames with two and three bays. The curves show the results of the nonlinear analysis for three observed frames, while the bilinear curves show solutions observed by a bilinear approximate equations (BA).

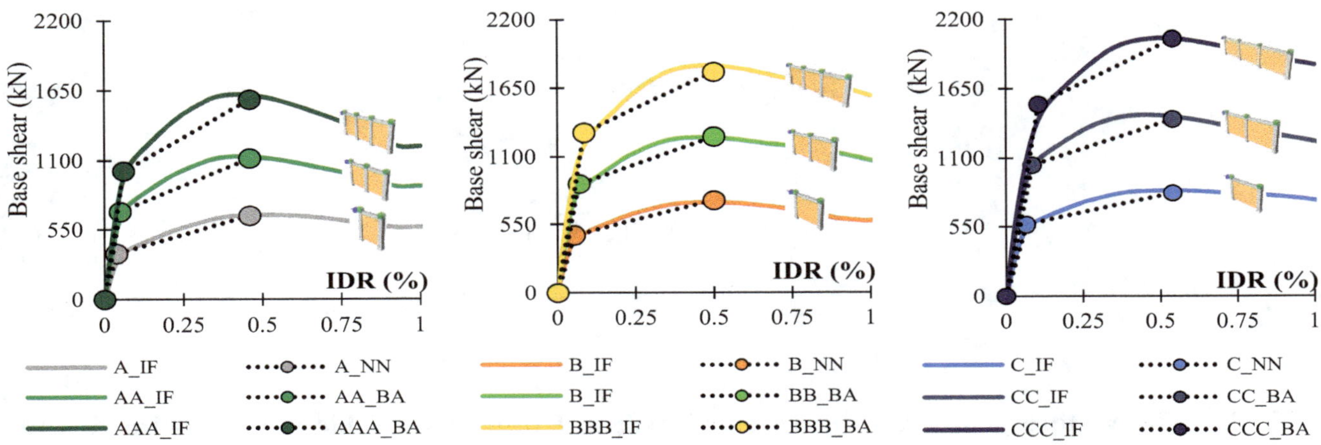

Figure 8. Capacity curves for multi-bay frames with same bay length.

Table 9. Evaluation of accuracy of proposed bilinear approximation equations for equal bay length.

Frame Type	Frame Combination	IDR_c (%)	IDR_m (%)	BS_c (kN)	BS_m (kN)
A	AA_IF	0.053	0.47	656.83	1129.91
	AAA_IF	0.067	0.47	996.97	1617.87
	AA_BA (3)-(6)	0.049{5}	0.46{1}	689.7{9}	1118.6 {2}
	AAA_BA (3)-(6)	0.061{10}	0.46{2}	1016.4 {2}	1579.2 {2}
B	BB_IF	0.08	0.48	846.43	1253.28
	BBB_IF	0.093	0.47	1219.77	1827.99
	BB_BA (3)-(6)	0.069{14}	0.5{4}	877.8{4}	1261.4{1}
	BBB_BA (3)-(6)	0.085{9}	0.5{6}	1293.6{6}	1780.8{3}
C	CC	0.08	0.55	1005.25	1386.18
	CCC	0.107	0.55	1405.98	2046.39
	CC_BA (3)-(6)	0.086{7}	0.54{2}	1045.25{4}	1406.96{2}
	CCC_BA (3)-(6)	0.1056{1}	0.54{2}	1525.5{9}	2045{1}

[1]—values in brackets { } represent relative error from data obtained by BA with respect to results from numerical analysis.

Relative errors for the multi-bay frames with equal bays are small. The average value of error for two and three bays for A frame is 4%, for B frame is 6% and for C frame is 4%. According to the observed results it can be concluded that the capacity curve for multi-bay frame with same bay length can be accurately calculated (94%) by the proposed bilinear approximation equations.

4.3. Application of Neural Networks on Multi-Bay Frames with Different Bay Length

The same idea from the previous section was applied to multi-bay frames with different bay lengths (Figure 9). The following equations are suggested:

$$\mathrm{IDR}_{\mathrm{cr,i}} = \mathrm{IDR}_{\mathrm{c1}}(\mathrm{NN}) + 0.3\sum_{j=2}^{n} \mathrm{IDR}_{\mathrm{cj}}(\mathrm{NN}) \tag{7}$$

$$\mathrm{IDR}_{\mathrm{mr,i}} = \frac{1}{n}\sum_{j=1}^{n} \mathrm{IDR}_{\mathrm{mj}}(\mathrm{NN}) \tag{8}$$

$$\mathrm{BS}_{\mathrm{cr,i}} = \mathrm{BS}_{\mathrm{c1}}(\mathrm{NN}) + 0.9\sum_{j=2}^{n} \mathrm{BS}_{\mathrm{cj}}(\mathrm{NN}) \tag{9}$$

$$\mathrm{BS}_{\mathrm{mr,i}} = \mathrm{BS}_{\mathrm{m1}}(\mathrm{NN}) + 0.7\sum_{j=2}^{n} \mathrm{BS}_{\mathrm{mj}}(\mathrm{NN}) \tag{10}$$

where:

- $\mathrm{IDR}_{cr,i}$—inter-storey drift ratio at cracking point of multi-bay infilled frame with different bay length,
- $\mathrm{IDR}_{mr,i}$—inter-storey drift ratio at maximum capacity point of multi-bay infilled frame with different bay length,
- $\mathrm{BS}_{cr,I}$—base shear at first cracking point of multi-bay infilled frame with different bay length,
- $\mathrm{BS}_{mr,i}$—base shear at maximum point of multi-bay infilled frame with different bay length,
- IDR_{cj} (NN)—inter-storey drift ratio at cracking point of one story one bay infilled frame obtained by NN
- IDR_{mj} (NN)—inter-storey drift ratio at maximum capacity point of one story one bay infilled frame obtained by NN
- BS_{cj} (NN)—base shear at cracking point of one story one bay infilled frame obtained by NN
- BS_{mj} (NN)—base shear at maximum point of one story one bay infilled frame obtained by NN
- i = 2, ..., n—number of bays of multi-bay frame
- j = 1, ..., n—ordinal number of bay in multi-bay frame

In Figure 10, the results for two-bay frames with combinations of A, B and C frames of different bays are shown. The curve shows the results from nonlinear FEM modelling of the two bay frames, while the bilinear curves show the initial results obtained by neural networks (ANN, BNN), and bilinear curves of approximating Equations (7)–(10), which are compared with the results of the nonlinear analysis (Table 10).

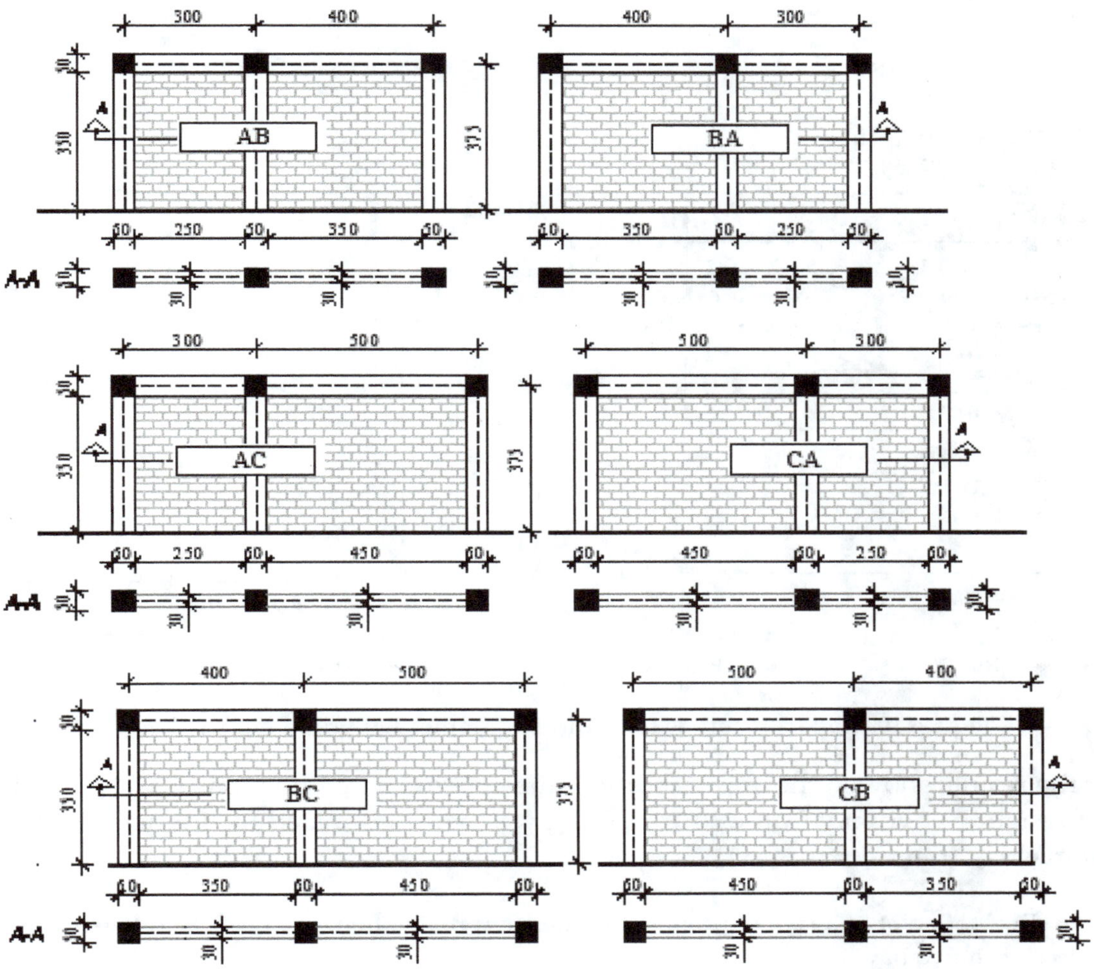

Figure 9. Infilled frame models—multi-bay models with different bay length.

For the AB and BA models, the greatest deviations are for IDRs with an average relative error of 9%. The average relative error for all four observed output values for the combination of AB and BA frames is 6% and 5%, respectively. As the results for frames with the same combination, but different frame distributions are not the same, it has been proved that the first frame is the one that contributes most to the behaviour of the multi-bay frame, with 100% value for both the yield area and the maximum capacity range, as it is assumed in bilinear approximation equations.

For the combination of frames A and C, the MRE is 7%, while for combination of the AC and CA frames MRE is 2%.

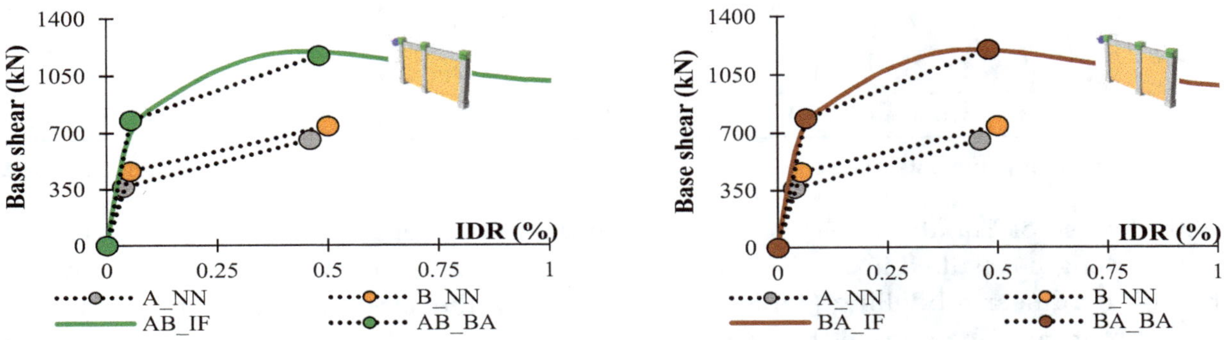

Figure 10. *Cont.*

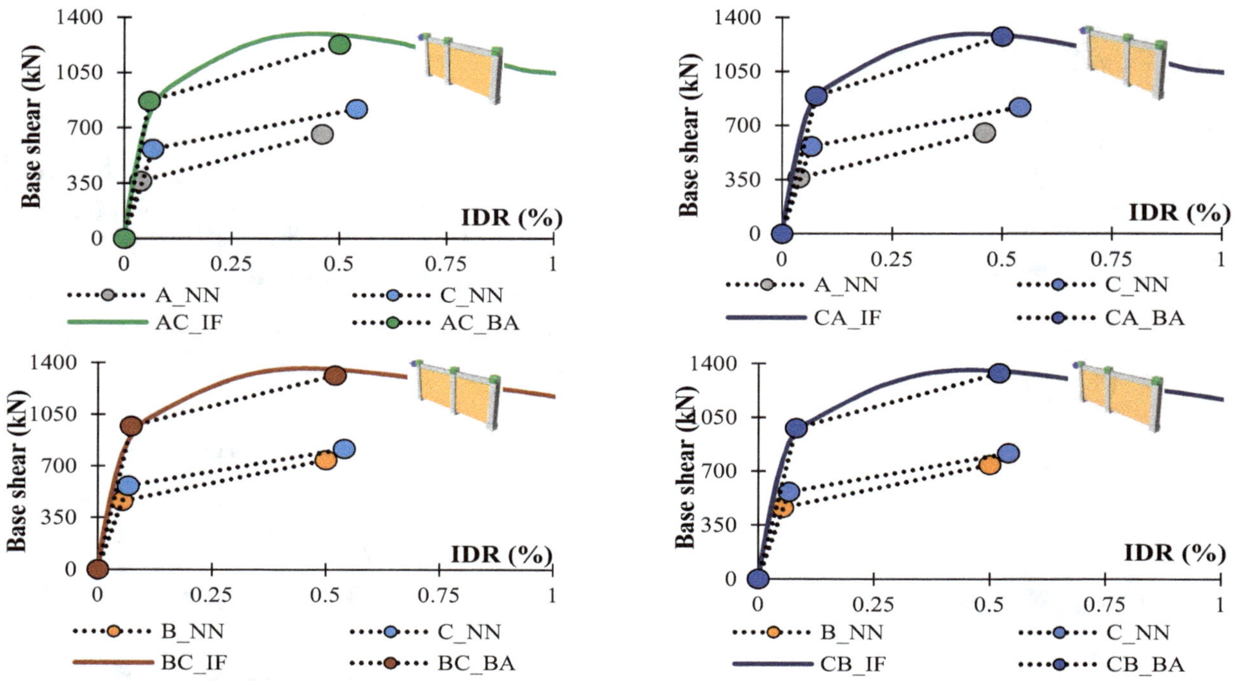

Figure 10. Capacity curves for multi-bay frames with different bay length.

After the results showed the applicability of the approximate Equations (7)–(10) for the different bay length ranges in two-bay frames, further analysis was performed for frames having a different length in three bay frames (Figure 11).

Table 10. Evaluation of accuracy of proposed bilinear approximation equations for different bay length for frames with two bays.

Frame Type	Frame Combination	IDR_c (%)	IDR_m (%)	BS_c (kN)	BS_m (kN)
A/B	AB_IF	0.06	0.44	764.33	1198.25
	BA_IF	0.06	0.44	765.62	1198.61
	AB_BA (7)-(10)	0.054{10}	0.48{9}	777.8{2}	1173.4{2}
	BA_BA (7)-(10)	0.064{7}	0.48{9}	787.8{3}	1199.8{0}
A/C	AC_IF	0.067	0.48	829.41	1292.75
	CA_IF	0.08	0.48	898.88	1291.81
	AC_BA (7)-(10)	0.058{13}	0.5{4}	870.5{5}	1226.6{5}
	CA_BA (7)-(10)	0.077{3}	0.5{4}	890.8{1}	1275.8{1}
B/C	BC_IF	0.08	0.49	934.02	1359.93
	CB_IF	0.08	0.49	986.58	1357.18
	BC_BA (7)-(10)	0.073{9}	0.52{6}	970.5{4}	1314.6{3}
	CB_BA (7)-(10)	0.082{2}	0.52{6}	980.8{1}	1337.4{1}

[1]—values in braces { } presents relative error from data obtained by BA in regard to results from numerical analysis.

With the use of Equations (7)–(10), the prediction of the behaviour of multi-bay frames with different bay lengths resulted in very small deviations (maximum error was 11%) in relation to values obtained by nonlinear modelling (Table 11). Accordingly, the acceptability of approximation and applicability for multi-bay frames has been demonstrated, regardless of the type (equal or different length of bay), number of bays and their distribution.

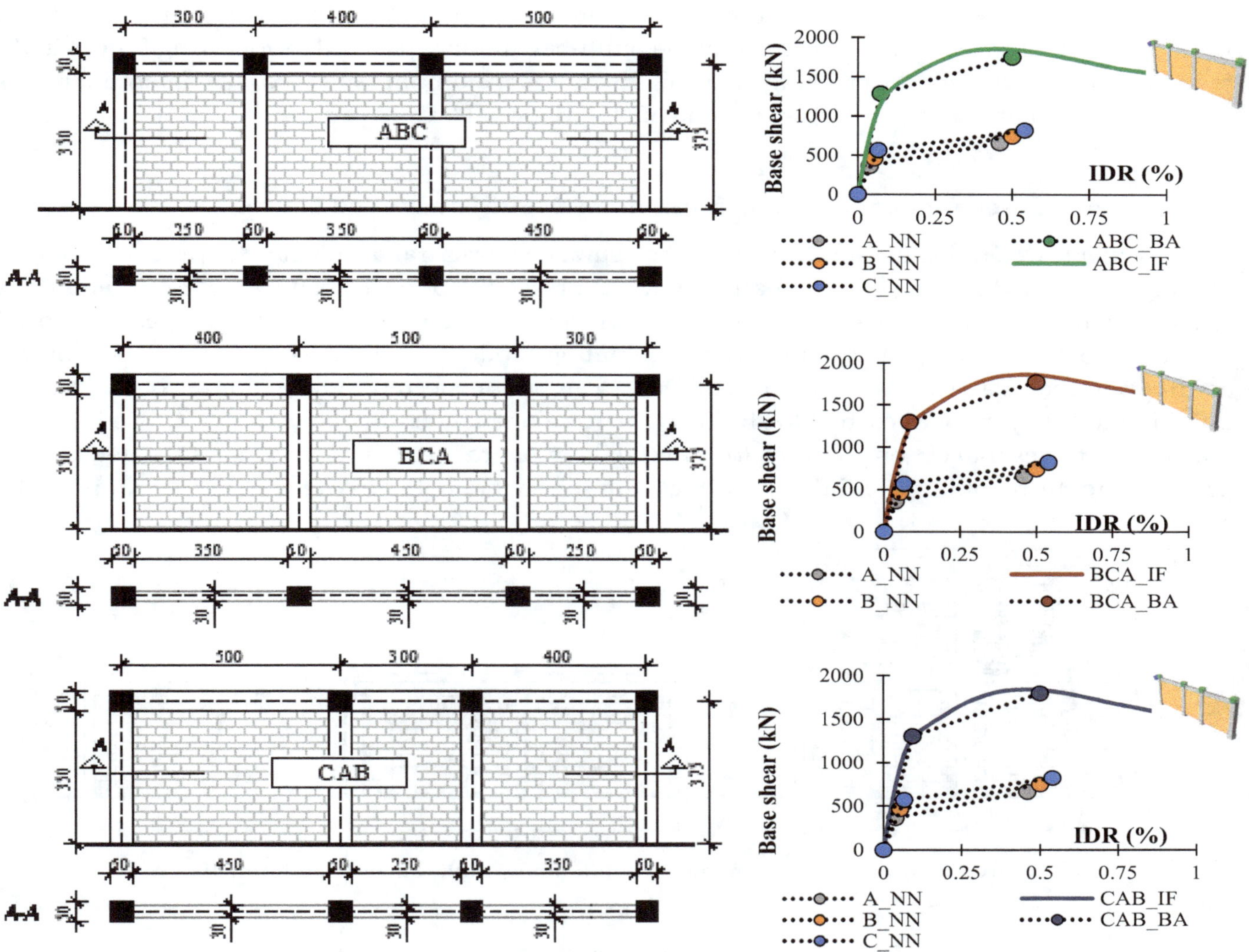

Figure 11. Multi-bay frame models and capacity curves for multi-bay frames with different bay length.

Table 11. Evaluation of accuracy of proposed bilinear approximation equations for different bay length for frames with three bays.

Frame Type	Frame Combination	IDR_c (%)	IDR_m (%)	BS_c (kN)	BS_m (kN)
A/B/C	ABC_IF	0.08	0.45	1197.65	1848.31
	ABC_BA (7)-(10)	0.073{8}	0.5{11}	1286.3{7}	1746{6}
	BCA_IF	0.085	0.47	1275.44	1857.63
	BCA_BA (7)-(10)	0.084{1}	0.5{6}	1296.3{2}	1772.4{5}
	CAB_IF	0.09	0.48	1285.47	1840.39
	CAB_BA (7)-(10)	0.093{4}	0.5{4}	1306.6{2}	1795.2{3}

[1]—values in braces { } presents relative error from data obtained by BA in regard to results from numerical analysis.

The primary curves of reinforced-concrete multi-bay frames with masonry infills are compared with the results of neural networks for one-bay frames and using the approximation Equations (3)–(10). From the comparison it is concluded that the behaviour of multi-bay frames can be predicted with an accuracy of at least 92%. The results of neural networks have shown that the prediction of behaviour with respect to the BS of reinforced frames with masonry infills is very realistically with a small average error of 5% for cracking capacity, and 4% for maximum capacity.

Equation suggestions used to determine the primary curves of the multi-bay frames include the rules that the first frame fully participates in the multi-bay frame capacity. The following frames in the sequence involve 90% for the cracking area, or 70% for the maximum range. For the determination of IDRs at the ultimate capacity, the mean value of all individual frames in a series of multi-bay frames is relevant.

5. Validation of Proposed Equations on Mult-Storey Multi-Bay Infilled Frames

In order to determine the applicability of the neural networks trained on the experimental database of masonry infilled frames, the evaluation of the expressions for multi-bay frames of different and the same lengths on multi-bay structures was performed. As neural networks emerged as a result of single-storey and single-bay frames, approximation models were developed. It is based on the assumption that masses and loads from all floors can be transferred to the columns of the ground floor. In multi-bay models, each of the bays is taken as a one-bay frame (Figures 12–14). Definition of the vertical forces that act on columns for input data in neural networks processing are determined according to the boundary conditions and the assumption that both forces are mutually equivalent.

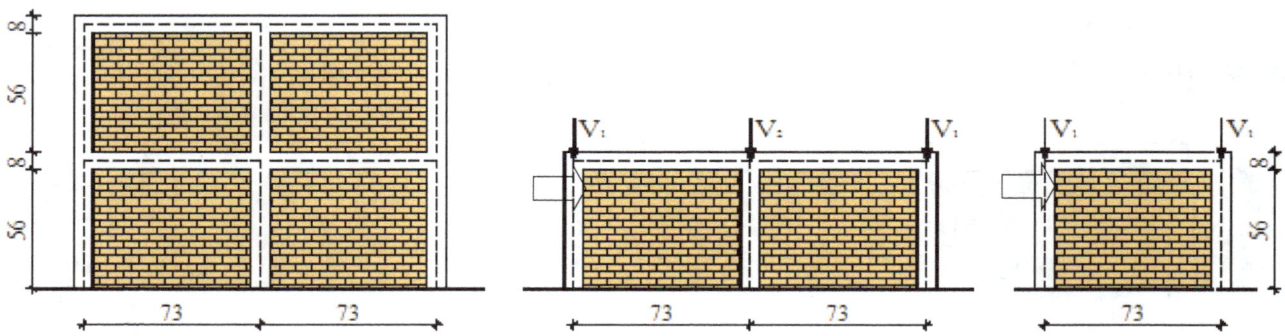

Figure 12. Approximation model for multi-storey multi-bay infilled frame IFS.

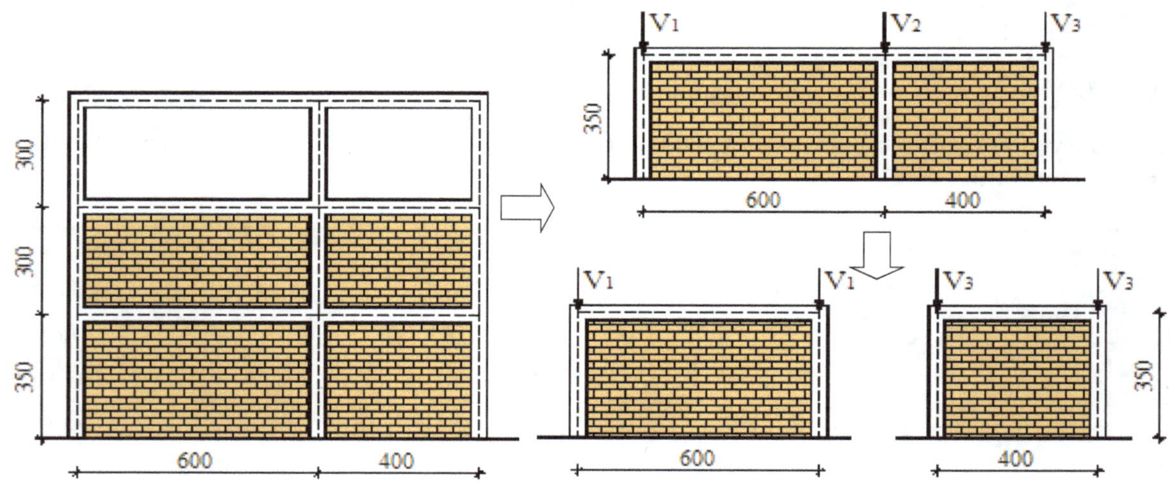

Figure 13. Approximation model for multi-storey multi-bay infilled frame Patras.

Validation was performed on experimental samples: two-storey IFS building [42], three-storey Patras building [43] and four-storey building designed according to EC2 and EC8 [44]. Since all of the frames that were used for validation are multi-storey structure, the gravitational load from the floors is modelled as a nodal force, acting on the columns of the ground floor. It represents its own weight of columns, beams, slabs and masonry, together with load of 2 kN/m^2. The Patras multi-storey muti-bay frame (Figure 13) consists of the three floors, of which only the ground floor and the first floor are filled with masonry infills (uneven distribution of masonry infill by height of the structure) with two bays. Since the bays have different bay length, for the application of approximation formulas it was necessary to use expressions (7–10). Although the second floor was a bare frame example, the same procedure was applied as in the IFS frame; floor masses are represented by a vertical load on the one-bay columns. The EC8 four-storey two-bay frame (Figure 14) was simplified by the use of approximate terms on a two one-storey one-bay frames of 4 and 6 m.

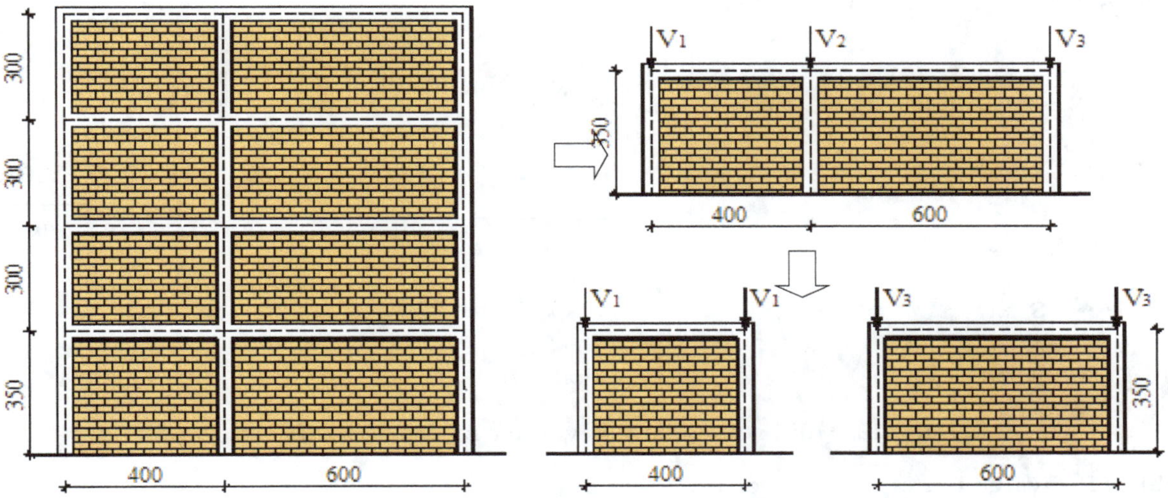

Figure 14. Approximation model for multi-storey multi-bay infilled frame EC8.

Neural network processing is done on previously trained neural networks with new input data from Table 12, calculated according to the loads from Table 13.

Table 12. Input data for neural network processing.

Buildings		a	b	g	r_c	f_y	λ_h	N
IFS		0.87	1	2.66	1.76	240	2.65	21.2
Patras	4m	0.875	0.76	3.57	2.00	555.0	2.22	194.9
	6m	0.583	0.76	3.57	2.00	555.0	2.12	146.7
EC8	4m	0.875	0.76	3.57	2.35	553.5	2.47	233.3
	6m	0.583	0.76	3.57	2.35	553.5	2.38	309.5

Table 13. Load on columns on one-story one bay approximated frames.

Buildings	V_1 (kN)	V_2 (kN)	V_3 (kN)
IFS	21.2	21.2	-
Patras	194.85	292.49	146.66
EC8	233.26	464.02	309.52

An approximation of IFS infilled frame consisted of a one-storey one-bay frame with the corresponding load, the vertical concentric forces representing the mass of the floor. As for the symmetrical system, equations for the equal length bays (3)–(6) were applied for the approximation. By comparing the results obtained from numerical modelling on the IFS frame and the application of approximate expressions of capacity curve, it can be concluded that the obtained values have sufficient acceptability that approves their application for multi-storey and multi-bay frames. The largest relative error in the IFS frame was 8%, while the mean relative error was 4.5% (Table 14, Figure 14).

The same principle was used for Patras and EC8 buildings. Obtained results are presented in Figure 15 and Table 14. It can be concluded that proposed bilinear Equations (7)–(10) can be used for both uniformly and non-uniformly distributed infilled frames. For the Patras frame, MRE was 8% and for EC8 frame error was 7%.

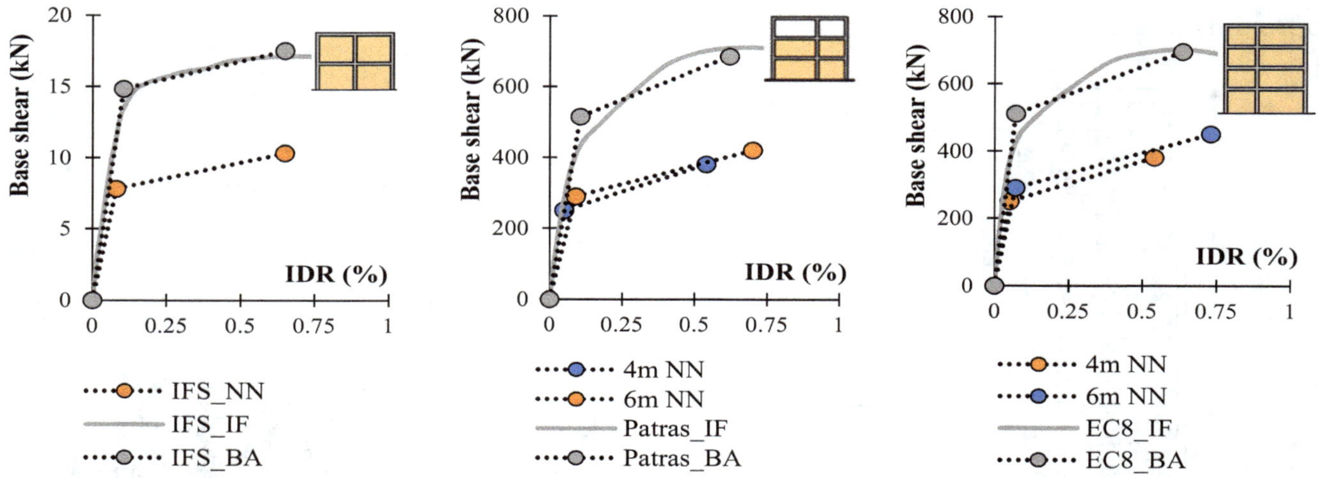

Figure 15. Multi-storey multi-bay frame models capacity curves.

Table 14. Evaluation of accuracy of proposed bilinear approximation equations for multi-storey multi-bay frames.

Buildings	Approximation	IDR_c (%)	IDR_m (%)	BS_c (kN)	BS_m (kN)
IFS	IFS_NN	0.08	0.65	7.8	10.3
	IFS_IF	0.104	0.65	14.82	17.51
	IFS_BA (7)-(10)	0.104{0}	0.67{3}	13.7{8}	17.13{2}
Patras	Patras 6m_NN	0.09	0.7	290	420
	Patras 4m_NN	0.05	0.54	250	380
	Patras_IF	0.11	0.69	446.24	711.25
	Patras_BA (7)-(10)	0.105{5}	0.62{9}	515{14}	686{4}
EC8	EC8 4m_NN	0.05	0.54	250	380
	EC8 6m_NN	0.07	0.73	290	450
	EC8_IF	0.071	0.635	511	695
	EC8_BA (7)-(10)	0.08{11}	0.615{3}	451.15{13}	701.65{1}

[1]—values in braces { } presents relative error from data obtained by BA in regard to results from numerical analysis.

6. Conclusions

The main idea of study was definition of the contribution of infill in masonry infilled frame response. According to the conducted experimental database and successfully processed neural networks results, further application in term of behaviour prediction was indispensable. The limit of direct use of neural network results was based on the fact that structural system which is used was a one-storey one-bay infilled frame. Therefore, the equations for definition of capacity curves for real buildings of infilled frame were proposed.

The attempt to simplify multi-bay infilled frames and the connection with obtained neural network results are presented by bilinear approximation equations for IDRs and BS for two behaviour ranges: cracking and maximum load. Equations were suggested for equal and different bay length.

In order to validate proposed method, multi-storey multi-bay frames with experimental results are used for evaluation. As it is presented the method was fully approved with the prediction of the characteristic values of the primary curve, displacement and forces, regardless of the number of floors, the difference of the bays and the unequal distribution of the masonry infills by the height of the structure with a mean accuracy of 92%.

It can be concluded and confirmed that approximate bilinear equations can be reliably applied for masonry infilled frame behaviour prediction. The contribution of masonry infill can be quantitatively summed for the prediction of the infilled frames capacity curve.

Direct contribution is based on the conclusion that the known response of a one-storey one-bay infilled frame can be used for the prediction of a multi-storey multi-bay infilled frame response. The initial response of one-storey one-bay frame can be defined by neural networks or by nonlinear FEM model, as it is approved that used FEM model is accurate and correct. According to this method the monolithic behaviour of masonry infilled frames is approved.

A performance based assessment can be implemented by using the proposed procedure in correspondence with the known performance levels for masonry infills based on IDR values.

Author Contributions: Conceptualization—T.K.Š., methodology—T.K.Š., software—T.K.Š., validation—T.K.Š. and K.S., writing—original draft preparation, T.K.Š..; writing—review and editing, T.K.Š. and K.S., visualization, T.K.Š. and K.S., supervision—T.K.Š.

References

1. Asteris, P.G.; Cotsovos, D.; Chrysostomou, C.; Mohebkhah, A.; Al-Chaar, G. Mathematical micromodelling of infilled frames: State of the art. *Eng. Struct.* **2013**, *56*, 1905–1921. [CrossRef]
2. Asteris, P.G.; Antoniou, S.T.; Spophianopoulos, D.S.; Chrysostomou, C.Z. Mathematical Macromodeling of Infilled Frames: State of the Art. *J. Struct. Eng.* **2011**, *137*, 1508–1517. [CrossRef]
3. Furtado, A.; Rodrigues, H.; Arêde, A.; Kalman Šipoš, T.; Varum, H. Masonry Infill Walls Participation in the Seismic Response of RC Structures. In *Masonry: Design, Materials and Techniques*; Nova Science Publishing, Inc.: New York, NY, USA, 2019; pp. 113–147.
4. Cascardi, A.; Micelli, F.; Aiello, M.A. An Artificial Neural Networks model for the prediction of the compressive strength of FRP-confined concrete circular columns. *Eng. Struct.* **2017**, *140*, 199–208. [CrossRef]
5. Cascardi, A.; Micelli, F.; Aiello, M.A. Analytical model based on artificial neural network for masonry shear walls strengthened with FRM systems. *Composites Part B* **2016**, *95*, 252–263. [CrossRef]
6. Asteris, P.G.; Repapis, C.C.; Repapi, E.V.; Cavaleri, L. Fundamental period of infilled reinforced concrete frame structures. *Struct. Infrastruct. Eng.* **2016**, *13*, 929–941. [CrossRef]
7. Kalman Šipoš, T.; Sigmund, V.; Hadzima-Nyarko, M. Earthquake performance of infilled frames using neural networks and experimental database. *Eng. Struct.* **2013**, *51*, 113–127. [CrossRef]
8. Asteris, P.G.; Nikoo, M. Artificial bee colony-based neural network for the prediction of the fundamental period of infilled frame structures. In *Neural Computing and Applications*; Springer: London, UK, 2019; pp. 1–11.
9. Marinilli, A.; Castilla, E. Experimental evaluation of confined masonry walls with several confining-columns. In Proceedings of the 13th World Conference on Earthquake Engineering, Vancouver, BC, Canada, 1–6 August 2004.
10. Combescure, D.; Pegon, P. Application of local to global approach to the study of infilled frame structures under seismic loading. *Nucl. Eng. Des.* **2000**, *196*, 17–40. [CrossRef]
11. Colangelo, F. *Pseudo-Dynamic Seismic Response and Phenomenological Models of Brick-Infilled RC Frames*; Report DISAT 1/99; University of L'Aquila: L'Aquila, Italy, 1999. (In Italian)
12. Cavaleri, L.; Fossetti, M.; Papia, M. Effect of vertical loads on lateral response of infilled frames. In Proceedings of the 13th World Conference on Earthquake Engineering, Vancouver, BC, Canada, 1–6 August 2004.
13. Lafuente, M.; Castilla, E.; Genatios, C. Experimental and Analytical Evaluation of the Seismic Resistant Behavior of Masonry Walls. *J. Br. Masonry Soc.* **1998**, *11*, 65–101.
14. Kakaletsis, D.J. Influence of masonry strength and rectangular spiral shear reinforcement on infilled RC frames under cyclic loading. *WIT Trans. Model. Simul.* **2007**, *46*, 643–653.

15. Dukuze, A. Behaviour of Reinforced Concrete Frames Infilled with Brick Masonry Panels. Ph.D. Thesis, The University of New Brunswick, Fredericton, NB, Canada, December 2000.
16. Žarnić, R. The Analysis of R/C Frames with Masonry Infill under Seismic Actions. Master's Thesis, FGG, Ljubljana, Slovene, 1985. (In Slovene).
17. Žarnić, R. Inelastic Response of r/c Frames with Masonry Infill. Ph.D. Thesis, University of Ljubljana, Ljubljana, Slovenia, December 1992.
18. Al-Chaar, G. Non-ductile behavior of reinforced concrete frames with masonry infill panels subjected to in-plane loading. Ph.D. Thesis, University of Illinois at Chicago, Chicago, IL, USA, December 1998.
19. Angel, R.; Abrams, D.; Shapiro, D.; Uzarski, J.; Webster, M. *Behavior of Reinforced Concrete Frames with Masonry Infills*; University of Illinois Engineering Experiment Station, College of Engineering, University of Illinois at Urbana-Champaign: Urbana, IL, USA, February 1994.
20. Mehrabi, A.B.; Shing, P.B.; Schuller, M.P.; Noland, J.L. *Performance of Masonry Infilled R/C Frames under in-Plane Lateral Loads*; National Science Foundation: Arlington, VA, USA, October 1994.
21. Crisafulli, F.J. Seismic Behavior of Reinforced Concrete Sructures with Masony Infills. Ph.D. Thesis, University of Canterbury, Christchurch, New Zealand, December 1997.
22. Fiorato, A.E.; Sozen, M.A.; Gamble, W.L. *Investigation of the Interaction of Reinforced Concrete Frames with Masonry Filler Walls*; University of Illinois Engineering Experiment Station, College of Engineering, University of Illinois at Urbana-Champaign: Urbana, IL, USA, November 1970.
23. Yorulmaz, M.; Sozen, M.A. *Behavior of Single-Story Reinforced Concrete Frames with Filler Walls*; University of Illinois Engineering Experiment Station, College of Engineering, University of Illinois at Urbana-Champaign: Urbana, IL, USA, May 1968.
24. Benjamin, J.R.; Williams, H.A. The Behavior of One-Story Brick Shear Walls. *J. Struct. Div.* **1957**, *83*, 1–49.
25. Zovkić, J.; Sigmund, V.; Guljaš, I. Cyclic testing of a single bay reinforced concrete frames with various types of masonry infill. *Earthquake Eng. Struct. Dyn.* **2012**, *41*, 41–60. [CrossRef]
26. Vafaei, M.; Alih, S.C.; Shad, H.; Falah, A.; Halim, N.H.F.A. Prediction of strain values in reinforcements and concrete of a RC frame using neural networks. *Int. J. Adv. Struct. Eng.* **2018**, *10*, 29–35. [CrossRef]
27. Yam, L.H.; Yan, Y.J.; Jiang, J.S. Vibration-based damage detection for composite structures using wavelet transform and neural network identification. *Compos. Struct.* **2013**, *60*, 403–412. [CrossRef]
28. Zang, C.; Imregun, M. Structural damage detection using artificial neural networks and measured FRF data reduced via principal component projection. *J. Sound Vib.* **2001**, *242*, 813–827. [CrossRef]
29. Gnana Sheela, K.; Deepa, S.N. Review on methods to fix number of hidden neurons in neural networks. *Math. Prob. Eng.* **2013**. [CrossRef]
30. Sonmez, H.; Gokceoglu, C.; Nefeslioglu, H.A.; Kayabasi, A. Estimation of rock modulus: For intact rocks with an artificial neural network and for rock masses with a new empirical equation. *Int. J. Rock Mech. Min. Sci.* **2006**, *43*, 224–235. [CrossRef]
31. Özturan, M.; Kutlu, B.; Özturan, T. Comparison of concrete strength prediction techniques with artificial neural network approach. *Build. Res. J.* **2008**, *56*, 23–36.
32. Seismosoft. SeismoStruct.—A Computer Program for Static and Dynamic Nonlinear Analysis of Framed Structures. Available online: http://www.seismosoft.com (accessed on 30 May 2018).
33. Kalman Šipoš, T.; Rodrigues, H.; Grubišić, M. Simple design of masonry infilled reinforced concrete frames for earthquake resistance. *Eng. Struct.* **2018**, *171*, 961–981. [CrossRef]
34. CEN Eurocode 8—Design of Structures for Earthquake Resistance. Part 3: Brussels. 2005. Available online: http://files.isec.pt/DOCUMENTOS/SERVICOS/BIBLIO/Documentos%20de%20acesso%20remoto/Eurocode-8-1-Earthquakes-general.pdf (accessed on 21 April 2018).
35. Hendry, A.W.; Malek, M.H. Characteristic compressive strength of brickwork walls from collected test results. *Masonry Int.* **1986**, *7*, 15–24.
36. Kaushik, H.B.; Rai, D.C.; Jain, S.K. Stress-strain characteristics of clay brick masonry under uniaxial compression. *J. Mater. Civ. Eng.* **2007**, *19*, 728–739. [CrossRef]
37. Mander, J.B.; Priestley, M.J.N.; Park, R. Theoretical stress-strain model for confined concrete. *J. Struct. Eng.* **1988**, *114*, 1804–1826. [CrossRef]
38. Menegotto, M.; Pinto, P.E. Method of Analysis for Cyclically Loaded Reinforced Concrete Plane Frames Including Changes in Geometry and Non-Elastic Behavior of Elements under Combined Normal Force and

Bending. In Proceedings of the IABSE Symposium on Resistance and Ultimate Deformability of Structures Acted on by Well Defined Repeated Loads, Lisbon, Portugal, 1973; pp. 15–22.

39. Stafford-Smith, B.; Carter, C. A method for the analysis of infilled frames. *Proc. Inst. Civ. Eng.* **1969**, *44*, 31–48. [CrossRef]

40. Decanini, L.D.; Fantin, G.E. Modelos simplificados de la mampostería incluida en porticos. Caracteristicas de stiffnessy resistencia lateral en estado limite. *Jornadas Argentinas de Ingeniería Estructural* **1986**, *2*, 817–836.

41. Kalman Šipoš, T.; Hadzima-Nyarko, M.; Miličević, I.; Grubišić, M. Structural performance levels for masonry infilled frames. In Proceedings of the 16th European Conference on Earthquake Engineering, Thessaloniki, Greece, 18–21 June 2018; pp. 1–12.

42. Žarnić, R.; Gostič, S.; Crewe, A.J.; Taylor, C.A. Shaking table tests of 1:4 reduced-scale models of masonry infilled reinforced concrete frame buildings. *Earthquake Eng. Struct. Dyn.* **2001**, *30*, 819–834. [CrossRef]

43. Fardis, M.N. *Experimental and Numerical Investigations on the Seismic Response of RC Infilled Frames and Recommendations for Code Provisions*; Laboratório Nacional de Engenharia Civil: Lisbon, Portugal, 1996.

44. Negro, P.; Pinto, A.V.; Verzeletti, G.; Magonette, G.E. PsD test on four-story RC building designed according to Eurocodes. *J. Struct. Eng.* **1996**, *122*, 1409–1417. [CrossRef]

6

Seismic Strengthening of the Bagh Durbar Heritage Building in Kathmandu following the Gorkha Earthquake Sequence

Rabindra Adhikari [1,2], Pratyush Jha [2,3], Dipendra Gautam [4,*] and Giovanni Fabbrocino [4,5]

1 Department of Civil Engineering, Cosmos College of Management and Technology, Lalitpur 44600, Nepal; rabindraadhi@cosmoscollege.edu.np
2 Interdisciplinary Research Institute for Sustainability, IRIS, Kathmandu 44600, Nepal; pratyush0119@gmail.com
3 Digicon Engineering Consult, Lalitpur 44600, Nepal
4 Structural and Geotechnical Dynamics Laboratory, StreGa, DiBT, University of Molise, Via F. de Sanctis, CB 86100 Campobasso, Italy; giovanni.fabbrocino@unimol.it
5 ITC-CNR, Construction Technologies Institute, National Research Council of Italy (CNR), L'Aquila Branch, Via Carducci, AQ 67100 L'Aquila, Italy
* Correspondence: dipendra.gautam@unimol.it

Abstract: The so-called Greco-Roman monuments, also known as neoclassical monuments, in Nepal represent unique construction systems. Although they are not native to Nepal, they are icons of the early 19th century in the Kathmandu valley. As such structures are located within the heritage sites and historical centers, preservation of Greco-Roman monuments is necessary. Since many buildings are in operation and accommodate public and critical functions, their seismic safety has gained attention in recent times, especially after the Gorkha earthquake. This paper first presents the background of the Bagh Durbar monument, reports the damage observations, and depicts some repair and retrofitting solutions. Attention is paid to the implementation of the different phases of the structural characterization of the building, the definition of reference material parameters, and finally, the structural analysis made by using finite element models. The aim of the contribution consists of comparison of the adequacy of the finite element model with the field observations and design of retrofitting solutions to assure adequate seismic safety for typical Greco-Roman buildings in Nepal. Thus, this paper sets out to provide rational strengthening solutions compatible with the existing guidelines rather than complex numerical analyses.

Keywords: seismic assessment; Greco-Roman construction; masonry building; seismic retrofitting; heritage construction; structural restoration

---

1. Introduction

On 25 April, 2015, a strong earthquake of magnitude 7.8 occurred at the Barpak neighborhood of the Gorkha district and affected central, eastern, and some western parts of Nepal. Per the National Planning Commission, as many as one million buildings were affected by the earthquake [1]. The earthquake caused 8790 fatalities and 22,300 injuries in 31 out of 75 districts in Nepal. At least 2900 cultural heritage buildings and monuments were damaged due to the main shock and the major aftershocks of 25 April (M_W 6.7), 26 April (M_W 6.9), and 12 May (M_W 7.3) of 2015. Gautam et al. [2] highlighted the vulnerability of Greco-Roman monuments in their recent contribution and placed them under European Macroseismic Scale (EMS-98) vulnerability class B. In Nepal, Greco-Roman heritage construction evolved in the 19th century and construction of such buildings was limited to the major urban as well as administrative centers of the country. Greco-Roman structures have relatively thicker brick masonry walls than other existing building forms in Nepal. They are primarily characterized by arches above the doors and windows and are multi-storied constructions. In most cases, multiple buildings are constructed as a jointed form to give rise to an aggregate system. Similarly, large structural, as well as false columns, are distinguishing features of Greco-Roman structural form. The name Greco-Roman is given as they possess the characteristics of both Greek and Roman monumental forms. The Gorkha earthquake damaged most of the Greco-Roman buildings in the Kathmandu valley including, for example, the main administrative center Singha Durbar, Administrative Staff College Jawalakhel, Babar Mahal, and Tangal Durbar, among others [3]. Although the construction age of such buildings was not similar neither comparable, damage occurrence was consistent in terms of the extent of damage in the massive walls.

Several historical earthquakes, such as 2003 Bam, 2009 L'Aquila, 2010 Chile, and 2016 Central Italy, among others, have reflected widespread damage to heritage structures (see e.g., [4–7]). The complex behavior of masonry structures is long discussed (e.g., [8–12], thus, the difficulties in seismic assessment are clear to the technical community. Existing literature on the numerical analysis of heritage structures by non-linear analyses documents well the complexity of the task and the uncertainties which remain unresolved in the interpretation of the seismic behavior of heritage masonry structures (see for e.g., [13–17]). The knowledge of structural features of heritage buildings, as well as of the material properties to be used in the safety assessment and retrofitting of existing structures, represents another complex task for engineers due to limitations arising from lack of destructive tests in heritage constructions [18,19]. Historical structures represent a particular time frame and the exact material properties and degradation cannot be adequately modeled.

Greco-Roman monuments are important buildings in Nepal due to their continued use in the present time and also due to recognition as archeologically important buildings by the Department of Archeology, government of Nepal, as most of them are more than 100 years old. Limited works related to Greco-Roman monuments can be found in Nepal [20,21], thus, to fulfill the research gap, seismic vulnerability analysis based on field observations and finite element modeling are needed. Despite the grave importance in understanding the vulnerability of Greco-Roman monuments, field tests, numerical modeling, detailed assessments, and retrofitting frameworks of existing Greco-Roman buildings are scarce, if not absent, in Nepal. Due to the heritage as well as administrative values of such buildings, it is important to assess the vulnerability and level of seismic safety to ensure at least a life safety performance level is in place in the case of future earthquakes. This paper presents the damage scenario observed in the Bagh Durbar monument after the Gorkha earthquake. The field tests and finite element analysis are presented and a retrofitting framework for each vulnerable/damaged component is proposed as well.

2. Materials and Methods

The original construction of Bagh Durbar (Tiger Palace) is believed to have occurred in 1805 by the then Prime Minister Bhimsen Thapa who was replacing a small house constructed by his father. The original construction covered an area of 61049 m^2; however, present-day Bagh Durbar is now limited to 2933 m^2. The palace is located (27°41′34.58″ N, 85°19′28.04″ E) in the central Kathmandu valley. The building was renovated by Bir Shamsher in 1885 and was reconstructed after the 1934 earthquake, which destroyed the monument [21]. Present-day Bagh Durbar (Figure 1), which is being used as the Kathmandu metropolitan city office, was constructed after the 1934 earthquake and was slightly altered from the Greco-Roman style by Juddha Shamsher and his son Hari Shamsher. The building has a rectangular combination of components with a central courtyard. The floors plans depict massive wall thickness and geometrically irregular constructions at several locations of the building.

Figure 1. (**a**) Location of Bagh Durbar (**b**) old Bagh Durbar (picture credit: Madan Pustakalay, Lalitpur), (**c**) changes in construction and neighborhood in Bagh Durbar area (picture credit: Madan Pustakalay, Lalitpur), (**d**) present day facade of Bagh Durbar.

The building is situated in the plain terrain of central Kathmandu and is a load-bearing masonry building. The building is a three-storied Greco-Roman construction; however, there are only two stories in some locations with higher story heights. The vertical configuration is generally regular, and the plan configuration has some projections. The total plinth area of the building is 8564 m^2. The thickness of most of the interior, as well as exterior walls, is 850 to 900 mm on the ground floor, 800 mm on the first floor, and 750 mm on the second floor, whereas the west extension of the building has a wall thickness of 650 mm. Several interior walls provided with timber and aluminum partitions in the building were added later, thus, the present-day form of Bagh Durbar is not the same as the one constructed in 1935. The walls were constructed as brick masonry walls in mud-mortar and the floor structure was mixed with timber joists, and a jack arch with a steel I-beam in the original construction. The foundation was provided up to the depth of ~3 m by the soling of brick bats in Surkhi mortar (mainly lime and brick powder). At the plinth level, damp proofing course of Surkhi, having 100 mm thickness, was observed during the excavation. However, a nearby deep borehole depicts that the upper 25 m depth is comprised of coarse-grained sand and clay mix. In between 25 m to 210 m depth, black carbonaceous clay (Kalimati formation) is present. From 210 m to 244 m, clayey sand is dominant in the area where the monument is located.

The building is constructed mostly from traditional local materials including some standard rolled steel sections and cast iron. The foundation of the building comprises stepped brick walls. The walls were mostly brick masonry in mud mortar; whereas, a few walls constructed later were brick masonry in cement mortar as well. The wall thickness varied per the story. Surkhi plaster was provided to the exterior walls of the monument and the interior walls were plastered with mud plaster. Cement plaster was also found in some locations that were repaired later. The corridors and a few rooms of the building were provided with a timber floor structure and brick tile finishing. Most of the rooms were provided with steel I-beams with jacked brick arches and brick tile finishing. Some of the rooms have a timber floor structure and timber plank finishing as well. The roof of the building was provided with corrugated iron sheets and timber trusses were provided to support the roof. The ceiling of the building has decorative metal sheets with cornices in the halls; whereas, some other rooms have a timber false ceiling. Although the structure is a load-bearing structure, some brick columns were found in the large halls, above which a timber-girder and flooring system was provided. During the field study, a timber flooring system, timber roofing system, and floor system of a jacked brick arch with steel I-beam, were observed at various locations of the monument.

Overview of the Damage

According to the Department of Urban Development and Building Construction (DUDBC), government of Nepal vulnerability guidelines [22], the expected damage is grade-4 in case of an intensity IX earthquake, which demands appropriate interventions to be provided to assure adequate seismic safety in the case of strong to major earthquakes. Some observed deficiencies leading to the vulnerability of the structure are summarized as follows:

- Irregular building shape in plan (courtyard with wings, unequal bay-width) and room shape (long-rectangular and inclined)
- Story height is more than 3 m, and the building is a three-storied brick in mud-mortar construction
- Outside walls/corridors are relatively long (>10m)
- Lack of vertical reinforcement in the walls, corners, and junctions
- Lack of horizontal bands, corner-stitch, and gable bands in the structure
- Flexible floors with variation in the floor-system and floor levels without effective bracing

These features of the building are reflected in the outcomes of the post Gorkha earthquake field reconnaissance reports. In fact, the monument suffered major structural and non-structural damages: most of the walls sustained slight structural damage; very few walls were heavily damaged without a risk of collapse of the whole structure; some floors were collapsed due to deteriorated timber joists; insignificant out-of-plane deformation was in very few walls, and some corners were separated. Similarly, the non-structural damage in the structure can be described as plaster delamination and collapse of some ceilings and floors (mainly due to deteriorated timbers).

The non-structural damage can be categorized as "Moderate" and structural damage can be depicted as "Slight" which yields the overall damage grade-2 (moderate) per the EMS-98 scale of damage grading [23]. Nevertheless, it is worth noting that intrinsic characteristics of the building and its value from an architectural standpoint made unrealistic the possibility of achieving immediate occupancy performance of the building under design as suggested by the Nepal Building Code (NBC)-105 [24], according to the seismic classification of the site. As the monument was constructed before the inception of design codes of practices, as well as by-laws, no seismic considerations were inputted during the construction, which led to the building being vulnerable in case of moderate to strong earthquakes, to which the codal provisions were formulated. Therefore, restoration of damaged components must be combined with seismic upgrading interventions capable of ensuring increased seismic performance and therefore preserving the monument. In this sense, the design process complies with relevant rules provided for architectural and historical heritage in some European countries (i.e., Italy), where specific guidelines have been released in recent years to combine property safety and conservation of monuments (e.g., [25]). Details of intervention and Finite element (FE) analysis after the introduction of retrofit design are presented in the following sections. Main damage locations are represented by uppercase letters in Figure 2 and the field evidence is presented in Figures 3–6. Figure 6a shows the damage led by the higher deflection which caused the corner separation and damage to the arch. Damages were mainly noted in terms of shear damage and corner separation.

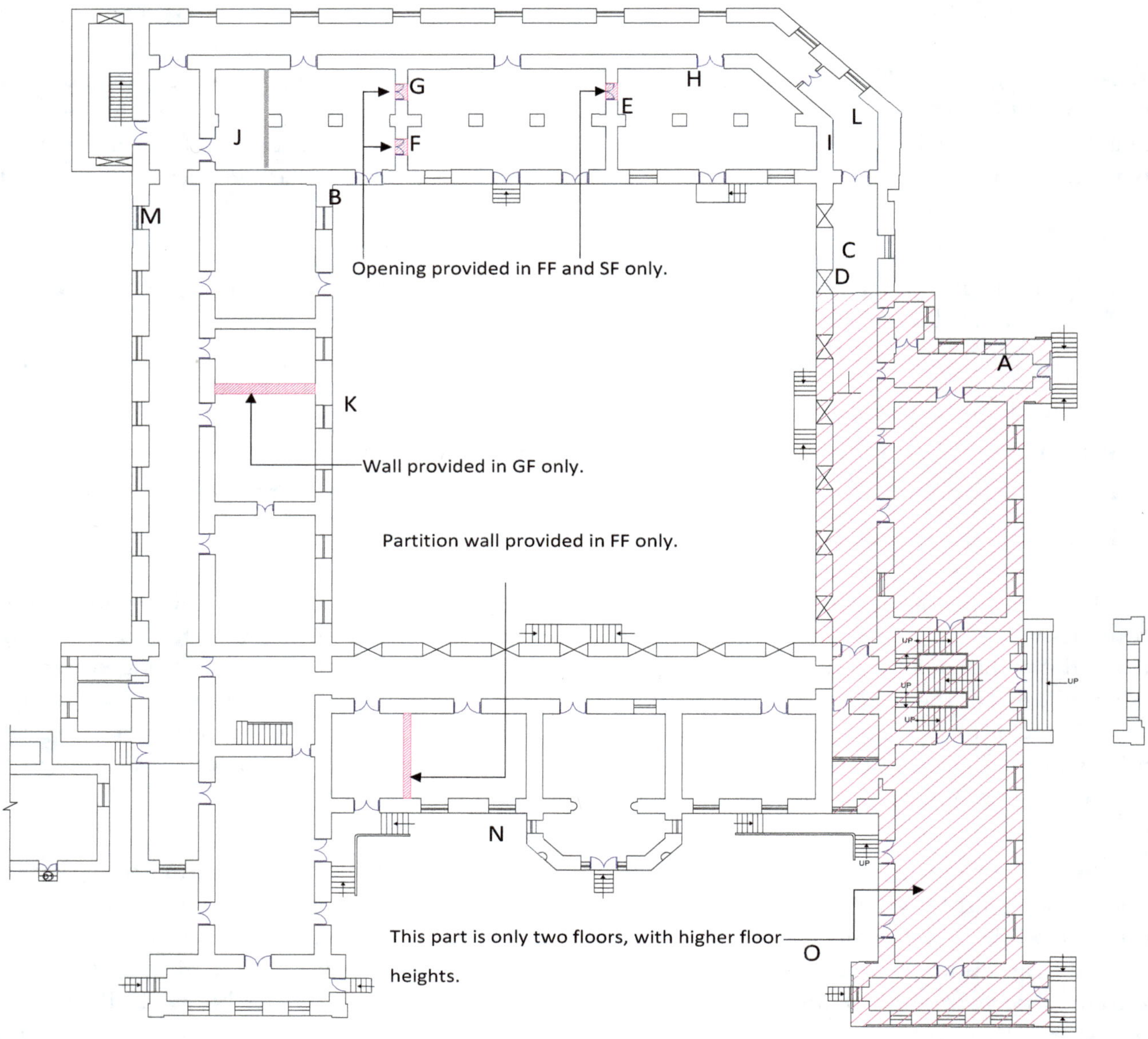

Figure 2. Major damage locations in the monument (capital letters represent the damage mechanisms as shown in Figures 3–6). In the figure, GF, FF, and SF respectively represent the ground floor, first floor, and the second floor.

Figure 3. Local damages in the monument (**a**) damage above the crown of arches, (**b**) corner damage at the junction of north and west lounge, (**c**) vertical crack, (**d**) crown damage (uppercase indications at the bottom of the structures are the location indications as shown in Figure 2).

Figure 4. Local damages in the monument (**a**) diagonal cracks on cross-wall, (**b**) damaged wall above the opening, (**c**) damage above the crown of the arch on cross wall, (**d–f**) damages around the openings (uppercase indications at the bottom of the structures are the location indications as shown in Figure 2).

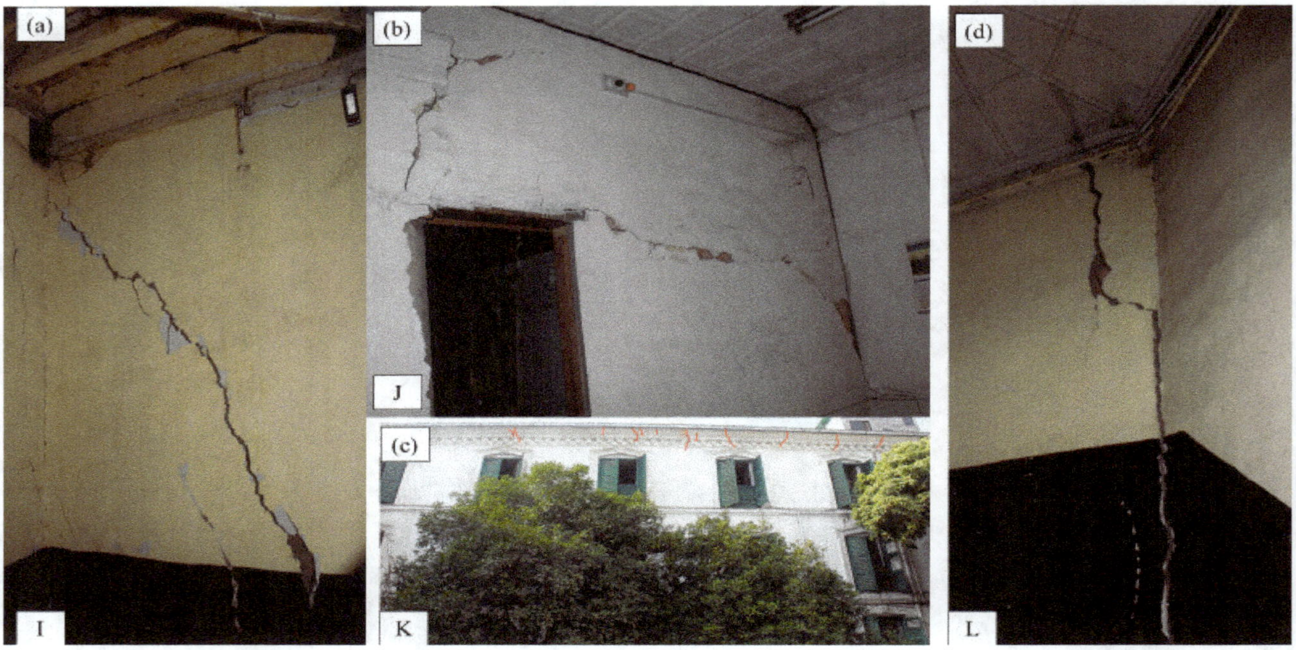

Figure 5. Local damages in the monument (**a**) diagonal crack on the third-floor wall, (**b**) horizontal and vertical cracks beside the opening, (**c**) exterior wall damage at the top floor, (**d**) corner damage (uppercase indications at the bottom of the structures are the location indications as shown in Figure 2).

Figure 6. Local damages in the monument (**a**) corner damage, (**b**) corner damage and crack propagation in the projection part, (**c**) cracks developed above the arch. In figure, uppercase indications at the bottom of the structures are the location indications as shown in Figure 2.

3. Structural Assessment

3.1. Qualitative Evaluation

Rapid visual screening identifies potential deficiencies in a building based on checks made with the help of standard checklists, and it often relies on the experience of the personnel in the same field. The qualitative evaluation was conducted using the Tier-1 evaluation of Federal Emergency

Management Agency (FEMA-310) [26] and qualitative structural assessment per the Department of Urban Development and Building Construction (DUDBC), Government of Nepal, guidelines [22]. Qualitative assessment of the building was conducted for the building system, lateral load resisting system, diaphragms, and geological site. For each evaluation aspect, compliance and non-compliance were noted based on the field observations. The shape of the building, proportion in the plan, story height, number of stories, unsupported wall length, openings in walls, and wall cracks were observed as non-compliant. Meanwhile, foundation, sloping ground, plumb line, wall thickness, wall core, height of the walls, position of openings, load path, vertical discontinuities, mass, masonry unit, and masonry lay-up were found to be compliant for the building system. Similarly, for the lateral force resisting system, vertical reinforcement, horizontal band, shear stress, and corner stitch were found to be non-compliant. In the case of diaphragms, all three components, diagonal bracings, lateral restraints, and unblocked diaphragms, were found to be non-compliant. For the geological site, area history and slope failure were compliant; whereas, liquefaction was non-compliant at the site. Based on the qualitative assessment, detailed seismic assessment of the monument was suggested. To accomplish this task, several in-situ tests were conducted along with the finite element analysis. As heritage and monuments are the types of buildings that require special attention during every intervention, understanding the state of the building has great importance [27–29]. To this end, field tests and investigations are of the utmost priority as reconstruction may not be a plausible solution in most of the cases. It is also very important to note that the properties of construction materials demand tests as there would be substantial changes in the manufacturing process in due course, which makes field investigations and in-situ, as well as laboratory, tests more pivotal.

3.2. Material Characterization

Owing to the fact that the allocation of material properties in modeling is likely to be suffered from the uncertainties, in-situ tests were performed to obtain the material properties when possible. Push shear tests were conducted in five different locations: one on the ground floor, two on the first floor, and two on the second floor. Using the FEMA-273 guidelines [26], the tests were carried out and then analyses were done. For further details regarding the push shear test procedure, the reader is directed to FEMA-273 [26]. The average shear strength at lintel-level of masonry wall on ground, first, and second floor was estimated to be 0.1, 0.1, and 0.08 MPa, respectively. Brick units from three different locations of the structure were also tested. Large scale intervention during the assessment was not possible. thus, only three brick units from the existing wall were tested. The brick samples were taken out carefully by scrapping the mortar around it. The samples were tested in the laboratory with the standard procedure. The dimensions of the brick samples were taken and averaged first. Thereafter, cement mortar was provided to smoothen the surface and curing was done as per standard practice. The load was applied in the direction of the depth in the universal testing machine. The load was gradually applied, and finally, the compressive strength was estimated with the help of the load and the area of the specimen. Three bricks considered for compressive strength test reflected the compressive strengths as 6.63, 1.14, and 5.30 MPa. One of the bricks showed relatively lower strength than the other specimen. This is probably due to random manufacturing process of that time and furthermore, the baking process might have been compromised too. Test results highlight that there is wide discrepancy in the strength of the brick units at various locations of the structure. Visually, the bricks were good, but the strength was not per the expectations and initial estimation. The average strength of the brick was found to be 4.3 MPa which can be classified as the lower bound standard of 3.5 MPa grade brick.

The mortar in most of the parts of the structure was found to be intact. The compressive strength of mud-mortar is generally very low, and estimation of exact compressive strength is challenging job; however, standard penetrometer test was conducted on mud-mortar at several locations to estimate and compare compressive strength of the mortar. The compressive strength was obtained in the range of 0.1 to 0.32 MPa with an average of 0.18 MPa for the structure. Brick bond was observed in the damaged portion and all locations without plaster. English bond was found in all tested and exposed locations of the building. Pit excavation test was also carried out to identify the properties of the footing. The test highlighted that there was no significant damage in the footing level, except hairline cracks in few brick units. The summary of the test results is presented in Table 1.

Table 1. Summary of material test results.

Parameter	Test Results	Remarks
Shear strength of masonry wall	0.1 MPa, 0.1 MPa, and 0.08 MPa for ground floor (GF), first floor (FF), and second floor (SF)	Two tests were carried out on each floor and the average value was taken. Due to lesser dead load, the value was low at the top floor, while because of slight damp conditions, the value did not increase on ground floor.
Compressive strength of brick	6.63 MPa, 1.14 MPa and 5.3 MPa for three units	There was restriction in obtaining more samples, thus, 3.5 MPa bricks were considered to match the categorization.
Compressive strength of mortar	Varied from 0.1 MPa to 0.32 MPa in 11 locations	Though not accurate, a penetrometer test was conducted that gave an average strength of 0.18 MPa.
Wall core and bonding type	All walls were solid English bond was found	All walls were identified to be solid brick walls as observed at the test and damage locations.
Foundation	Masonry strip footing with base width of 1700 mm	Excavation was done in two locations. Stepping was found to be done to increase the width of footing at the base, increasing 50 mm in width at each side at every 150-mm depth.

3.3. *Analytical Modeling*

A three-dimensional finite element model was created in SAP2000 v.20 [30] based on the concept of homogenized wall materials assuming the building to be supported at the plinth level by hinges as shown in Figure 7. The hinge support assumption allows capturing the redistributed stress after formation of tension cracks near plinth, which is more realistic than the assumption of the fixed base analysis. Thin shell elements (iso-parametric 4-nodded and 3-nodded area elements) were used to model the structural masonry walls. Two-nodded frame elements (line elements) were used to model beams and timber joists in the floor and roof. Timber floors were modeled with thin shells equivalent to the provided timber planks that provided partial in-plane rigidity. The building has over 10,500 square meters of wall and floor area in the model that demands very high input in modeling and analysis of the structure. Further, the accurate assessment of material properties, accurate constitutive model of masonry element, and rigorous non-linear analysis are important for rational performance evaluation. Owing to the fact that nonlinearity characteristics of masonry play a dominant role in discrepancies that arise during nonlinear dynamic analysis, a detailed study regarding nonlinear properties of materials is needed. However, to do so, sophisticated facilities are needed, and such facilities are not available in Nepal. Thus, because of accurate damping characteristics, natural periods, Poisson's ratio, friction between different materials, and other parameters, a simplified modeling technique was adopted using a homogenized macro element for the wall modeling.

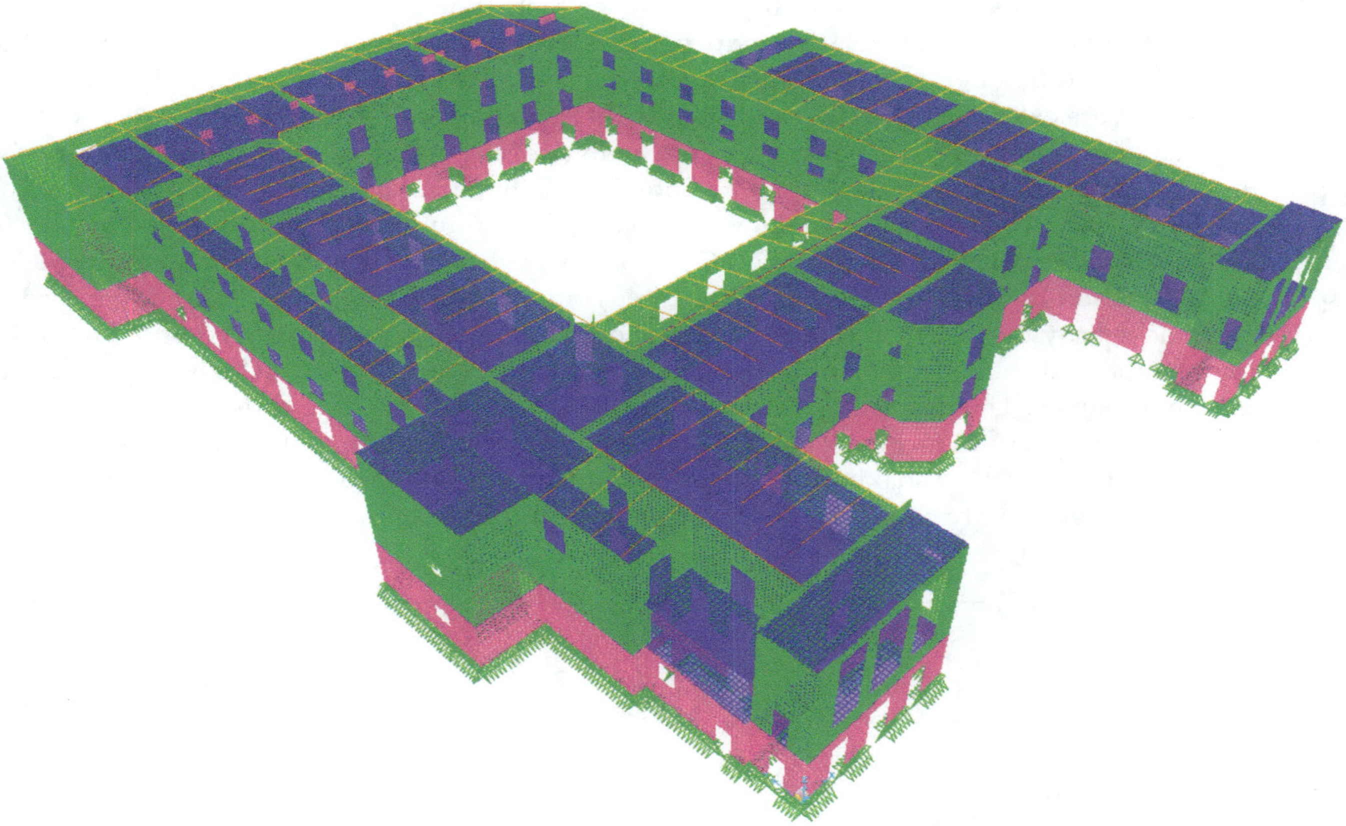

Figure 7. Three-dimensional finite element model of Bagh Durbar.

The mechanical characteristics of materials for modeling are based on laboratory tests, and related previous studies. As the minimum basic strength of masonry per the identified brick strength from tests and weakest mortar according to Indian Standard (IS) 1905:1987 is 0.25 MPa, that corresponds to the average compressive strength of 1 MPa, and the same strength was adopted in this study. Three specimen masonry prisms tested in Nepal (personal communication with Ram Udar Yadav) for similar construction material affirmed the adopted value, as the test results showed a compressive strength of nearly 1 MPa. The Young's modulus of masonry was taken as 550 MPa, which is calculated as 550 times the average compressive strength of masonry as recommended by Indian Standard 1893 (part 1) and 2016. The Poison's ratio of masonry was taken as 0.2. In the case study building, the height to width ratio of brick units is less than 0.75; slenderness ratio of walls is below 6; and no portion of the wall has an area lower than 0.2 m^2; thus, per the IS 1905:1987 [31] recommendations, the permissible stress in masonry for design equals basic compressive strength of masonry, which is one fourth of the average compressive strength of masonry (1 MPa). Several studies (e.g., [27–29]) highlight the necessity of non-destructive tests and material identification in order to predict the actual behavior of complex aggregate buildings and historical constructions. Thus, laboratory tests are required to characterize such properties.

For performance assessment, the building was evaluated against standard requirements as suggested by the Nepal Building Code (NBC). A live load of 2.5 KPa to 4 KPa was considered as per IS 875 (part 2). The analysis was carried out for the design earthquake considering the base shear coefficient of 0.384. Similarly, the base shear coefficient for the retrofitted model was adopted as 0.301, as recommended by the NBC-105 [28]. The values of basic seismic coefficient (C) (corresponding to a low natural period of about 0.12 s), seismic zone factor (Z) (for Kathmandu), and importance factor (I) (considering large number of occupants), were taken as 0.08, 1.0, and 1.2, respectively. The value of structural performance factor (K) was taken as 4 and 3.14 for existing and retrofitted models corresponding to unreinforced masonry and masonry with horizontal and vertical bands. As the

NBC recommends the value of structural performance factor for unreinforced masonry only, the structural performance factor for relatively ductile masonry with horizontal and vertical confining members was considered per common practice taking a slightly higher value, owing to the fact that the confinements are externally provided and slightly inferior compared to monolithically constructed confining elements. To this end, response spectrum analysis was performed considering the response spectrum as suggested by the Indian Standard Code [32]. Different response modification factors as suggested by the Indian Standard Code [32] were adopted for non-retrofitted and retrofitted models to match the base shear recommended by the NBC. A comparative plot of the Indian Standard response spectrum for soft soil is presented along with the three components response spectra for the recorded accelerograms nearby the building (Figure 8). As shown in Figure 8, the vertical component of the accelerogram showed that a higher response was observed at a low period, which surpassed the horizontal components. This is a unique response spectrum for the far-field records, thus, some of the damage to the non-structural members like the partition walls and damage to the upper stories might be attributed to strong vertical shaking. As highlighted by Figure 8, the horizontal components of the response spectra were clearly below the design spectrum for the structural period of 0.12 s.

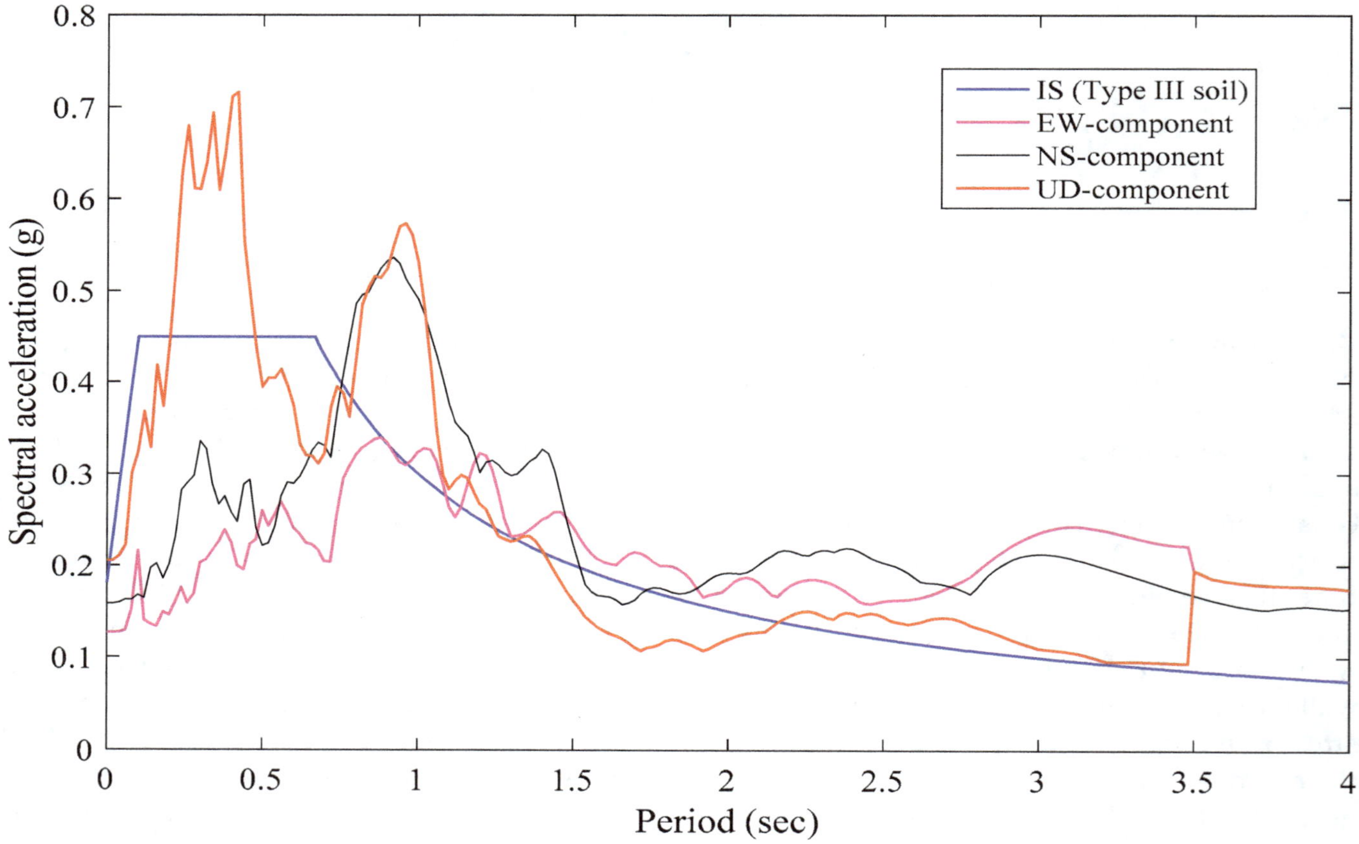

Figure 8. Response spectra of the recorded accelerogram at the Department of Mines and Geology plotted along with Indian Standard code response spectrum for soft soil. In the figure, EW indicates the east-west component of acceleration, NS indicates the north-south component of acceleration, and UD indicates the vertical acceleration component.

Before the detailed analysis of the structure, average stress on the wall at the ground floor was checked. The simplified average stress check showed that the average shear stress demand is about two times greater than the in-plane shear capacity of the mud-mortar walls, which indicates the need for strengthening of the structure. Also, the vertical stress for the dead load and full live load is found to exceed the permissible compressive strength of 0.25 MPa, but was found within the average compressive strength of 1 MPa. This fact indicates that even at minor shaking or disturbance, some parts of the masonry may be damaged. Tensile and compressive stresses due to rocking of the building were found to be negligible in the structure due to the large plan area. However, rocking of individual walls induced additional tension and compression, which were investigated in the detailed analyses. Stresses in individual wall piers of the structure were analyzed by finite element modeling. The performance of each masonry wall was evaluated for the most critical combination of loads per the Nepal Building Code.

The modal analysis highlighted that the first four modes contributed significantly, leading to 60% mass participation in the x-direction and 50% in the y-direction. Mass participation of 85% was achieved by processing 44 modes. The period of the structure was 0.295, 0.292, 0.276, and 0.272 s, respectively, in the first four modes. After the seventh mode, mass participation was less than 5% for both directions in any individual mode except the 12th mode which showed 6% mass participation in the y-direction. A summary of significant mass participation, along with the period of each mode, is presented in Table 2. Figure 9 shows the mode shapes for the first four modes. The first mode reflects that the walls in the western portion experienced higher deformation in the first mode (Figure 9a). The second mode shows that higher deformation was concentrated in the east side of the aggregate in the narrow corridor (Figure 9b). Similar damage occurrence was also noted in the walls of the east side of the aggregate as shown in Figure 5. The arching of the north side of the aggregate should have caused in-plane damages to the cross-walls as shown in Figure 5. The damage mechanisms are represented in Figure 4 by uppercase letters E, F, and G on the third floor and, E′, F′G′, and H on the second floor. The third mode depicted that the projection in the south-east direction experienced the largest deformation as shown in Figure 9c. Consistently, in-plane damage as shown in Figure 6c was observed due to the earthquake which is indicated by the uppercase letter 'O' in Figure 2. As shown in Figure 9d, the fourth mode shows higher deformation in the south-west projection of the structure. Consistently, the damage also occurred in the same wall due to the Gorkha earthquake.

Table 2. Summary of modal mass participation in x- and y-direction in the first 15 modes.

Mode	Period (sec)	Modal Mass Participation Ratio in x-direction	Modal Mass Participation Ratio in y-direction
1	0.295	0.450	0.071
2	0.292	0.054	0.046
3	0.276	0.087	0.170
4	0.272	0.000	0.190
5	0.249	0.025	0.038
6	0.242	0.014	0.051
7	0.228	0.001	0.060
8	0.222	0.008	0.001
9	0.200	0.050	0.006
10	0.199	0.002	0.001
11	0.198	0.001	0.007
12	0.198	0.000	0.060
13	0.187	0.011	0.015
14	0.181	0.011	0.007
15	0.167	0.001	0.008

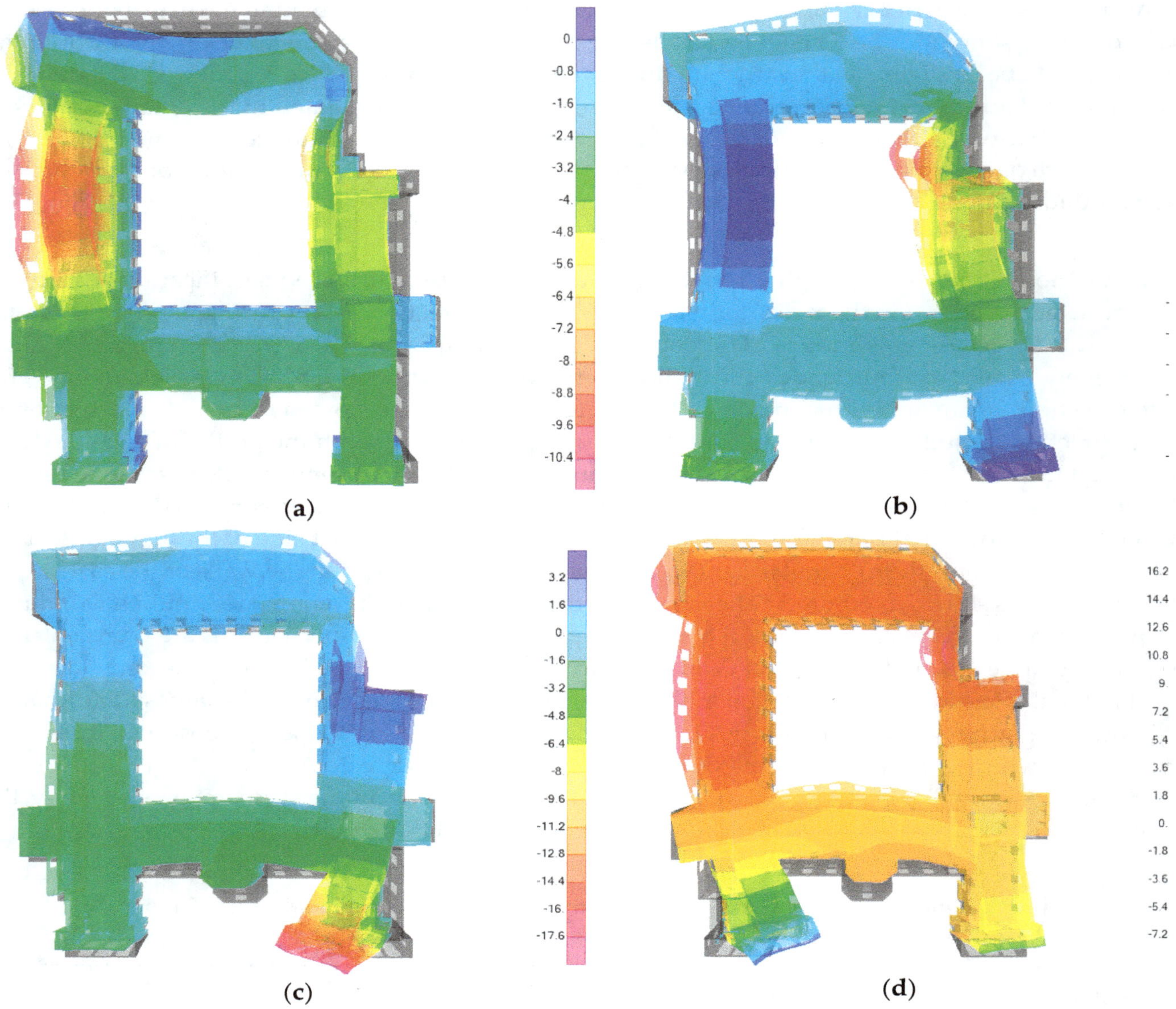

Figure 9. Deformed shape (**a**) mode 1 at the period of 0.295 s, (**b**) mode 2 at the period of 0.292 s, (**c**) mode 3 at the period of 0.276 s, and (**d**) mode 4 at the period of 0.272 s.

Stresses were checked for the wall sections accounting for different load combinations per NBC-105 [24] recommendations. For each wall, vertical and horizontal loads, as well as the moments in the two directions, were used to analyze the stresses. Compliance test results agree that the damage due to the Gorkha earthquake as the response spectrum analysis of the as-built model showed that 49% of the walls were non-compliant, indicating some level of direct tension. Similarly, 40% of the walls depicted non-compliance with permissible compression. However, 97% of the walls were found to be deficient in shear. As almost every wall demanded ductility enhancement and confinement, retrofitting was proposed for the monument as outlined in the following section.

4. Design of Structural Restoration and Seismic Strengthening

Method selection is a crucial step in providing interventions to monumental constructions. Based on the building typology and expected level of performance following the retrofitting, compliance with existing codes of practices, functionality requirements, cost constraints, availability and compatibility of retrofitting solutions, safety and serviceability of the structure, aesthetics, and cultural values, among others, were kept in mind during the selection of the method. With the aforementioned considerations, different strengthening options were compared. After several expert level consultations, retrofitting using reversible and traditional construction materials with steel plates were considered as the most rational and viable option, and analysis was done for the same. Traditional materials (or previously used materials) like Surkhi, lime, timber, and steel were selected for the interventions in most of the locations. In the meantime, reinforced concrete was not used in the design to preserve the morphology of the building. Herein, appropriateness in the use of steel sections was due to ease in installation as well as in removal or replacement of the retrofitted portions when required.

The retrofitting framework developed for the Bagh Durbar monument is presented in Figure 10. The details of the bands, diagonal bracings, and timber joist end connections are respectively presented in Figures 11–13. As shown in Figure 11, the flexible floors are provided with floor level bracings to increase the floor rigidity. With a revised base shear after the interventions, which decreases significantly because of an increase in ductility of structure, the compliance of individual walls against shear resistance was increased appreciably and satisfied the design criteria as well. The global shear resistance of the structure, including provided sections, is at least two times greater than the overall demand which caters for localized high-stress demands. After retrofitting, a part of compression (under critical load) would be shared by vertical steels as well.

As the design is based on permissible stress, average stress capacity of the existing wall and added steel elements would be higher, which provides the necessary safety margin.

A horizontal band of 300 × 8 mm was designed to suffice the out-of-plane bending of the wall at the lintel level. Although the existing thickness of the wall makes it less vulnerable to out-of-plane instability, the bands simultaneously improve the integrity and ductility of the wall and the structure. Horizontal bands were also provided at the floor level using plates on the exterior face and ISMC-250 on the interior side. All horizontal bands were welded to vertical bands using a single bevel butt weld and extra lapping plate. Horizontal bands on both faces of the walls were connected by bolts and anchored to the wall. The total moment of resistance of the wall, together with bands, exceeded the out-of-plane moment of the wall from analysis under critical load combination; thus, the provided steel section is sufficient. Both I-beam with jack arch floors and wooden joist floors were made more rigid against in-plane deformation. For an I-beam with jack arch floor, ISMC-250 was placed at the interior face of all rooms and joined together as a box. Diagonal strips were welded to the underside of existing I-sections for satisfactory in-plane rigidity. The I-sections were connected to the floor band ISMC-250 as detailed in drawings.

For timber floors, since most of the visible timber in the floors had deteriorated, all flooring timber was replaced by seasoned timber with a strip of plastic damp-proof course underneath and related protection work, especially at the embedded length. Wooden floors were stiffened in-plane by the provision of two layers of timber plank flooring oriented in the perpendicular direction. Furthermore, wooden joists were connected to supporting walls by steel strips welded to the horizontal band on the exterior side of the wall for proper anchorage as detailed in Figure 13. The roof truss was observed to be displaced in some locations and some timber portions were damaged and deteriorated. The entire roof required dismantling and reconstruction. On average, 40% of the existing timber needed to be replaced and the rest could be reused with necessary protection works. Vertical bands were required to be extended up to the roof level. Gable band and roof band consisting of wooden joists were required to be placed and the roof truss required to be reconstructed per the existing design and detail. Then diagonal braces were needed between rafters and the rafters needed to be connected to the roof band using steel strips. Further, the roofing sheets required replacement by new corrugated galvanized iron (CGI) sheets and were anchored firmly to the roof truss.

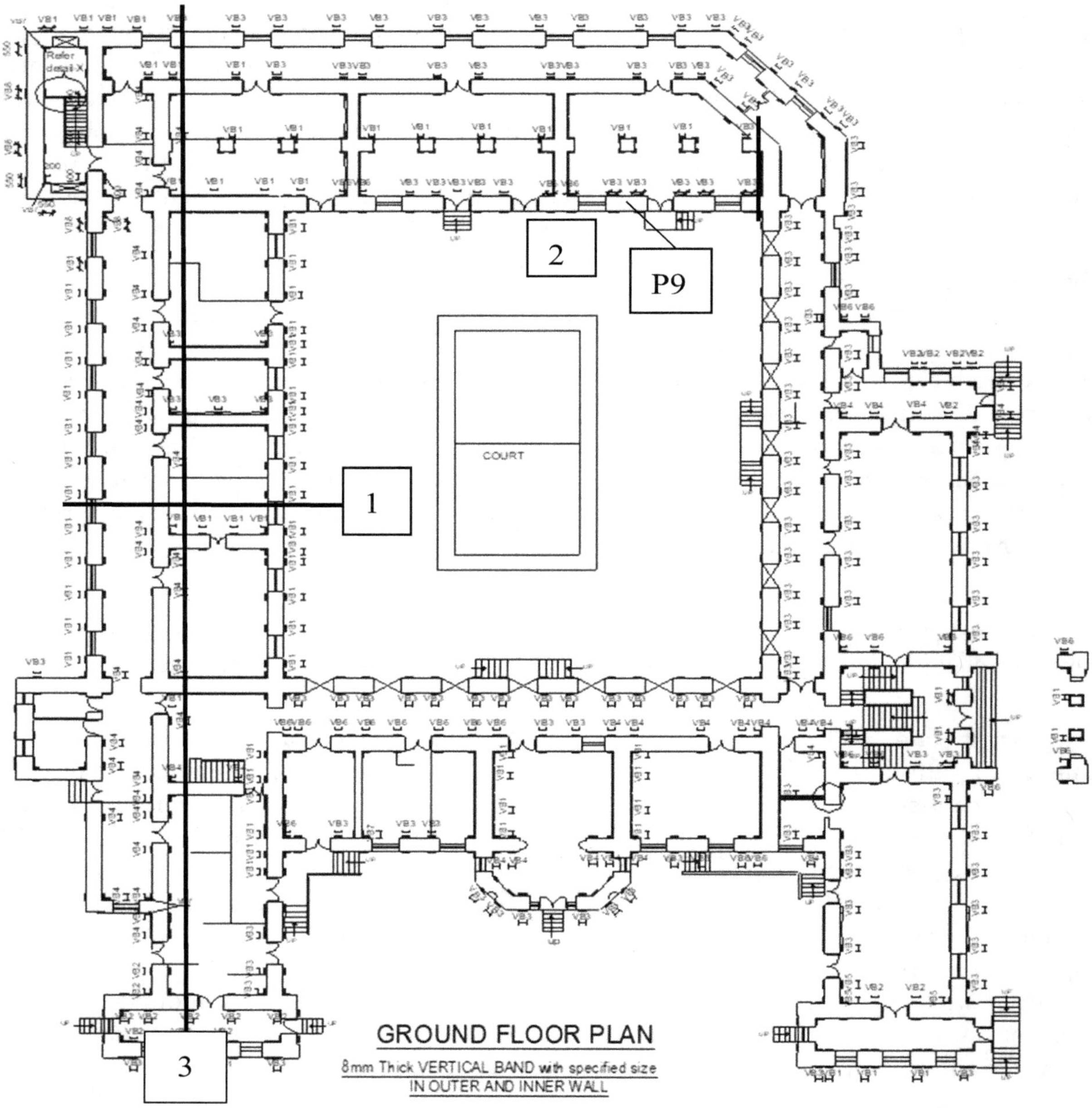

LEGENDS: VERTICAL BAND DESCRIPTION

SN	SYBMOL	DESCRIPTION
1	VB1	VERTICAL BAND- 8mmx100mm
2	VB2	VERTICAL BAND- 8mmx125mm
3	VB3	VERTICAL BAND- 8mmx150mm
4	VB4	VERTICAL BAND- 8mmx200mm
5	VB5	VERTICAL BAND- 8mmx225mm
6	VB6	VERTICAL BAND- 8mmx250mm
7	VB7	VERTICAL BAND- 8mmx300mm
8	VB8	VERTICAL BAND- 8mmx400mm
9	VB9	VERTICAL BAND- 8mmx500mm
10	VB10	VERTICAL BAND- 8mmx550mm

Figure 10. Retrofitting measures adopted for the monument (line and text in box indicates section-cut number and pier number).

In flat roofs, coats of water proofing materials are required to prevent seepage. The timber in existing staircases was deteriorated. Hence, all staircases needed to be replaced by new, but similar, ones. The staircase projecting in the north-west corner was found to be especially vulnerable from analysis. The staircase of the exterior walls needed to be stiffened and anchored by the addition of ISMC-250 sections underneath the landing. Secure connections and ductility of the exterior architectural columns and porch were necessary for life safety. Decorative masonry columns were confined with vertical steel bars embedded around the columns by cutting the groves, and the entrance porch was tied with the main structure using horizontal bands at two levels. Provision of vertical bands around all openings and wide horizontal bands above openings help to redistribute the stress concentrated around openings. Arches were observed to be especially vulnerable in lateral shaking during past earthquakes, hence steel strips were used around arches to hold them to the nearby vertical band and horizontal lintel band.

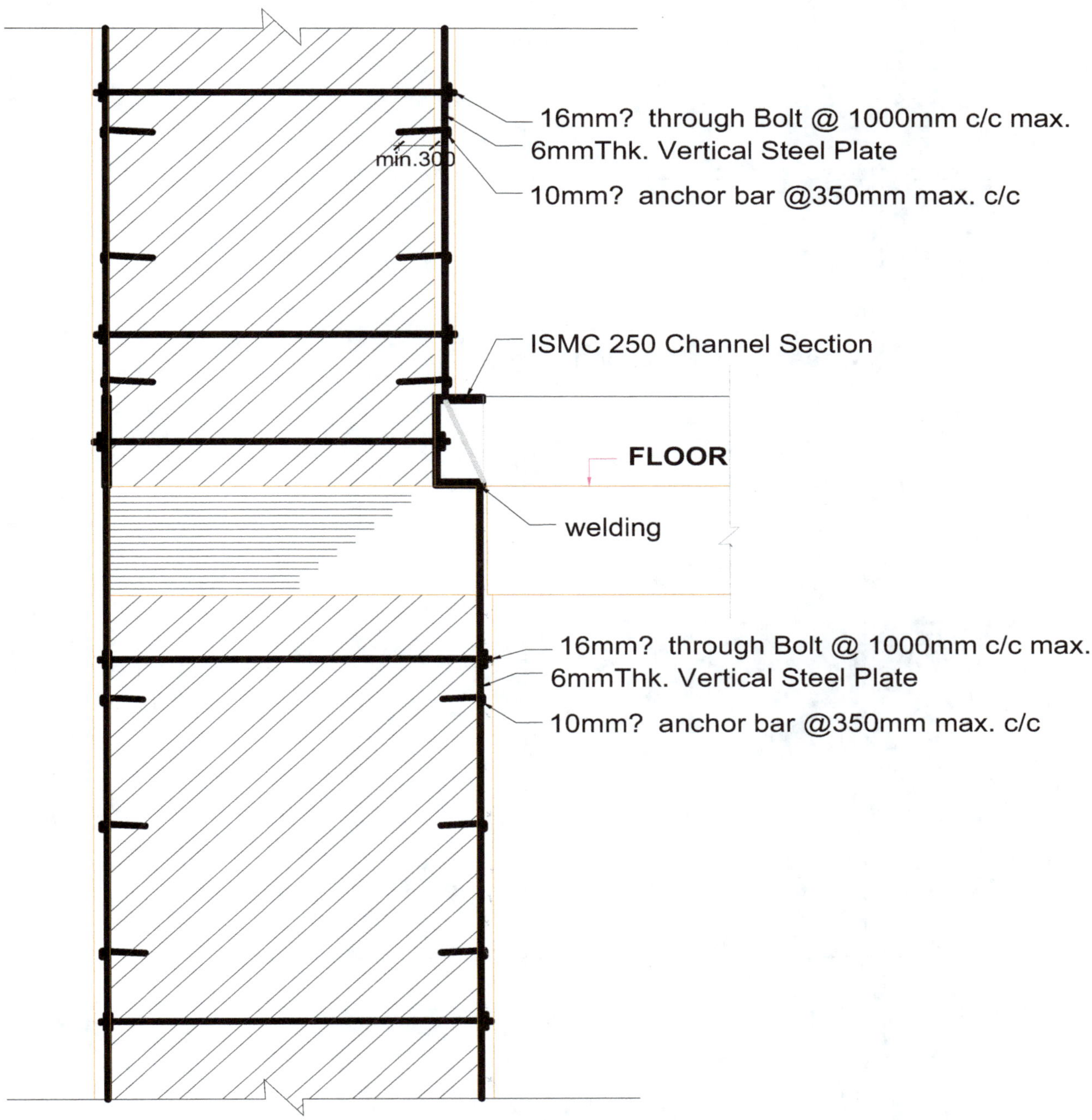

Figure 11. Details of band per floor.

The primary intervention system is composed of vertical and horizontal steel members which integrates and confines the masonry system; all such interventions would be hidden under the plaster to preserve the morphology of the monument. Strengthening of floors is done by providing floor braces and continuous beams at the floor levels. The shear strength of walls is enhanced by providing vertical steel members at the ends with horizontal bands and diagonal braces on walls at the selected locations. The sizing, position, and their connections are based on the analysis results. These elements are designed to resist the stress beyond the capacity of the existing wall. The stresses in the wall were computed as the ratio of internal forces per meter to the thickness of the corresponding wall as shown in Figures 14–16. Although examples are shown for different load combinations, the most critical cases were considered to perform necessary analysis and retrofit design.

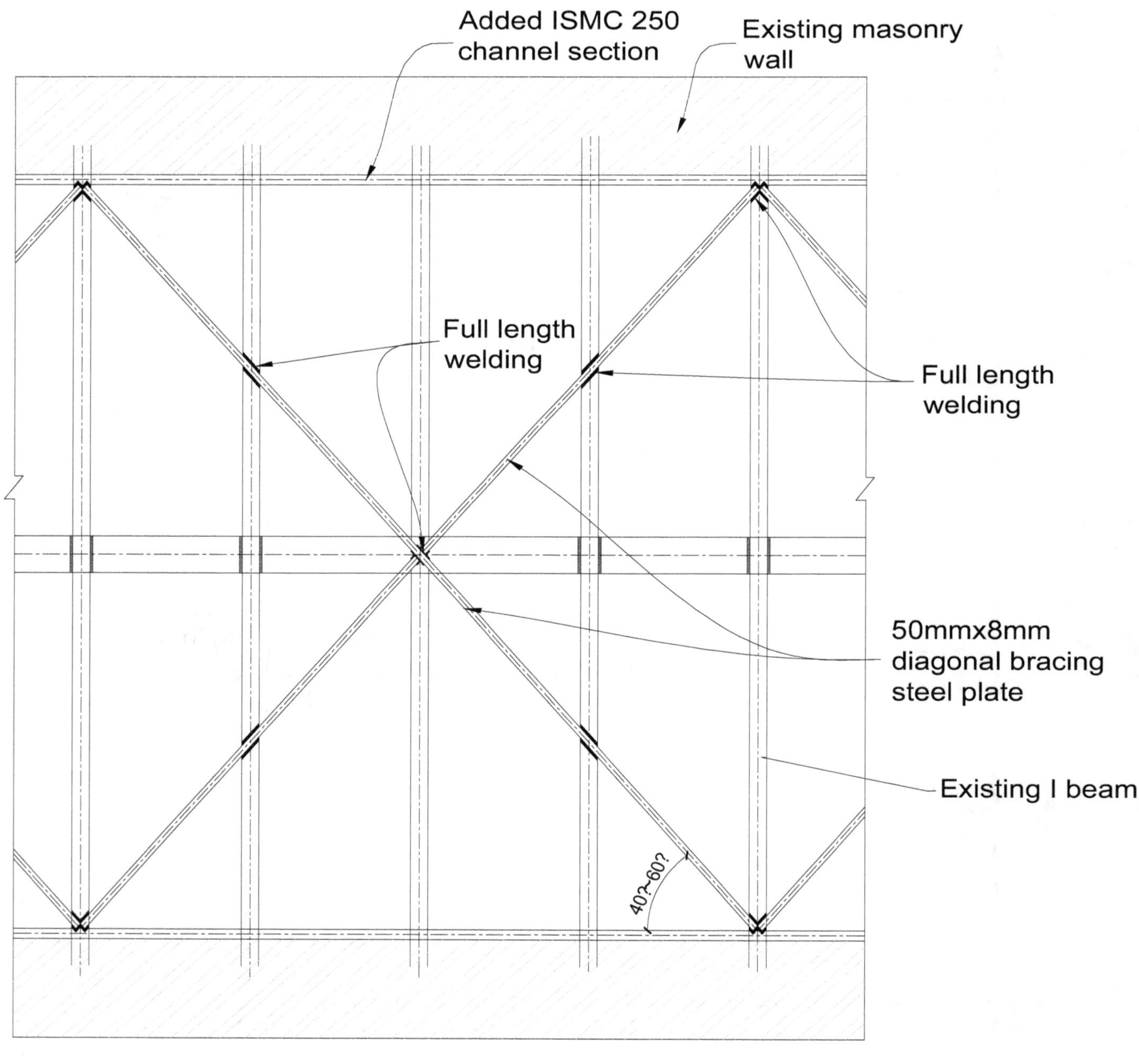

Figure 12. Details of bracing.

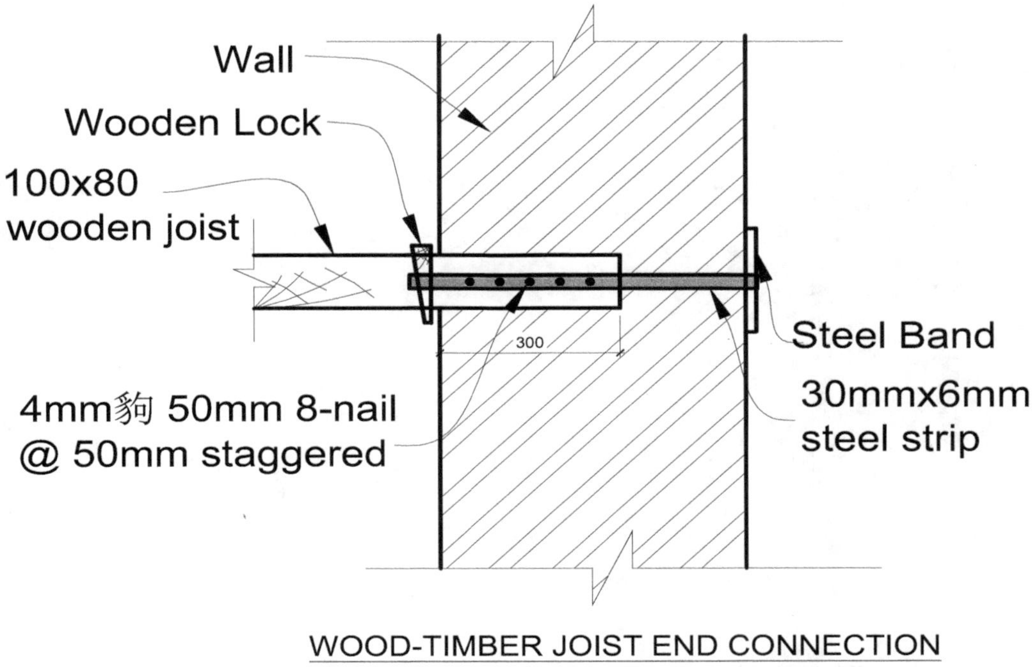

Figure 13. Details of timber joist end connection.

Tensile stresses were concentrated around opening and pier edges only due to the in-plane action of the earthquake load as indicated by Figure 14. There was an uplift in some portion of the piers, but due to development of tensile cracks, stresses were redistributed. Vertical steel plates are provided along the edges of each pier to take these stresses. Hence, tensile cracks are not allowed to propagate and subsequently prevent the masonry degradation in the cyclic load. Compressive stresses were also concentrated around opening, corners, and pier edges (see Figures 14–16). Although average compressive stress was obtained to be less than the average compressive strength of the walls; it was still higher than the permissible design stress. The provided steel plates that are stiffly anchored to walls take part in the compression of the wall to keep the stress within safe limits. Further, the confinement with vertical and horizontal steel bands increases the ductility and compressive strength of the wall, which is not considered in the calculation, and regarded as an additional factor of safety. Most masonry piers experienced higher shear stress than their design capacity. However, the design capacity of the wall was identified as ~0.1 MPa, but because of high damping in mud-mortar structures, the actual damages would be generally lesser.

Improving the structure with ductile bands assure advantages in two ways. Firstly, it improves the ductility of the overall structure, requiring lesser design force, and secondly, these bands also increase the capacity of the structure. The sizing of these members is done based on the demand observed from response spectrum analysis in the FEM model. Engineering judgments were used in predicting the actual non-linear behavior of the structure and providing suitable adjustments in the required interventions. Figure 17a,b show the displacements in the x- and y-directions, respectively, before retrofitting. The maximum displacements in the x- and y-directions were noted as 40.65 mm and 33.54 mm, respectively. The maximum displacement along the x-axis was obtained in the east wall which was the most damaged wall. Similarly, the maximum displacement along the y-axis was obtained in the southern part of the monument where damage was also significant. Figure 17c,d reflect the displacements along the x- and y-axis after retrofitting. The inter-story drift was limited to 0.26% which indicates satisfactory performance level (immediate occupancy level) during the earthquake, provided the integrity of all the components is maintained. The maximum deformations along the x-axis were 22.6 mm on the west wall, whereas, the same for the y-axis was 16.9 mm on the south wall of the monument.

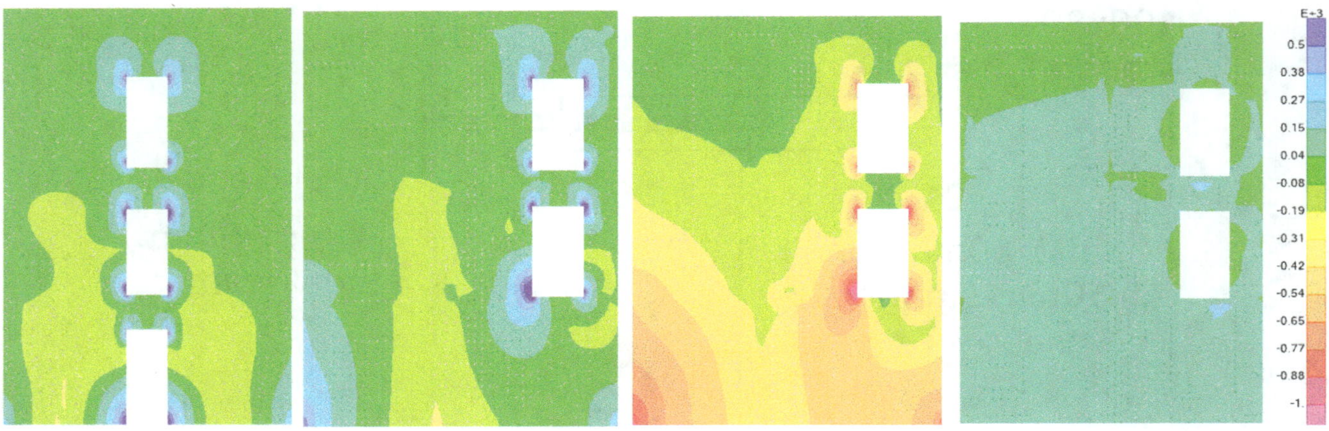

(a) (b) (c) (d)

Figure 14. Force per meter under load combination dead + live + response spectrum-x (**a**) tensile force at section cut-1, (**b**) tensile force at section cut-2, (**c**) compressive force at section cut-2, and (**d**) shearing force at section cut-2. (See Figure 10 for section cuts).

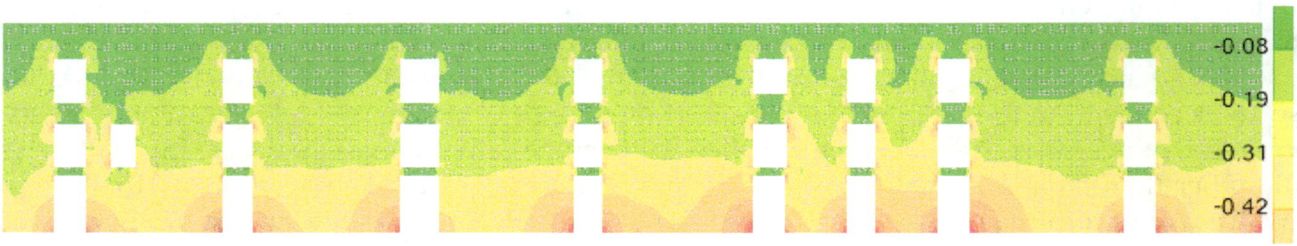

Figure 15. Compressive force per meter at section cut-3 (Figure 10) under load combination dead + live + response spectrum-x.

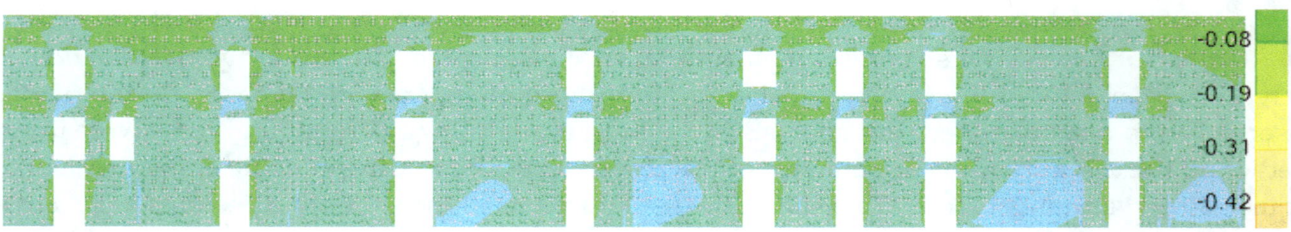

Figure 16. Shearing force per meter at section cut-3 (Figure 10) under load combination dead + live + response spectrum-x.

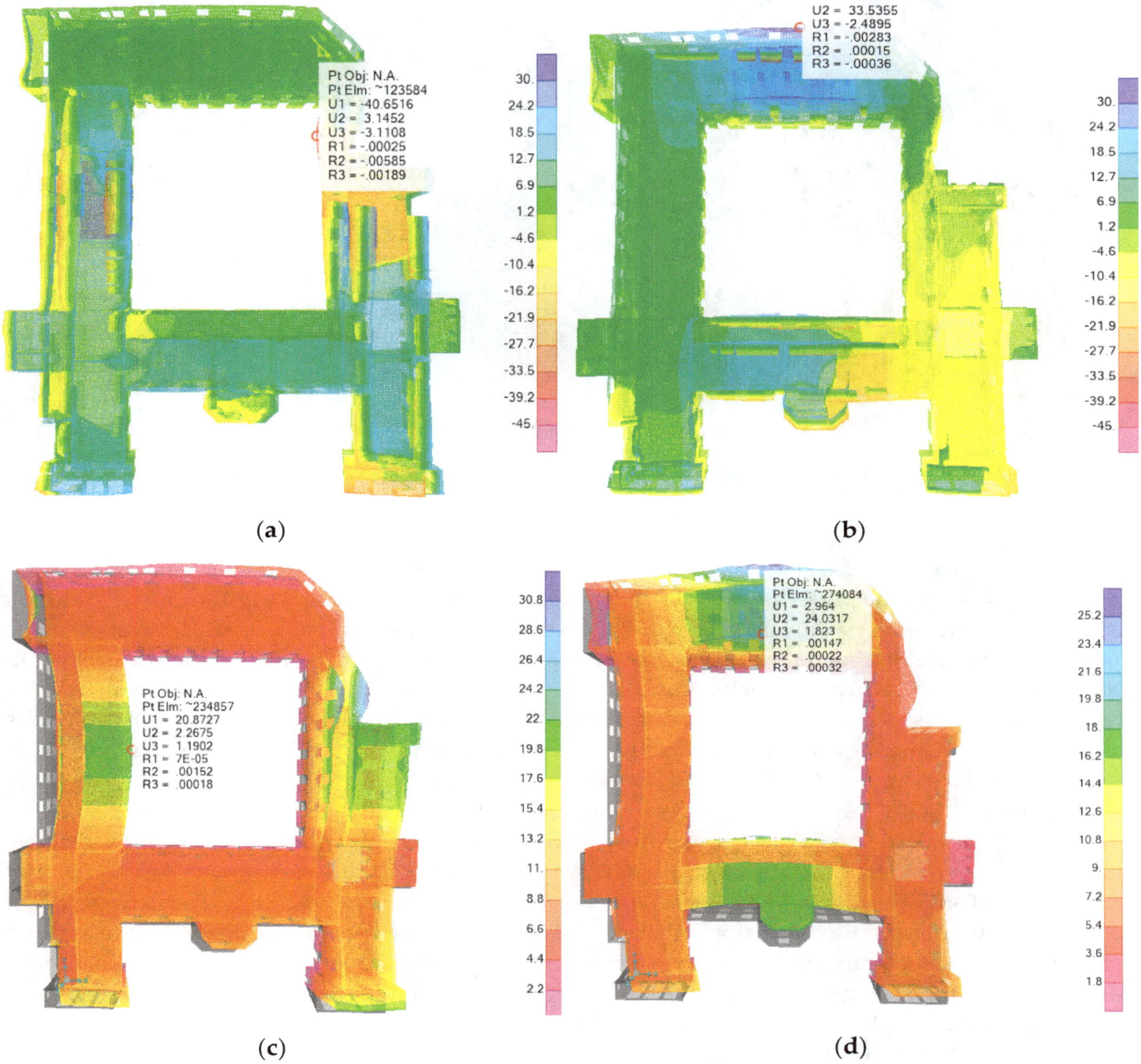

Figure 17. Deformation of the monument (**a**) before retrofitting along the x-direction, (**b**) before retrofitting along the y-direction, (**c**) after retrofitting along the x-direction, and (**d**) after retrofitting in the y-direction.

Figure 18 shows the three-dimensional view of the pier P9 indicated in Figure 10. The section cut force from the analysis for P9 shows an axial, shear, and moment demand of 743 KN, 181 KN, 332 KN-m respectively. Meanwhile, there was no direct tension on the pier. The analysis further showed that the pier was deficient in compression capacity and rocking shear capacity. To this end, 100 × 8 mm vertical steel plates were provided at pier ends on both faces in addition to the horizontal plates at lintel and sill levels. Analysis including the strength of steel plates showed that the aggregate capacity exceeds the demand forces. The same approach was used for all piers and safety checks were performed.

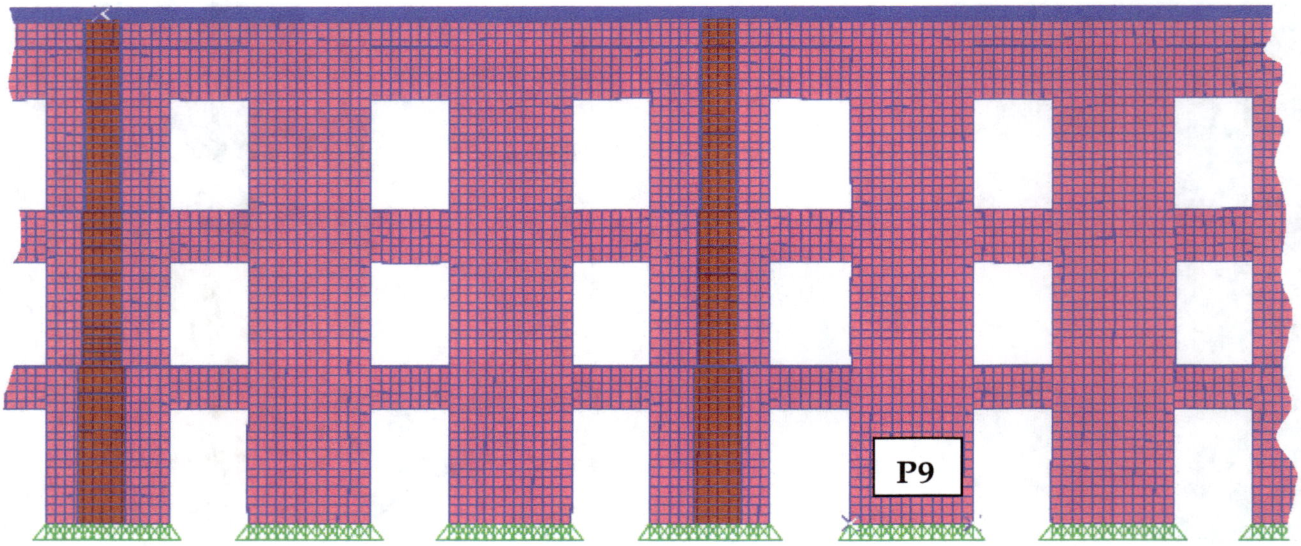

Figure 18. Elevation of wall showing the considered pier P9 (see Figure 10).

5. Conclusions

Restoration of Nepalese heritage buildings is a relevant topic in relation to their monumental and functional values. In the present paper, Bagh Durbar monumental building, a 19th-century landmark construction, was taken as a case study. As Greco-Roman constructions are still in considerable number and have monumental value, assessment and strengthening is a pertinent topic in Nepal. To achieve this objective, structural analyses and design of strengthening solutions were performed in this study. Cultural preservation and seismic safety were considered vis-a-vis and strengthening solutions were proposed in such a way that no morphological discrepancies would take place. Observation of the building conditions after the Gorkha earthquake and the characterization of the damage state confirmed a number of vulnerable features of the structural system. In compliance with national and international standards on existing structures, a two-phase process was adopted in this study. The first phase was associated with the comprehensive knowledge of the structure, including material properties and structural detailing. The second one consisted of the definition of a structural model to be loaded by means of conventional actions: the performance level in the current state and the performance level after implementation of specific interventions aimed at removing critical issues and increasing the strength and ductility of different components. It is shown that the localized non-compliances which occurred in the as-built model are significantly improved by the proposed interventions. This is one of the main outcomes of the paper, in the sense that a rational and virtuous process can preserve Greco-Roman monuments that are still widely used for administrative purposes, especially in the Kathmandu valley, by increasing at the same time their seismic safety. Apart from their administrative value, they have monumental values too; thus, field investigation, numerical analysis, and subsequent retrofit designs would be important to preserve them. The same retrofit option may be applicable to other monuments in the Kathmandu valley as the construction system and age of almost all Greco-Roman buildings matches.

In conclusion, it is worth noting that the level of the structural analysis discussed in this paper represents a point of balance between the ease-of-use of the representation of the seismic action (response spectrum), and the complexity of the two-dimensional representation of the walls and spandrels in the masonry building. Nevertheless, this is not necessarily a limitation of the approach that can be improved by including additional levels of analysis. In such a case, it is also clear that the detail of the structural and material knowledge needs to be increased to cover the requirements of refined and detailed constitutive laws of the materials and of the main structural components. Furthermore, one of the limitations of this work is that strengthening solutions are explored based on basic engineering principles to assure compliance with the existing guidelines that are practiced in Nepal (i.e., Nepali and Indian building codes only). So future studies may incorporate other international standards and practices to improve the outcomes.

Author Contributions: Conceptualization, R.A. and P.J.; methodology, G.F.; software, R.A., P.J., and D.G.; validation, G.F. formal analysis, R.A., and D.G.; investigation, P.J.; resources, P.J.; writing—original draft preparation, D.G. and R.A.; writing—review and editing, G.F., D.G. and R.A.; supervision, G.F.

Acknowledgments: The authors are grateful to Digicon Engineering Consult, Kupandole for the arrangements. For the laboratory tests, Multi Lab and NS Engineering and Geotechnical Services Pvt. Ltd. are acknowledged. The support from various people at Kathmandu Metropolitan City Office is also greatly acknowledged. Suggestions from Professor Filippo Santucci de Magistris are also warmly acknowledged.

References

1. National Planning Commission. *Post-Earthquake Damage Assessment*; Government of Nepal, National Planning Commission: Kathmandu, Nepal, 2015; Volume A and B.
2. Gautam, D.; Fabbrocino, G.; Santucci de Magistris, F. Derive empirical fragility functions for Nepali residential buildings. *Eng. Struct.* **2018**, *171*, 617–628. [CrossRef]
3. Pandit, A.K.; Yadav, R.; Jha, S.K.; Adhikari, R. Seismic vulnerability assessment of masonry buildings in Kathmandu Valley after Gorkha Earthquake 2015: A case study of Administrative Staff College building. In Proceedings of the International Conference on Earthquake Engineering and Post Disaster Reconstruction Planning, Bhaktapur, Nepal, 24–26 April 2016; pp. 244–251.
4. Langenbach, R. Performance of the earthen Arg-e-Bam (Bam Citadel) during the 2003 Bam, Iran, earthquake. *Earthq. Spectra* **2005**, *21*, S345–S374. [CrossRef]
5. D'Ayala, D.; Benzoni, G. Historic and traditional structures during the 2010 Chile earthquake: Observations, codes, and conservation strategies. *Earthq. Spectra* **2012**, *28*, S425–S451. [CrossRef]
6. Lucibello, G.; Brandonisio, G.; Mele, E.; De Luca, A. Seismic damage and performance of Palazzo Centi after L'Aquila earthquake: A paradigmatic case study of effectiveness of mechanical steel ties. *Eng. Fail. Anal.* **2013**, *34*, 407–430. [CrossRef]
7. Mazzoni, S.; Castori, G.; Galasso, C.; Calvi, P.; Dreyer, R.; Fischer, E.; Fulco, A.; Sorrentino, L.; Wilson, J.; Penna, A.; et al. 2016–17 Central Italy Earthquake Sequence: Seismic retrofit policy and effectiveness. *Earthq. Spectra* **2018**. [CrossRef]
8. Ferreira, T.M.; Vicente, R.; Mendes da Silva, J.A.R.; Varum, H.; Costa, A. Seismic vulnerability assessment of historical urban center in Seixal, Portugal. *Bull. Earthq. Eng.* **2013**, *11*, 1753–1773. [CrossRef]
9. Vicente, R.; Rodrigues, H.; Varum, H.; Mendes Da Silva, J.A.R. Evaluation of strengthening techniques of traditional masonry buildings: Case study of four-building aggregate. *J. Perform. Constr. Facil.* **2011**, *25*, 202–216. [CrossRef]
10. Lamego, P.; Lourenco, P.B.; Sousa, M.L.; Marques, R. Seismic vulnerability and risk analysis of the old building stock at urban scale: Application to a neighborhood in Lisbon. *Bull. Earthq. Eng.* **2017**, *15*, 2901–2937. [CrossRef]
11. Asteris, P.G.; Chronopoulos, M.P.; Chrysostomou, C.Z.; Varum, H.; Plevris, V.; Kyriakides, N.; Silva, V. Seismic vulnerability assessment of historical masonry structural systems. *Eng. Struct.* **2014**, *62–63*, 118–134. [CrossRef]
12. de Felice, G.; De Santis, S.; Lourenco, P.B.; Mendes, N. Methods and challenges for the seismic assessment of historic masonry structures. *Int. J. Archit. Herit.* **2017**, *11*, 143–160. [CrossRef]
13. Casapulla, C.; Maione, A.; Argiento, L.C. Seismic analysis of an existing masonry building according to the multi-level approach of the Italian guidelines on cultural heritage. *Ingegneria Sismica* **2017**, *34*, 40–59.
14. Clementi, F.; Gazzani, V.; Poiani, M.; Mezzapelle, P.A.; Lenci, S. Seismic assessment of a monumental building through nonlinear analyses of a 3D solid model. *J. Earthq. Eng.* **2018**, *22*, 35–61. [CrossRef]
15. Lagomarsino, S.; Cattari, S. PERPETUATE guidelines for seismic performance-based assessment of cultural heritage masonry structures. *Bull. Earthq. Eng.* **2015**, *13*, 13–47. [CrossRef]
16. Milani, G.; Venturini, G. Safety assessment of four masonry churches by a plate and shell FE nonlinear approach. *J. Perform. Constr. Facil.* **2013**, *27*, 27–42. [CrossRef]

17. Rossi, M.; Cattari, S.; Lagomarsino, S. Performance-based assessment of the great mosque of Algiers. *Bull. Earthq. Eng.* **2015**, *13*, 369–388. [CrossRef]
18. Marra, A.; Brigante, D.; Rainieri, C.; Fabbrocino, G. Structural characterization and performance assessment of the Villa d'Este Palace in Tivoli. In Proceedings of the 16th International Brick and Block Masonry Conference "Masonry in a world of challenges", Padua, Italy, 26–30 June 2016.
19. Fabbrocino, G.; Marra, A.; Savorra, M.; Fabbrocino, S.; Santucci de Magistris, F.; Rainieri, C.; Brigante, D.; Celiento, A. Increasing the resilience of cultural heritage to earthquakes by knowledge enhancement: The lesson of the Carthusian Monastery in Trisulti. In *Resilienza Delle città D'arte ai Terremoti*; Atti Dei Convegni Lincei; Accademia Nazionale dei Lincei: Roma, Italy, 2016; Volume 306, pp. 553–566.
20. Sonda, D.; Miyamoto, K.; Kast, S.; Khanal, A. The restoration and seismic strengthening of the earthquake-damaged UNESCO heritage palace in Kathmandu. *Int. J. Archit. Herit.* **2018**. [CrossRef]
21. Gautam, D. Seismic performance of world heritage sites in Kathmandu valley during Gorkha seismic sequence of April-May 2015. *J. Perform. Constr. Facil.* **2017**. [CrossRef]
22. DUDBC (Department of Urban Development and Building Construction). *Seismic Vulnerability Evaluation of Guideline for Private and Public Buildings*; Government of Nepal, Department of Urban Development and Building Construction: Kathmandu, Nepal, 2011.
23. Grunthal, G. *European Macroseismic Scale 1988 (EMS-1988)*; Centre Européen de Géodynamique et de Séismologie: Luxemburg, 1998.
24. NBC (Nepal Building Code). *Seismic Design of Buildings in Nepal (NBC-105)*; Government of Nepal, Department of Urban Development and Building Construction: Kathmandu, NBC, Nepal, 1994.
25. Recommendations PCM. *Guidelines for the Assessment and the Mitigation of Seismic Risk of Cultural Heritage with Reference to Italian NTC2008*; Directive of the Prime Minister: Rome, Italy, 2011. (In Italian)
26. FEMA (Federal Emergency Management Agency). *NEHRP Guidelines for Seismic Rehabilitation of Buildings*; Federal Emergency Management Agency Report (FEMA-273); FEMA: Washington, DC, USA, 1997.
27. Rainieri, C.; Fabbrocino, G.; Verderame, G.M. Non-destructive characterization and dynamic identification of a modern heritage building for serviceability seismic analyses. *NDT E Int.* **2013**, *60*, 17–31. [CrossRef]
28. Rainieri, C.; Fabbrocino, G. Development and validation of an automated operational modal analysis algorithm for vibration-based monitoring and tensile load estimation. *Mech. Syst. Signal Process.* **2015**, *60–61*, 512–534. [CrossRef]
29. Rainieri, C.; Marra, A.; Rainieri, G.M.; Gargaro, D.; Pepe, M.; Fabbrocino, G. Integrated non-destructive assessment of relevant structural elements of an Italian heritage site: The Carthusian monastery of Trisulti. *J. Phys. Conf. Ser.* **2015**, *628*, 012018. [CrossRef]
30. CSI (Computer and Structure Inc). *SAP: Integrated Software for Structural Analysis and Design*; v. 20; CSI: Walnut Creek, CA, USA, 2000.
31. Bureau of Indian Standards (BIS). *Indian Standard Criteria for Earthquake Resistant Design of Structures: Part 1 General Provisions and Buildings (Fifth Revision)*; IS 456 (Part 1); BIS: New Delhi, India, 2016.
32. Bureau of Indian Standards (BIS). *Indian Standard Code of Practice for Structural Use of Unreinforced Masonry*; Bureau of Indian Standard: New Delhi, India, 1987.

7

Deflection Prediction of No-Fines Lightweight Concrete Wall using Neural Network Caused Dynamic Loads

Ridho Bayuaji [1,*] and Totok Ruki Biyanto [2]

1 Department of Civil Infrastructure Engineering, Faculty of Vocations, Institut Teknologi Sepuluh Nopember, Surabaya 60116, Indonesia
2 Department of Engineering Physic, Faculty of Industrial Technologi, Institut Teknologi Sepuluh Nopember, Surabaya 60116, Indonesia; trbiyanto@gmail.com
* Correspondence: bayuaji@ce.its.ac.id

Abstract: No-fines lightweight concrete wall with horizontal reinforcement refers to an alternative material for wall construction with an aim of improving the wall quality towards horizontal loads. This study is focused on artificial neural network (ANN) application to predicting the deflection deformation caused by dynamic loads. The ANN method is able to capture the complex interactions among input/output variables in a system without any knowledge of interaction nature and without any explicit assumption to model form. This paper explains the existing data research, data selection and process of ANN modelling training process and validation. The results of this research show that the deformation can be predicted more accurately, simply and quickly due to the alternating horizontal loads.

Keywords: wall; hysteresis; dynamic; no-fines lightweight concrete; artificial neural network

---

1. Introduction

Indonesian territory is prone to earthquake disasters occurring as a consequence of its position. As it is the point of the interaction of four tectonic plates, including Australian, Eurasian, Pacific, and Philippine, moreover, it is the place in which two primary earthquake courses (Circum-Pacific Earthquake Belt and Trans Asiatic Earthquake Belt) pass through. Over the last five decades, dozens of large earthquakes occurred in Indonesia and they have led to many casualties. It is worth noting here that even the conventional force-based seismic design approach is strongly connected to the deformation capacity parameter through the force-reduction factor which is used in the estimation of design force of structures in the force-based seismic design approach. Furthermore, the deformation capacity plays a crucial role in seismic assessment and retrofitting of existing structures, which has become one of the main research topics in structural engineering.

This phenomenon can be minimized through awareness of unreinforced masonry building, even for the non-engineered building construction. A simple house construction commonly uses a brick masonry wall which, when a horizontal force (earthquake) occurs, can lead to total and immediate collapse starting from the damage and the collapse in the part of masonry wall followed by other building structures. This happens since the masonry wall in a simple house is used as a structural part supporting the lateral loads (limited friction) due to the earthquake load. Seismic behaviour of masonry walls was specifically discussed for its shear capacity drives in several papers [1,2].

The displacement capacity is a key parameter in the seismic design and assessment of structures. Unfortunately, our current state of knowledge of the displacement capacity of masonry walls is limited. On the one hand, the available experimental data has pronounced variability, so it is not possible to identify rational values for the displacement capacity of masonry walls based only on such experimental data and, on the other hand, there are no reliable analytical models for either the displacement capacity or the force–displacement relationship of masonry walls [3].

In general, the displacement capacity of masonry walls is a very complex value; it is influenced not only by the failure mode but also by many other factors such as the constituent materials, geometry, boundary conditions, and pre-compression level. Currently, we are still not able to take into account properly the influences of all factors affecting the displacement capacity of masonry walls due to inhomogeneous experimental data and a lack of reliable analytical models.

The assessment of the nonlinear seismic behaviour of masonry walls represents a subject of great importance, however it is rather difficult to solve.

The behavior of unreinforced masonry wall panels subject to dynamic excitation such as seismic loading is very complex and—up to today—not yet entirely investigated and understood. There are neither instructions for the determination of the seismic input onto the wall nor a concept for the approximation of the dynamic stability leaving it up to the structural engineer to find a practicable method for the seismic verification. Quite often the seismic input is hence determined by using the definition for non-structural components that may be an acceptable approximation in many cases. However, there are no guidelines specifically for vital non-structural components and for components that represent a potential risk. Including a seismic verification using a realistic model and realistic response spectra that take into account the filtering effects of the structure is required. They are not explained clearly by Eurocode 8 [4].

The deformation occurring under lateral loading on unreinforced masonry wall was explained Bourzam [5] which the element's stiffness depends on the mechanical properties of constituent material, the geometry and boundary conditions. When subjected to a lateral load V, a confined masonry wall generates a horizontal deflection δ. This lateral displacement is the sum of the deflection due to flexure and the deformation due to shear as defined in the Equations (1)–(5). The detail correlation lateral load (V) and horizontal deflection as deformation of unreinforced masonry wall under lateral loading is figured on Figure 1.

$$\delta = \frac{Vh^3}{\alpha E_{eqv} I_w} + \frac{1.2Vh}{G_{eqv} A_w} \tag{1}$$

$$V = K_e\, \delta \tag{2}$$

where:

K_e = the effective stiffness of confined wall
I_w = $tl^3/12$, moment inertia of the wall's cross section
A_w = area of the wall's horizontal
1.2 = the shear coefficient for rectangular cross-section
α = coefficient depends on the boundary condition, α =3 for cantilever panel and α =12 in case of fixed-ended wall
$E_{Equation}$ = modulus of elasticity equivalent
$G_{Equation}$ = shear modulus equivalent

$$E_{eqv} = \frac{E_m A_m + 2E_c A_c}{A_m + 2A_c} \tag{3}$$

$$G_{eqv} = \frac{G_m A_m + 2G_c A_c}{A_m + 2A_c} \tag{4}$$

After substituting the value of V from Equation (2) into Equation (1) and rearranging its different terms, the general equation for the effective stiffness of confined masonry wall in the elastic domain is obtained and expressed as follows:

$$K_e = \frac{G_{eqv} A_w}{1.2h\left[1 + \mu \frac{G_{eqv}}{E_{eqv}} \left(\frac{h}{l}\right)^2\right]} \tag{5}$$

μ = coefficient describes the applied restraint conditions of the wall, μ = 3.33 for cantilever walls and μ = 0.83 in case of fixed-ended walls.

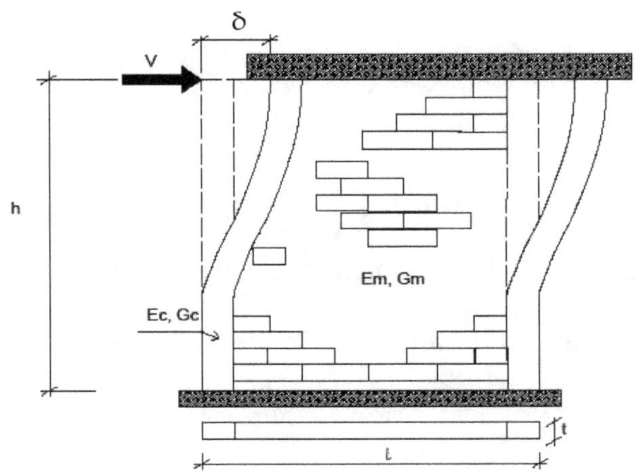

Figure 1. Deformation of unreinforced masonry wall under lateral loading.

This clearly shows that further research into this field of seismic engineering is inevitable. The outcome of the research should be a new verification concept that allows for a quick and easy verification based on limit values and also provides guidance for a more advanced investigation of the dynamic stability using displacement-based approaches. It is interesting to observe the model of an artificial neural network [6–9] of dynamic capacity of no-fines lightweight concrete wall with or without any horizontal reinforcement. As the name implies, no-fines lightweight concrete is a conventional lightweight concrete which eliminates the engagement of fine aggregate, in order to achieve some advantages. For example, it won't segregate and is cheap.

Modelling with an artificial neural network (ANN) [10] refers to the Black Box modelling in which the input is installed with a proper output. This model consists of connections and processing elements (neurons).

In common, the structure of the ANN is multilayer perceptron (MLP). Figure 2 illustrates the MLP structure consisting of input, hidden and output layer.

Cybenko [11] explained that the ANN model using the function of tangent hyperbolic activation in a hidden layer and linear function in its layer output is able to predict the accuracy of all modelled systems. Because of that, Equations (6)–(19) used to arrange the ANN modelling.

To determine the weight, for example by connecting $\hat{y}_i$ output to φ_i input, needs an attempt called training/learning. In training, the weight is adjusted to obtain the network output suitable with the process or target output. This learning algorithm will continuously adjust the weight until the target desired is reached. MLP can be written mathematically as follows:

$$y_i = F_i\left[\sum_{j=1}^{n_h} W_{i,j}.f_j\left(\sum_{l=1}^{n_\phi} w_{j,l}\phi_l + w_{j,0}\right) + W_{i,0}\right] \tag{6}$$

The learning algorithm used in this research was Levenberg Marquardt Algorithm. Though being more complex in comparison with the back-propagation algorithm, this algorithm is able to provide a better result. The derivation of this algorithm can be seen in Norgaard [12] and can be explained as follows.

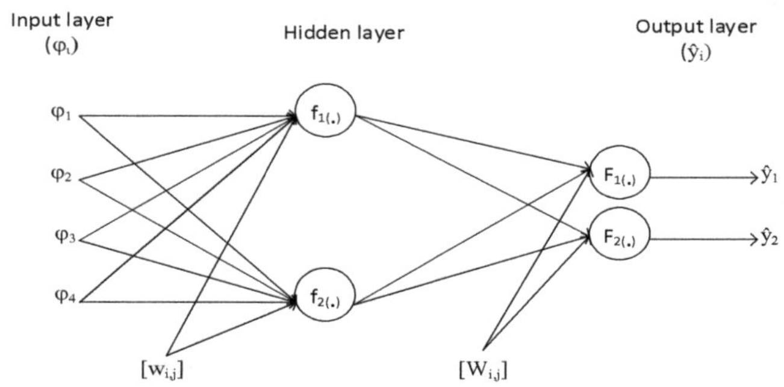

Figure 2. Structure of multilayer perceptron.

Training Data refers to a set of inputs $u(k)$ paired with the desired output of $y(k)$ as given below:

$$Z^N = \{[u(k), y(k)] \mid k = 1, \ldots, N\} \tag{7}$$

The objective of this learning is to determine the weight that might be from the data pair given.

$$Z^N \rightarrow w \tag{8}$$

Thus, the network can result in an estimation of output of $y(k)$ that is equal or closer to the output of $y(k)$. The error estimation will be approached using mean square error criterion:

$$\begin{aligned} V_N(w, Z^N) &= L^{(i)}(w) \\ &= \frac{1}{2N} \sum [y(k) - \hat{y}(k|w)]^T [y(k) - \hat{y}(k|w] \end{aligned} \tag{9}$$

The weight obtained:

$$w = \arg\min_{w} V_N(w, Z^N) \tag{10}$$

$$w^{(i+1)} = w^{(i)} + \mu^{(i)} f^{(i)} \tag{11}$$

w^i refers to the recent weight, $f^{(i)}$ denotes the direction of searching and $\mu^{(i)}$ is the extent of step. Levenberg Marquardt is a standard method for the minimization of mean square error criterion. This algorithm has λ parameter to maintain convergence. The value of λ is controlled by the ratio between the decrease of actual value and the prediction value.

$$r^{(i)} = \frac{V_N(w^{(i)}, Z^N) - V_N(w^{(i)} + f^{(i)}, Z^N)}{V_N(w^{(i)}, Z^N) - L^{(i)}(w^{(i)} + f^{(i)})} \tag{12}$$

where:

$$L(w^{(i)} + f) = \sum_{k=1}^{N} \left(y(k) - \hat{y}(k|w) - f^T \frac{\partial \hat{y}(k|w)}{\partial w} \right)^2 = V_N(w^{(i)}, Z^N) + f^T G(w^{(i)}) + \frac{1}{2} f^T R(w^{(i)} f \tag{13}$$

G shows a gradient of criteria by referring to the weight and R refers to the approach of Hessian. If the ratio gets closer to one, $L(i)$ $(w^{(i)} + f)$ approaches V_N, and λ should be reduced through some factors. Conversely, if the ratio is little or negative, λ should be added. The Levenberg Marquardt algorithm can be summarized as follows.

1. Select the vector of initial weight of $w^{(0)}$ and the initial value of $\lambda^{(0)}$ where w refers to the weight and λ is given the initial value.

2. Determine the direction of searching in which I refers to the matrix of identity.

$$\left[R(w^{(i)} + \lambda^{(i)} I \right] f^{(i)} = -G(w^{(i)}) \tag{14}$$

Thus, f is obtained and put into

$$w = \arg\min_{w} V_N(w, Z^N) \tag{15}$$

$$w^{(i+1)} = w^{(i)} + \mu^{(i)} f^{(i)} \tag{16}$$

If the objective function in the current iteration is less than the previous iteration or V_N $(w^{(i)} + f^{(i)}, Z^N) < V^N$ $(w^{(i)}, Z^N)$; thus, the current weight has been added to be a new weight $w^{(i+1)} = w^{(i)} + f^{(i)}$. For this, the new searching direction of searching is the old searching direction $\lambda^{(i+1)} = \lambda^{(i)}$. If not, finding new λ must be found from the r value.

$$r^{(i)} = \frac{V_N(w^{(i)}, Z^N) - V_N(w^{(i)} + f^{(i)}, Z^N)}{V_N(w^{(i)}, Z^N) - L^{(i)}(w^{(i)} + f^{(i)})} \tag{17}$$

If $r^{(i)} > 0.75$ thus $\lambda^{(i)} = \lambda^{(i)}/2$.
If $r^{(i)} < 0.25$ thus $\lambda^{(i)} = 2\ \lambda^{(i)}$.
where V is calculated from the equation of Levenberg Marquard L

$$\begin{aligned} V_N(w, Z^N) &= L^{(i)}(w) \\ &= \tfrac{1}{2N} \Sigma\, [y(k) - \hat{y}(k|w)]^T [y(k) - \hat{y}(k|w] \end{aligned} \tag{18}$$

$$L^{(i)}(w^{(i)} + f^{(i)}) = (\lambda^{(i)} f^{(i)\mathrm{T}} f^{(i)}) - (f^{(i)\mathrm{T}}\,\mathrm{G}) \tag{19}$$

3. If the criteria are achieved, the calculation is terminated. Conversely, if the criteria are not achieved yet, it must be started from step 2.

2. Research Method

Table 1 shows the specimen of no-fines lightweight concrete wall was made into 2 (two) variations: no-fines lightweight concrete wall without horizontal reinforcement (lightweight concrete wall, LCW) and no-fines lightweight concrete wall with horizontal reinforcement with 6 mm-diameter with the distance of 200-mm inter-reinforcements (lightweight concrete wall with horizontal bars, LCWHB-200) as shown in Figures 3 and 4.

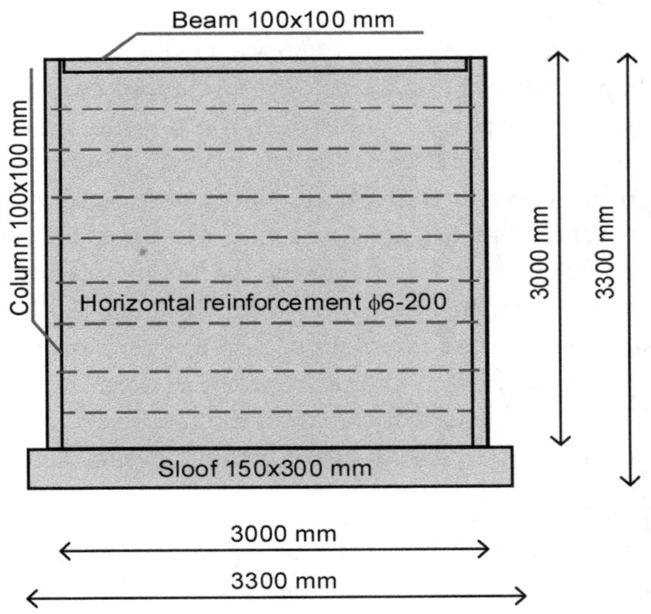

Figure 3. LCWHB-200.

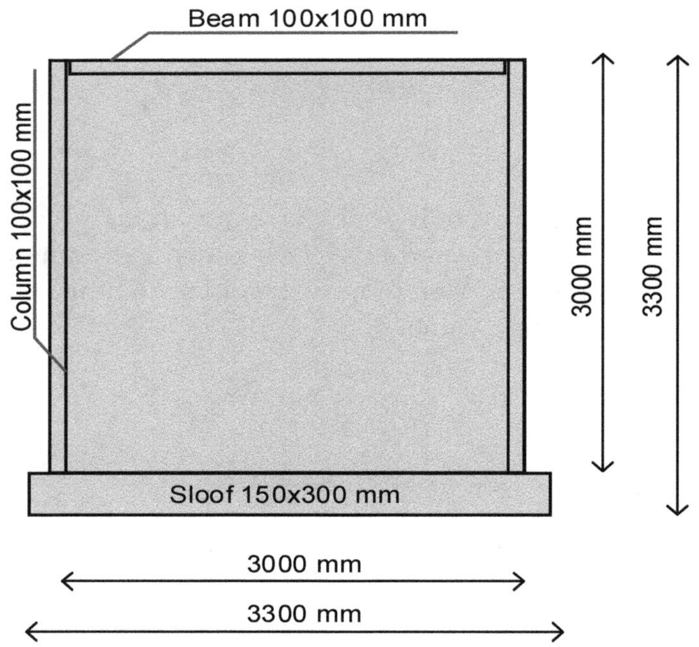

Figure 4. LCW.

Table 1. Dimension of the specimen of no-fine lightweight concrete wall.

Code		LCW	LCWHB-200
Size (mm)	Column	100 × 100	100 × 100
	Beam	100 × 100	100 × 100
Number of Steel reinforcements		4 Ø8	4 Ø8
Confinement distance		Ø6-150	Ø6-150
Horizontal reinforcement		-	Ø6-200

The constituent materials of no-fines lightweight concrete wall in this research was structured from the binding material of 50-kg Portland cement type I (Indocement), PDAM water at Structure Laboratory, Department of Civil Engineering of Gadjah Mada University, 10–20 mm diameter aggregate originated from Kemiri Village, Pakem Sleman. PDAM has acronym Perusahaan Daerah Air Minum, Indonesia regional water utility company. In addition, the reinforcement steel used in forming the reinforced concrete frame was the ISPAT steel, the company was set up as a 60,000 tpa Greenfield project, for rolling from Surabaya with 8-mm plain steel reinforcement (fy = 341.62 MPa) as the column reinforcement and practical beam. The 6-mm reinforcement (fy = 263.06 MPa) was used as the shear reinforcing and horizontal reinforcement on the wall. The no-fines lightweight concrete wall was planned with the stress force on the average of 4.45 Mpa, elasticity modulus on the average of 1632.75 Mpa, splitting tensile was on the average of 0.248 Mpa, and adhesive reinforcement in no-fine concrete at 1.661 Mpa (34.62% of the adhesive reinforcement in normal concrete). No-fines lightweight concrete has a weight of 1572 kN/m^2, 94.3% lower than the red brick. The ready-mix normal concrete with fc′ = 25.90 MPa for sloof beam with the ratio of Portland Cement:Sand:Coarse Aggregate = 1:2:3.

The equipment used in the research is included loading frame, load cell, crane, rigid floor, hydraulic jack and hydraulic pump machine with the capacity of 50 tons, data lodger, and computer. 6 units of 50-mm Linear Variable Differential Transformer (LVDT) with the accuracy of 0.01 mm and all was connected to data lodger were used to measure the level of deflection. Set up of the test on those three specimens can be seen in Figure 5.

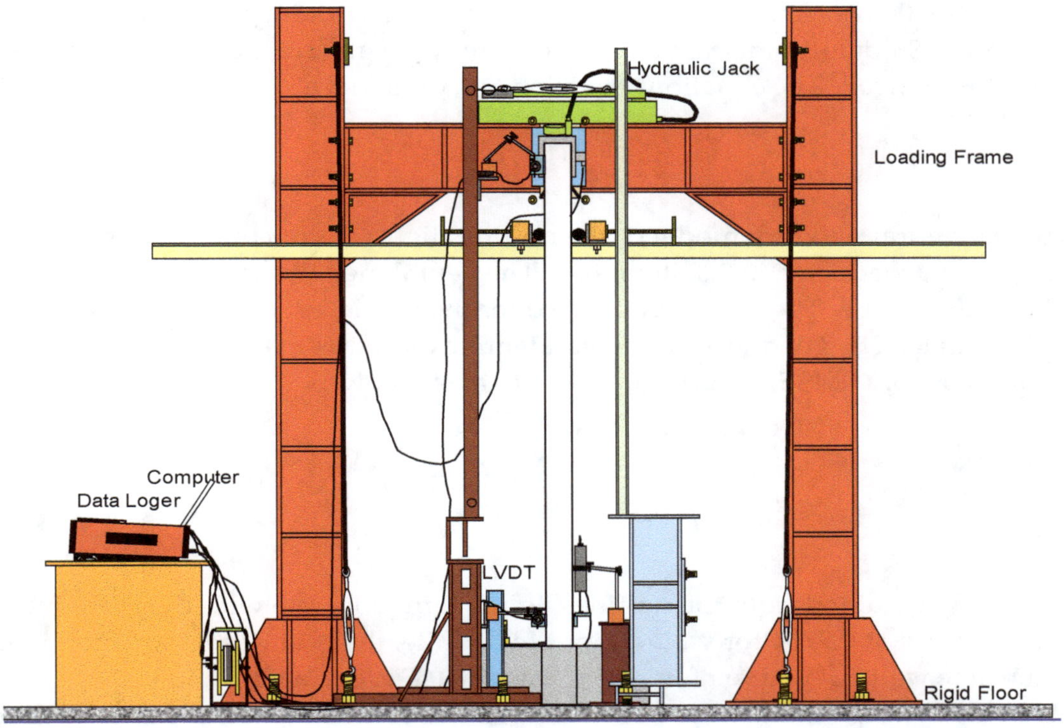

Figure 5. Test Setting.

Cyclic lateral load was given as unidirectional with the axis of wall strength referring to American Society for Testing Materials (ASTM) E 2126-02a Method B (amplitudes of the Reversed Cycles). The lateral load was given to the specimen with alternating direction until reaching the yield in which it was then continued to reach a collapse (failure) and loading was based on the displacement control.

Meanwhile, data obtained from the test above was used to make an ANN-based model that can be used to predict the horizontal deflection by entering the input condition that would be predicted. ANN will predict in an interpolation and extrapolation of horizontal deflection. Here, the model of ANN proposed to predict the horizontal deflection (force and deviation) was developed by means of MATLAB R2013 (TechSource Systems Pte Ltd, Singapore) MATLAB 7.1. Figure 6 presents the structure of the artificial neural network model in which the force was the input data and deflection was the output.

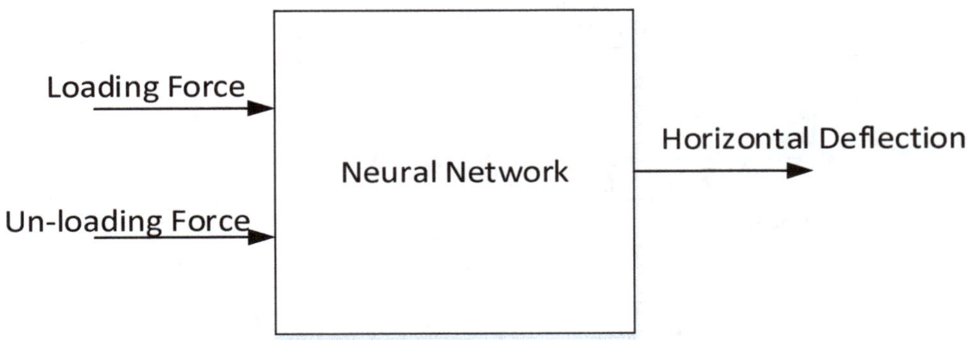

Figure 6. The Proposed model of artificial neural network (ANN) for the force and deflection.

3. Results and Analysis

This neural network model used a FIR (Finite Impulse Response) structure. It was characterized by placing the variable of input model from the input itself. The input data used in this study was

1217 data points for the no-fines lightweight concrete wall without the horizontal reinforcement unreinforced and 2688 data points for the no-fines lightweight concrete wall with the horizontal reinforcement unreinforced. The equation, Equation (20) of model output can be presented as follows:

$$\hat{Y} = f(U_1, U_2, U_3) \quad (20)$$

The artificial neural network model of the no-fines lightweight concrete used MLP; trained by Marquardt algorithm conducted in 200 iterations. The neural model was trained with training data to determine the weight value. Then, the weight used for the validation of neural models by using other input and output data. The goodness of system identification was measured by using Equation (21), Root Mean Square Error (RMSE), which can be written as follows:

$$RMSE = \sqrt{\frac{\sum_{i=1}^{N} (y_i - \hat{y}_i)^2}{N}} \quad (21)$$

The best RMSE obtained from the no-fines lightweight concrete with the reinforcement was at 0.2629 for training and at 0.7701 for validation. Meanwhile, the best RMSE obtained in the no-fines lightweight concrete wall without reinforcement was at 0.0431 for training and at 0.0462 for validation. The trained neural network model was also validated in a set of data that was not used for network training (Figures 7 and 8). Using fixed values of the weights that obtained in training phase, the neural networks should produce the predicted output from the new input data. Figure 9 and 10 illustrate the validation phase.

These results confirm that no-fines lightweight concrete with the reinforcement provides better results than without bar steel horizontal. The mechanical properties of horizontal reinforcement improved the strength behavior of no-fine lightweight concrete as the wall element. This is shown in the RMSE results and images on training and validation. Figure 7 shows the training for no-fines lightweight concrete wall with the horizontal reinforcement showing a more orderly and non-random pattern, as well as Figure 9 as a validation image showing similar trends. These are different with Figures 8 and 10. They were figured as training and validation for no-fines lightweight concrete without the reinforcement.

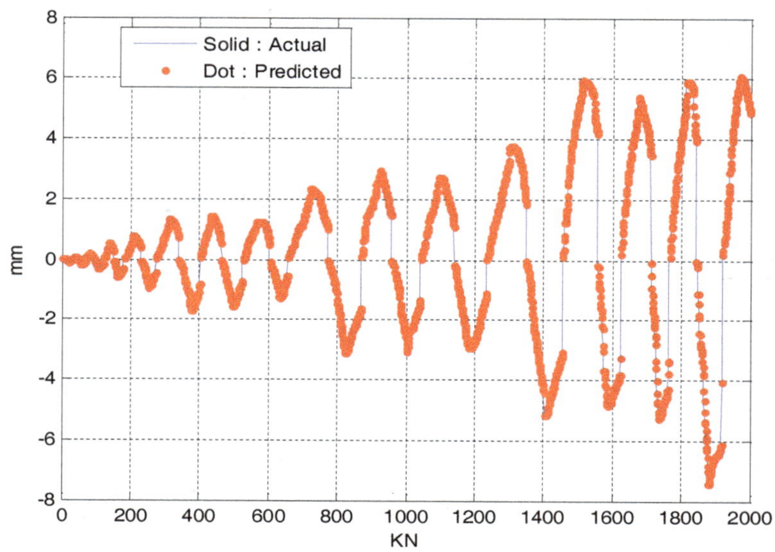

Figure 7. Training for no-fines lightweight concrete wall with the horizontal reinforcement.

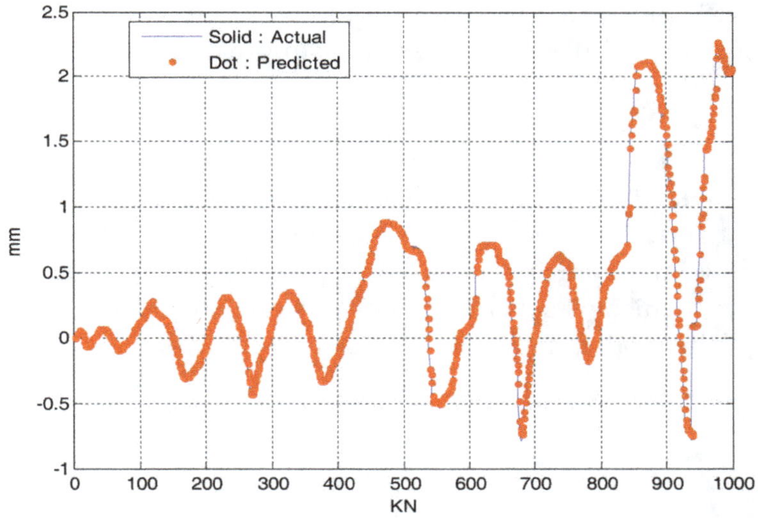

Figure 8. Training for no-fines lightweight concrete wall without the horizontal reinforcement.

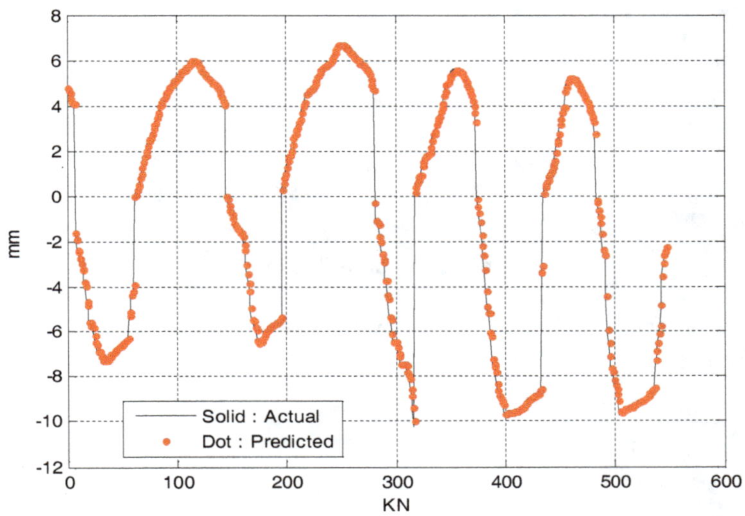

Figure 9. Validation for no-fines lightweight concrete wall with the horizontal reinforcement.

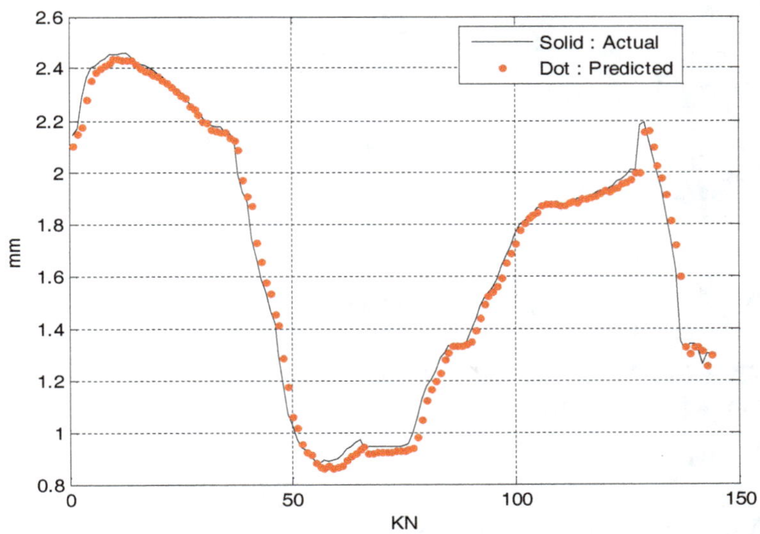

Figure 10. Validation for no-fines lightweight concrete wall without the horizontal reinforcement.

The correlation coefficient R that shows at Equation (22) what proportion of the variation of the predicted values can be attributed to the linear relationship with the actual values is given by the formula:

$$R = \frac{S_{xy}}{\sqrt{S_{xx} \times S_{yy}}} \quad (22)$$

The experimental deflection of the no-fines lightweight concrete wall due to cyclic load is illustrated in Figures 11 and 12. The correlation coefficients are 0.9927 for deflection of no-fines lightweight concrete wall with reinforcement and 0.9955 for deflection of no-fines lightweight concrete wall without reinforcement which was distributed evenly on both sides of the line, which indicates an excellent performance of the model.

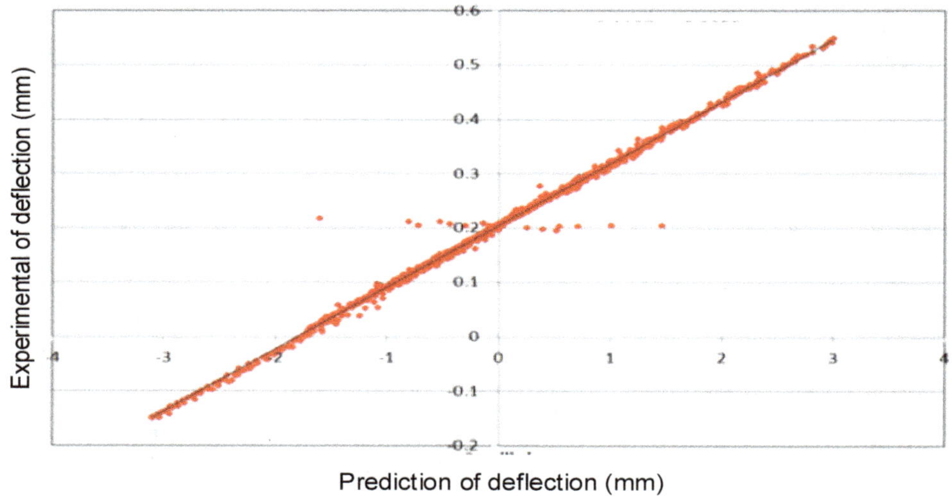

Figure 11. Predicting the accuracy of neural-network system for no-fines lightweight concrete wall with the horizontal reinforcement.

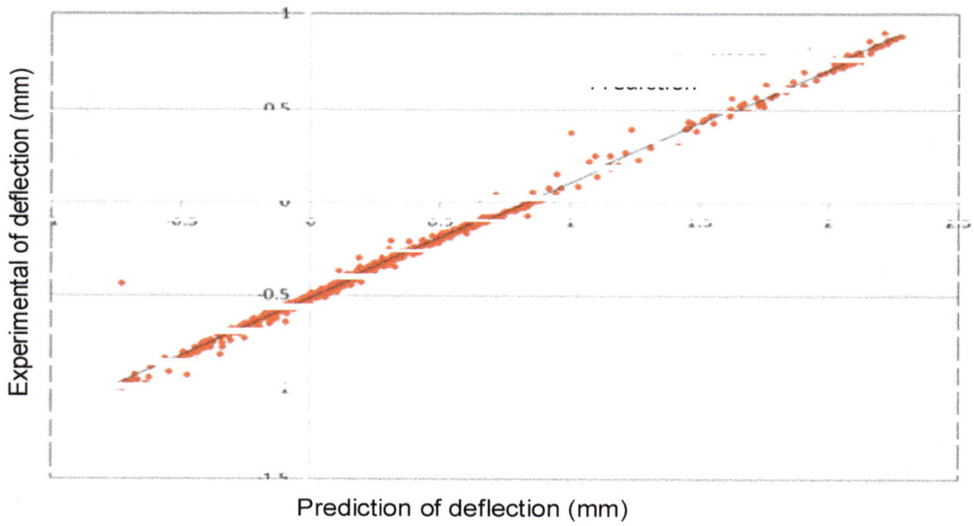

Figure 12. Predicting the accuracy of neural-network system for no-fines lightweight concrete wall without the horizontal reinforcement.

4. Conclusions

The development of a model using ANN to predict the deflection due to the dynamic force in no-fines lightweight concrete wall comprising 2 input variables and 1 output variable has been presented in this paper. The model structure was referred to as perceptron multilayer. In general, the neural network is designed to be able to anticipate the nonlinear and complex interactions between input/output variables and no-fines lightweight concrete wall. Therefore, the ANN model can be used as an alternative model to predict the deformation more accurately, simply and quickly due to the alternating horizontal loads on a no-fines lightweight concrete wall.

Author Contributions: Ridho Bayuaji conducted preparing and maintaining facility test conditions, the collection data, data analysis and wrote the paper. Totok Ruki Biyanto conducted the neural network modelling and carried out data analysis.

Acknowledgments: The author thank to Ministry of Research, Technology and Higher Education of the Republic of Indonesia and LPPM ITS for funded this research and publication.

References

1. Bousabbah, L.; Bruneau, M. Review of the Seismic Performance of Unreinforced Masonry Wall. In Proceedings of the 10th World Conference on Earthquake Engineering, Madrid, Spain, 19–24 July 1992; pp. 4537–4540.
2. Calderón, S.; Sandoval, C.; Arnau, O. Shear Response of Partially-Grouted Reinforced Masonry Walls with a Central Opening: Testing and Detailed Micro-Modelling. *Mater. Des.* **2017**, *118*, 122–137. [CrossRef]
3. Salmanpour, A.H.; Mojsilović, N.; Schwartz, J. Displacement Capacity of Contemporary Unreinforced Masonry Walls: An Experimental Study. *Eng. Struct.* **2015**, *89*, 1–16. [CrossRef]
4. European-Standard-En1996-1-1. *Euro Code 6: Design of Masonry Structures*; Comité Européen de Normalisation: Brussels, Belgium, 2005.
5. Bourzam, A.; Goto, T.; Miyajima, M. Shear Capacity Prediction of Confined Masonry Walls Subjected To Cyclic Lateral Loading. *Doboku Gakkai Ronbunshuu A* **2008**, *25*, 47–59.
6. Bayuaji, R.; Biyanto, T.R. Model Jaringan Saraf Tiruan Kuat Tekan Beton Porus Dengan Material Pengisi Pasir. *J. Civ. Eng.* **2013**, *20*, 23–32.
7. Nuruddin, M.; Diah, A.; Biyanto, T.; Darmawan, M.S.; Bayuaji, R. Concrete Structures Life Span Based on Carbonation Rate Using Artificial Neural Network. *J. Aplikasi* **2011**, *9*, 41–47. [CrossRef]
8. Mardiyono, M.; Suryanita, R.; Adnan, A. Intelligent Monitoring System on Prediction of Building Damage Index Using Neural-Network. *Telecommun. Comput. Electron. Cont.* **2012**, *10*, 155–164. [CrossRef]
9. Xiaoping, Y.; Wenhui, Z.; Yamin, F. Neural Network Adaptive Control for X-Y Position Platform with Uncertainty. *Telkomnika* **2014**, *12*, 79–86. [CrossRef]
10. Murray, R.; Neumerkel, D.; Sbarbaro, D. Neural Networks for Modeling and Control of a Non-Linear Dynamic System. In Proceedings of the 1992 IEEE International Symposium on Intelligent Control, Glasgow, Scotland, 11–13 August 1992; pp. 404–409.
11. Cybenko, G. Approximation by Superpositions of a Sigmoidal Function. *Math. Cont. Signal Syst.* **1989**, *2*, 303–314. [CrossRef]
12. Norgaard, M.; Ravn, O.; Poulsen, N.K.; Hansen, L.K. *Neural Network for Modelling and Control of Dynamic Systems*; Springer: New York, NY, USA, 1999.

8

Study of the Seismic Response on the Infill Masonry Walls of a 15-Storey Reinforced Concrete Structure in Nepal

André Furtado *, Nelson Vila-Pouca, Humberto Varum and António Arêde

Construct-Lese, Departamento de Engenharia Civil, Faculdade de Engenharia da Universidade do Porto, 4200-001 Porto, Portugal; nelsonvp@fe.up.pt (N.V.-P.); hvarum@fe.up.pt (H.V.); aarede@fe.up.pt (A.A.)
* Correspondence: afurtado@fe.up.pt

Abstract: Following the strong earthquake on April 25, 2015 in Nepal, a team from the University of Porto, in collaboration with other international institutions, made a field study on some of the most affected areas in the capital region of Kathmandu. One of the tasks was the study of a high-rise settle of buildings that were damaged following the earthquake sequence. A survey damage assessment was performed to a 15-storey infilled reinforced concrete structure, which will be detailed in the manuscript. Moreover, ambient vibration tests were carried out to determine the natural frequencies and corresponding vibration modes of the structure. The main aim of this manuscript is to present a numerical study concerning the influence of the masonry infill walls in the structure seismic response. For this, three numerical models were built discriminating the situations with and without damage and nondamaged infill walls. Validation and calibration of the numerical model was ensured by comparing the numerical frequencies with those obtained from ambient vibration tests. In addition, linear elastic analyses were carried out, using real accelerograms from the Gorkha earthquake to assess and quantify the major differences between the models in terms of inter-storey drifts ratios, inter-storey shear forces and seismic loadings.

Keywords: Nepal earthquake; high-rise reinforced concrete structure; masonry infill walls; ambient vibration test; survey damage assessment; numerical modelling

---

1. Introduction

Recent earthquakes demonstrated that the masonry infill walls have an important contribution in the seismic response of the reinforced concrete (RC) structures. Most of the structural and seismic codes consider the infill panels as nonstructural elements, which according to post-earthquake damages report and to several experimental studies is not correct. As a result, even light to moderate earthquake shaking/acceleration or drift levels can cause damage to the infill walls and this damage may result in life safety hazards, immediate evacuation and loss of function of buildings, limiting the use of internal spaces. In many cases, the influence of the infill panels showed to be the reason of extensive damages or even the buildings collapses [1,2].

Three failure mechanisms were observed in most of the analysed cases along the last major earthquake events and also after the Gorkha earthquake. The first is associated with cases where masonry walls do not extend towards all the inter-storey height for openings, leaving a short portion of the columns clear, creating a short-column mechanism. The second is associated also with the short-column mechanism but induced by the stair-slabs connected to the column. In both mechanisms, the non-consideration of the nonstructural infill panels, or of the secondary elements (as the staircases) in the design, may not represent the real behaviour of the columns, underestimating the column stiffness and, consequently, of the forces attracted, leading to unexpected shear failure [3]. The failure of several structural elements was observed after the Gorkha earthquake due to this specific mechanism, according to some authors. Finally, the third and last one is related to the vertical stiffness irregularity due to the irregular distribution of the infill panels, which can lead to the concentration of the deformation in the storeys with less presence of these elements [4]. A high reduction of the infill panels on the ground floor is quite common in Nepal for commercial purposes or garages, which increase their seismic vulnerability. Several collapses were observed during the Gorkha earthquake. Finally, the local failure of the infill panel, characterized by the detachment of the panel from the envelope frame, diagonal cracking, shear sliding was also reported following the Gorkha earthquake, as well was observed in the last major earthquakes in Europe.

During the past three decades, the number of RC buildings in Nepal has increased considerably, mostly built by the landowners or local builders, too dependent on previous knowledge and experience which has led to insufficient detailing, bad quality of materials or lack of proper design rules and practice. The construction of RC buildings in Nepal presents several weaknesses on the quality control of materials (improper vibration of concrete, improper size of the aggregates and steel bars with insufficient ductility) and reduced construction quality (reinforcement detailing and provisions, and insufficient percentage of reinforcement), which have a direct impact on the bearing capacity as well as the deformation capacity of the structural elements. Another important issue regarding the seismic vulnerability of these types of structures is related to decreasing number of masonry walls on the ground floor, leading to the high potential to develop soft-storey mechanisms, and subsequent partial/total collapse of some buildings. However, it should be mentioned that the major part of the collapses and extensive damages that occurred in the Gorkha earthquake were related to masonry buildings and historic constructions [5].

Regarding the high-rise RC buildings' (with 10–18 storeys) performance from the earthquake of April 25, 2015, it was observed that most of the damage was related to the infill walls [6–8]. Figure 1 shows the structure damage of a 14-storey RC and from the observation of the vertical profile of the building, the damages are concentrated the first seven floors of the structure, as can be seen in Figure 1a. The damages are composed by detachment of the walls from the envelope frame (Figure 1b), diagonal cracking (Figure 1c) and some slight out-of-plane detachment of the wall from the frame. The reduced level of damage suits the proper seismic behaviour of these structures, which were designed as per the Indian National rules and standards. Even though, in many cases, damage was limited to the infill panels, it is worth noting that in many of these structures, the occupants had to be moved for temporary shelter (for some, more than a year) due to safety issues related to possible failure of the panels and until all repairs are complete. In some cases, the high costs associated with the repair of the buildings exceeded the structural costs of construction.

a) b) c)

Figure 1. Fourteen-storey reinforced concrete structure damaged after the Gorkha earthquake: (**a**) distribution of the damages from the first to the seventh storey; (**b**) detachment of the walls from the envelope frame; (**c**) diagonal cracking with slight out-of-plane detachment of the wall.

The main aim of the present manuscript is to study the effect of infill panels in the seismic response of a 15-storey infilled RC structure. For this, three numerical models were built discriminating the situations with and without damage and nondamaged infill walls. Comparing the numerical frequencies with those obtained from the ambient vibration tests ensured validation and calibration of the numerical model. In addition, linear elastic analyses were carried out, using real accelerograms from the Gorkha earthquake to assess and quantify the major differences between the models in terms of inter-storey drifts ratio, inter-storey shear forces and seismic loadings.

2. Case Study

2.1. Introduction

The study of the infill masonry walls influence in the seismic response of a RC structure was carried out based on the study of an existent structure in Nepal. The RC building is located in Kathmandu, is a 15-storey height, and is dated from 2012. Throughout the present section, the structure will be deeply described such in architectural or structural components. Posteriorly, the infill masonry walls characteristics and disposition will be detailed. An extensive damage survey assessment performed after the Gorkha earthquake focusing on the infill masonry walls' damages will be also included. Finally, ambient vibration tests were carried out to achieve the building structure vibration modes and corresponding natural frequencies. The major results of the testing campaign will be included within this section.

2.2. General Description

The building is a residential structure, located in the small town of Satdobato and belongs to a large luxury development composed of small houses and four high-rise structures (Figure 2a). The building is located at the left side of those grouped together. The structure is composed of two underground floors and 15 storeys (Figure 2b). The ground-floor height is 4 m and the others are 3.2 m high—a total of 48.8 m high.

a) b)

Figure 2. Global overview of the high-rise structure under study: (**a**) location of the houses and the high-rise buildings; and (**b**) front view.

The tower is composed by RC frames filled with masonry infill walls made with solid clay bricks aligned according to the longitudinal and transverse direction and two stiff RC cores destined to the elevator boxes. The RC frames are composed by beam-column elements with spans relatively small. The beams are typically composed by two cross-sections, namely 300 × 600 mm and 230 × 600 mm. On the other hand, the columns are composed of 8 types of cross-sections, which are constant in height and very robust, as can be observed in Figure 3 and is described in Table 1. The longitudinal and transverse reinforcement is different among the different storeys. Complete detailing of all the cross-sections and reinforcements can be found in [9].

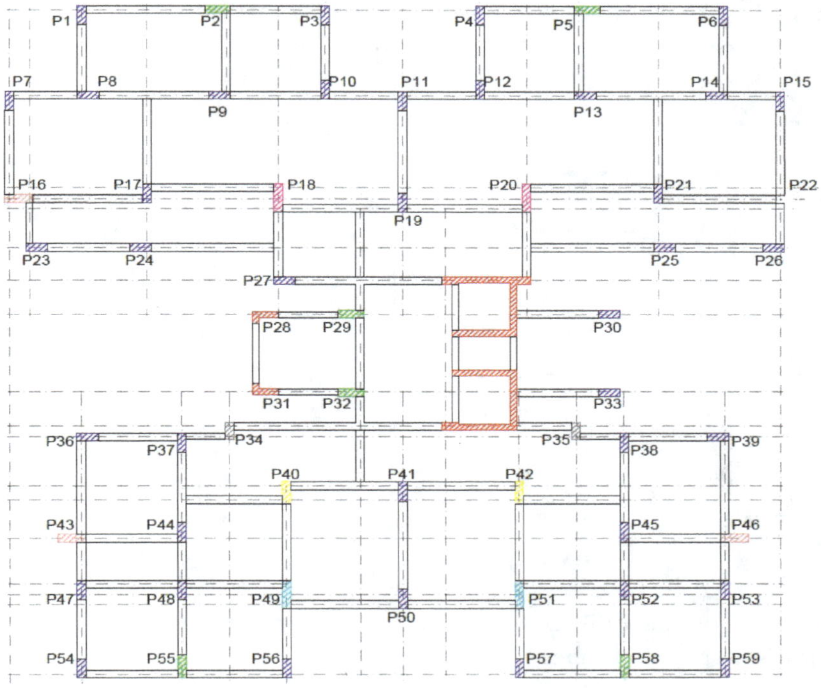

Figure 3. Case study: columns cross-section.

Table 1. Case study: columns cross-section and nomenclature.

Dimensions [mm]	Color	Columns Nomenclatures
900 × 300	Fluorescent green	P2 I P5 I P29 I P32 I P55 I P58
970 × 300	Pink	P16 I P43 I P46
1125 × 300	Hot pink	P18 I P20
1100 × 300	Baby blue	P49 I P51
855 × 300	Yellow	P40 I P42
800 × 300	Grey	P22
645 × 300	Gray	P34 I P35
750 × 300	Blue	Remaining columns

The solid RC slabs are supported by RC frame resisting system and have two different thicknesses: 125 mm and 110 mm. Two stiff RC cores destined to the elevator boxes are distributed in the middle alignment of the structure (red in Figure 3).

Concerning infill masonry walls, two different types of walls were noted from the visual inspection, namely: (i) Façade walls made with two leaf-panels, where each two rows of bricks are connected through one single row disposed perpendicularly, for a total thickness of 230 mm–250 mm; (ii) internal partition walls made with single rows of solid bricks with a total thickness of 150 mm. The solid bricks dimensions are 240 × 115 × 57 mm (length × thickness × height). A cement mortar was used for the horizontal and vertical bed joints, as well as for the plaster, which was approximately 2–3 cm thick.

Regarding the plan distribution of the infill panels, it was found the distribution was different between the ground floor and the remaining floors. At the ground floor, the major part of the walls did not exist to allow for the passage of people (Figure 4a). Through the observation of the plan disposition of the infill walls, some asymmetry is visible, which could introduce some torsion effect in the structure dynamic response. However, this could not be analysed without also considering the vertical structural elements, which also presents some asymmetry. Concerning the vertical disposition of the infill walls seems to present also some asymmetry, since there is a high number of infill panels was observed in the top storeys, as can be seen in Figure 4b.

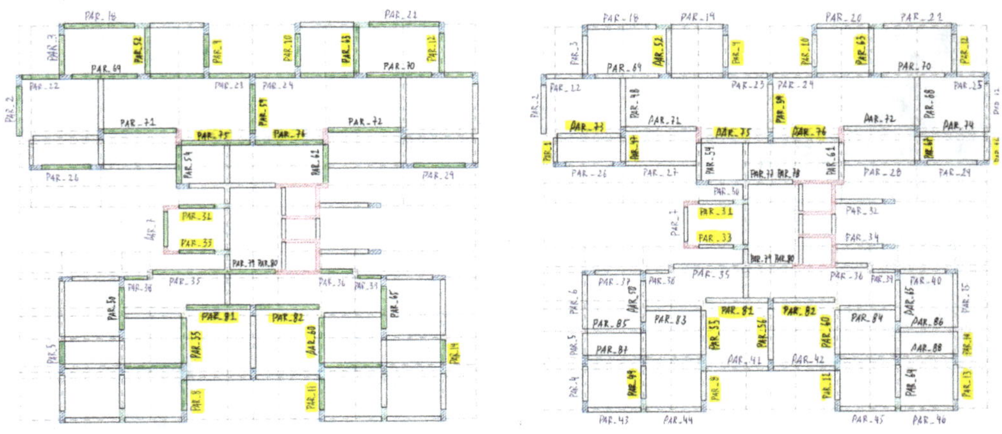

a) b)

Figure 4. Distribution of the infill masonry walls: (**a**) ground floor; and (**b**) remaining floors. Where Par_€ means infill panel number €.

2.3. Damage Survey Assessment

Globally, the structural seismic performance was positive since the structural damages observed were slight to moderated. Figure 5 shows the vertical profiles of the structure from two different orientations, from which it becomes clear that the damage extension is higher in the bottom storeys (until 7th storey). Regarding the damages observed throughout the building's façades, only a few RC columns located in the ground floor suffered cracking and spalling of the concrete. It was not observed any column affected in the top storeys of the structure. Otherwise, the infill masonry walls suffered extensive damages. Diagonal cracking occurred in most of the cases followed by the detachment from the envelope frame. Out-of-plane collapse along the buildings' façade was not observed, which is justified by the robustness of the façade walls.

From the survey assessment of the interior of the building, it was observed that several panels detached from the surrounding frame and suffered diagonal cracking (Figure 6a,b). Due to the high stiffness of these infill panels, and a consequently reduced capacity to accommodate lateral distortions, shear sliding cracking was visible in some situations (Figure 6c). This damage was followed by the detachment of the panel from the surrounding envelope frame. Lastly, it was observed that some nonconfined interior partition walls suffered out-of-plane collapse (Figure 6d).

a) b)

Figure 5. Survey damage assessment of the building: (**a**) north façade; (**b**) south-east façade.

Figure 6. Survey damage assessment of the interior of the building: (**a**) diagonal cracking in façade wall; (**b**) diagonal cracking in interior partition wall; (**c**) sliding cracking in interior partition wall of storey 1; and (**d**) out-of-plane collapse of interior partition wall.

2.4. Modal Identification Through Ambient Vibration Tests

The modal identification combines experimental techniques with analytical methods characterization of the dynamic properties of a structure. It is often used to support the inspection and assessment of structures, when it is aimed to study the structural behaviour due to dynamic loadings such as wind or earthquake, or when it planned to determine the structural properties such as the lateral stiffness [10].

With the aim of obtaining the natural frequencies and the corresponding vibration modes of the damaged structure, ambient vibration tests were carried out. For this, three seismometers GeoSIG (GSR-18) were used. Each seismometer allows to recording acceleration signals of three orthogonal directions and assuming specific conditions of trigger, reading and sampling rates. Series of 900 s duration were considered, with defining sampling rates of 250 Hz. The adopted time series lengths for each setup were essentially limited by restrictions for the tests' duration; still, the presented results show that they were adequate for identifying of the most relevant natural frequencies. Each seismometer was set to work independently, avoiding the use of cables and minimizing the work associated with test preparation.

This was made possible by resorting to internal clock synchronization. The tests consisted in successive vibration measurements in building points, corresponding to different locations in the building plan (at the top story) and along the building height. The vibration data was collected through four different test setups, in which the seismometer number 3 (here designated S03) was the reference. This reference seismometer was placed in the center of the building plan, close to the elevators core at the 16th storey during all setups. In the first setup, seismographs S01 and S02 were placed on floor 16 in the left block of the building to capture the local torsion mode. The second test setup was comprised by two seismometers placed at two opposite corners (16th storey) aiming to capture the global torsion mode of the structure. Finally, in the third and fourth setup all seismometers were in the same plan location of the reference seismometer. The difference among the test setups was the position of the seismometers along the vertical height of the structure, namely in the setup 3 and 4, seismograph S01 and S02 were placed at the 10th storey and 13th storey, and the 4th storey and 7th storey, respectively. Figure 7 shows the floor layout for all setups.

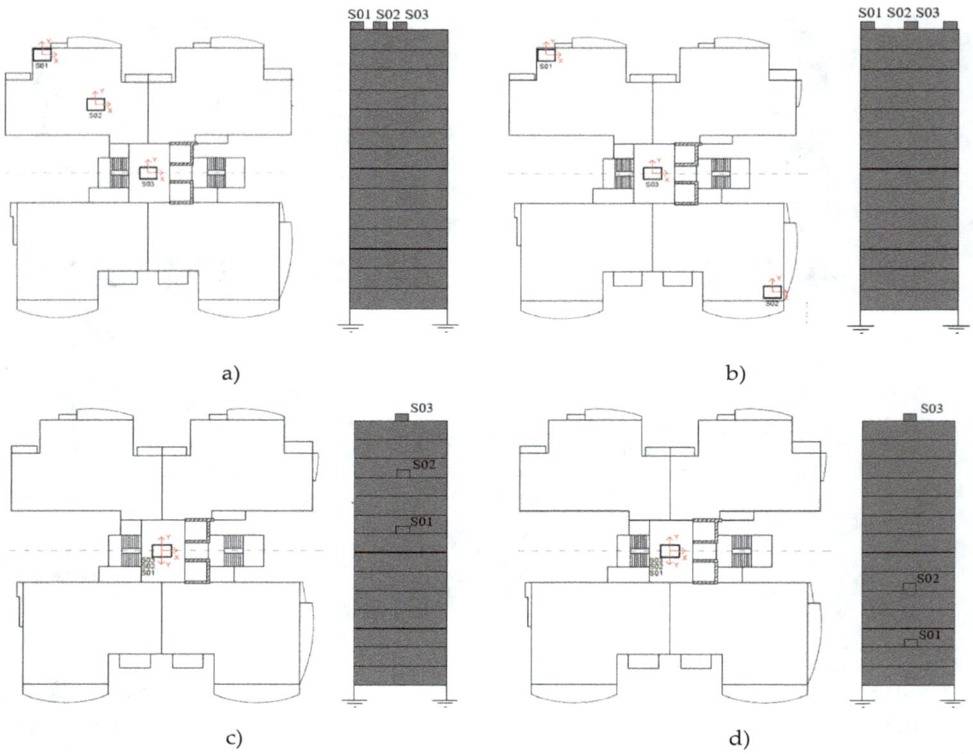

Figure 7. Ambient vibration tests: schematic layout of the seismometers disposition (**a**) Setup 1; (**b**) Setup 2; (**c**) Setup 3, and (**d**) Setup 4.

The determination of the natural frequencies and corresponding vibration mode shapes of the building is based on the acquisition of the acceleration measurements. For this purpose, the ARTeMIS [11] software for analysis and signal processing was used. The peak picking and the frequency domain decomposition (FDD) methods were used. Concerning the peak picking method, natural frequencies were identified from the peaks of the normalized average power spectra of the measured accelerations in each section, if the dynamic output in resonance is due only to one vibration mode shape. The modal identification was performed through the application of the enhanced frequency domain composition method (EFDD). The EFDD technique theory can be found in [12]. The spectral density matrices obtained from the analysis is plotted in Figure 8. From the plot 6 points are indicated which correspond to the structure natual frequencies. The 5-point red star is related to the first mode (translational mode along direction X—Figure 9a) equal to 0.61 Hz, and the blue one is related to the second mode (translational mode along direction Y—Figure 9b) equal to 0.75 Hz. Regarding the diamond scatter, it starts with the green one and it corresponds to a natural frequency equal to 1 Hz and is related to a torsional mode (Figure 9c). The gold diamond scatter corresponds to a second order vibration mode equal to 2.39 Hz (Figure 9d). Finally the gray and yellow diamond scatters correspond also to second order vibration modes equal to a natural frequency of 2.46 Hz and 2.78 Hz, respectively.

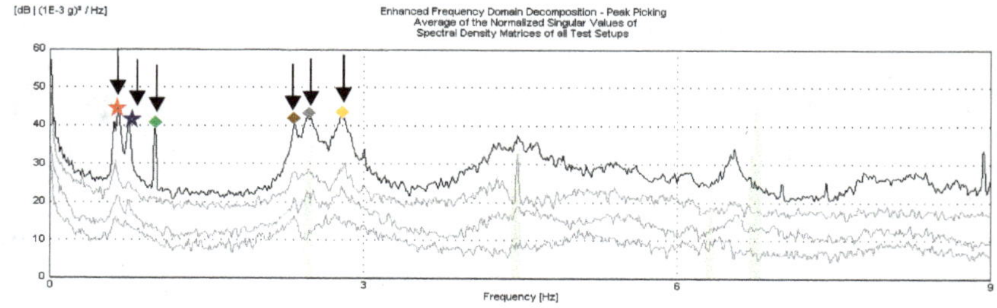

Figure 8. Ambient vibration test: normalized single values of spectral density matrices.

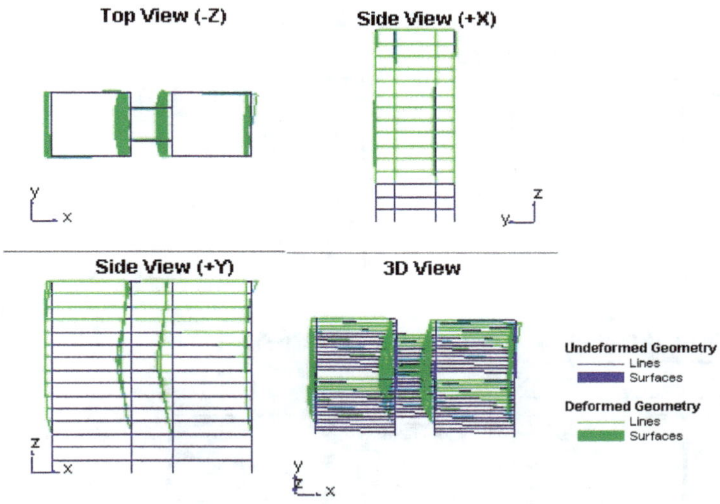

f_1=0.61Hz

a)

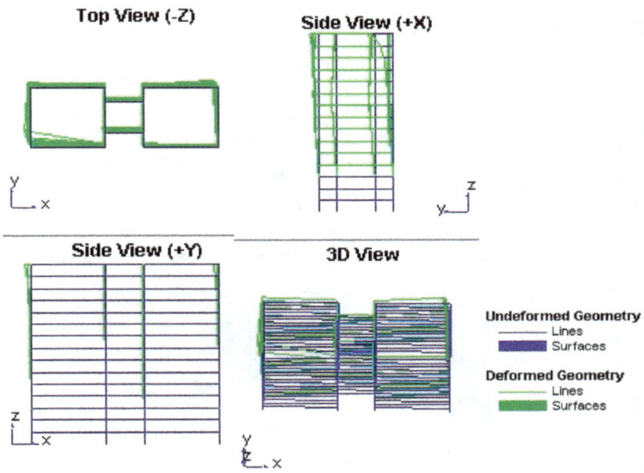

f_2=0.75Hz

b)

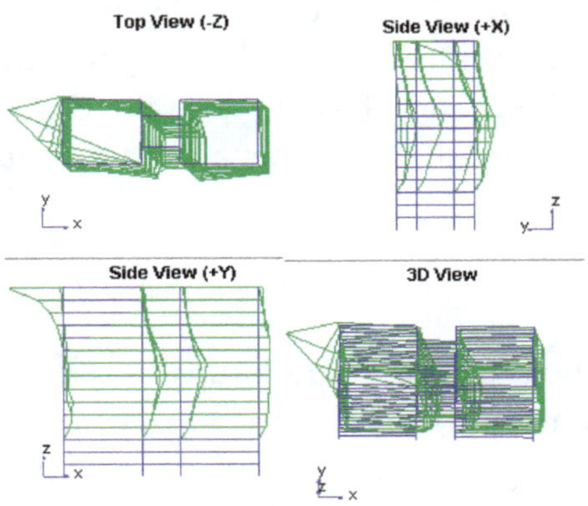

f_3=1.00Hz

c)

Figure 9. *Cont.*

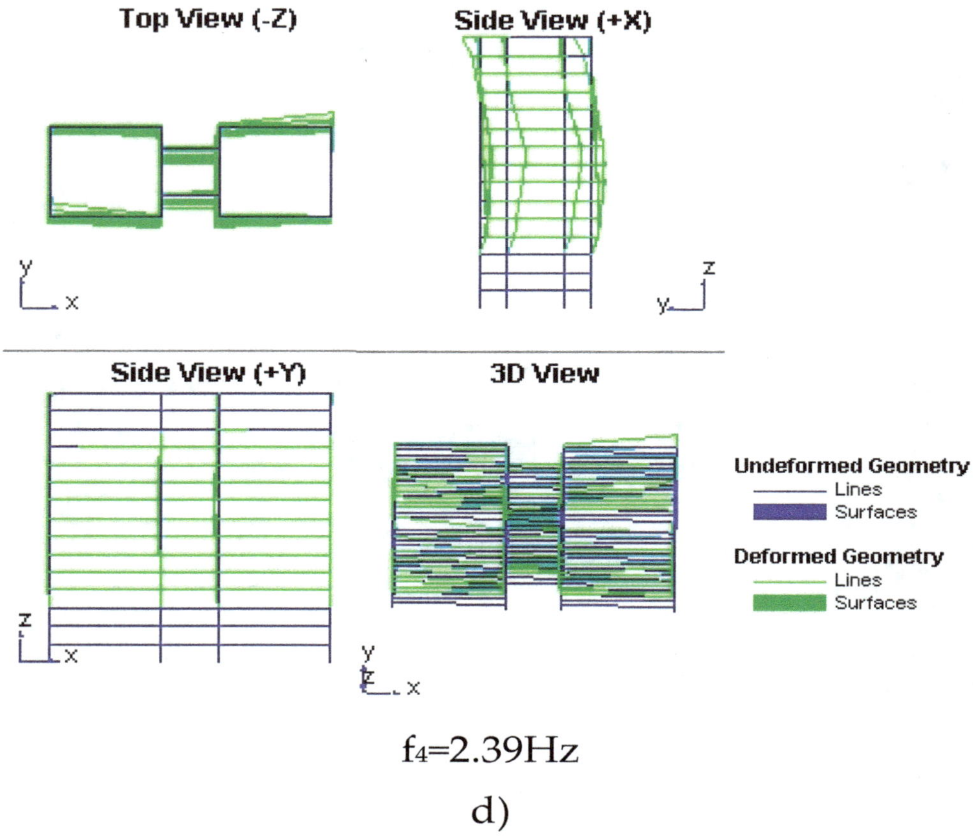

Figure 9. Ambient vibration test results: vibration modes corresponding to (**a**) 1st frequency; (**b**) 2nd frequency, (**c**) 3rd frequency and (**d**) 4th frequency.

3. Numerical Modelling

3.1. Introduction

The present section aims to detail the modelling strategies adopted within this work. Thus, it will start with the description of RC structural members modelling and then the modelling of the nonstructural elements. 3D numerical models were built in the software SAP2000 [13]. Three different numerical models were built considering different strategies related to the infill masonry walls, namely: (1) structure without infill masonry walls—Model 1; (2) structure with infill masonry walls (not considering existent damage)—Model 2; and (3) structure with damaged infill masonry walls—Model 3. The input material properties considered, as well as all the remaining modelling assumptions, will be discussed.

3.2. RC Structure Modelling

The numerical modelling of the building under study started by considering the plan and vertical disposition of the structural elements according to the structural drawings. Beams and columns were modelled through bar elements. Regarding the modeling of the two central RC cores, both were modelled using finite elements (plate elements) with dimensions that correspond to each thickness (Figure 10a). The slab modelling was performed by considering the rigid diaphragm at the storey levels, meaning each one behaved as rigid body (Figure 10b). Differences between infill masonry walls and rigid diaphragm concerning the dynamic action can be found in [14]. The modelling of the staircases was neglected since their contribution in terms of stiffness was considerably lower than the RC core's stiffness. However, the staircases' mass contribution was considered. A 3D view of the bare frame numerical model is shown in Figure 10c.

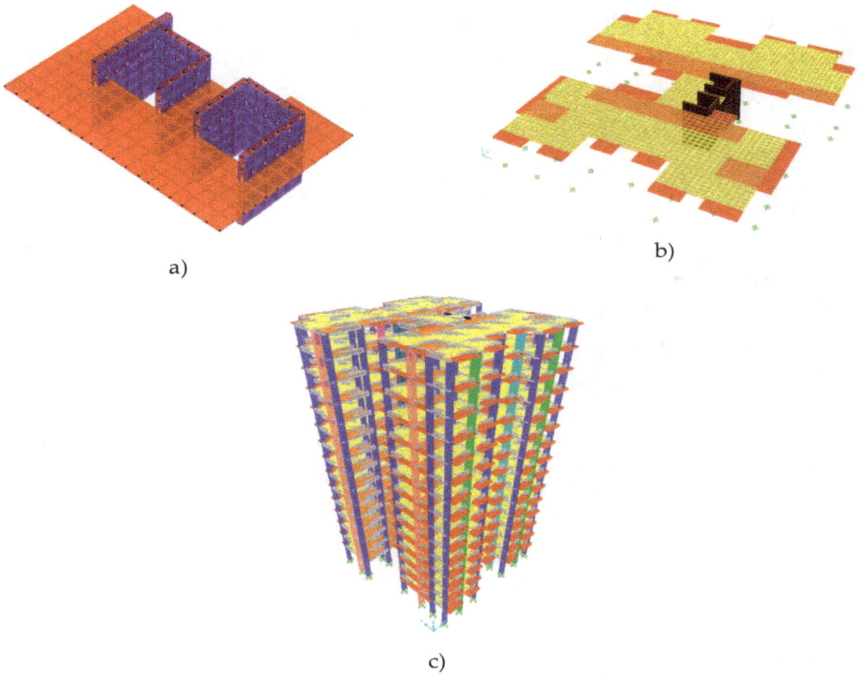

Figure 10. Numerical modelling of RC structure: (**a**) detail of the RC cores and slabs connection; (**b**) plan view; (**c**) 3D view.

Regarding the global vertical load, it was assumed a value of 5.09 kN/m^2 plus a variable load of 2.0 kN/m^2. The concrete compressive strength was assumed to be 20 MPa. A concrete elasticity modulus of 33 GPa and a tensile strength of 2.9 MPa was adopted.

3.3. Infill Masonry Walls Modelling

Many different approaches can be assumed to simulate the infill masonry walls seismic behaviour, starting from strut model concept [15–19] to detailed micro-modelling approaches [20]. Concerning the infill masonry walls modeling, the one-strut model approach was adopted, proposed by Al-Chaar [21] (Figure 11), which basically simulate the stiffness and strength contribution of the infills to the RC frame by the connection of the strut to the beam-column joints. Concerning the strut modelling parameters, the equivalent strut width, w, calculated for each infill panel was calculated according the Paulay and Priestley [22] proposal, which is given by Equation (1). The consideration of the infill panels' openings and quantification of the existent damage was achieved by the application of reduction factors according to the Al-Chaar [21] proposal. The authors suggested the application of the reduction factors R_1 and R_2 that affect the equivalent strut width. Thus, the effective reduced strut width, w_{red}, was calculated according to Equation (2). The reduction factor R_1 is related to the openings dimension, which is only applied if the panel area is lower than 60% of the panel area. For infill panels in which the openings area is higher than 60% of the panel area, some authors suggest to not consider the contribution of the wall in terms of strength and stiffness. The reduction factor R_1 was determined according to Equation (3).

$$w = 0.25 \times d \tag{1}$$

$$w_{red} = w \times R_1 \times R_2 \tag{2}$$

$$R_1 = 0.6 \times \left(\frac{A_{opening}}{A_{panel}}\right)^2 - 1.6 \times \left(\frac{A_{opening}}{A_{panel}}\right)^2 + 1 \tag{3}$$

Regarding the consideration of a reduction factor that considers the level damage, Al-Chaar [21] suggests the use of R_2, which can assume different values depending on the damage severity.

For example, in the case of a panel without damage, the coefficient R_2 is assumed as 1. Table 2 summarizes the R_2 reduction factor according to the panel geometry and level of damages.

Both simulate the infill masonry wall's in-plane behaviour through one equivalent strut (Figure 11).

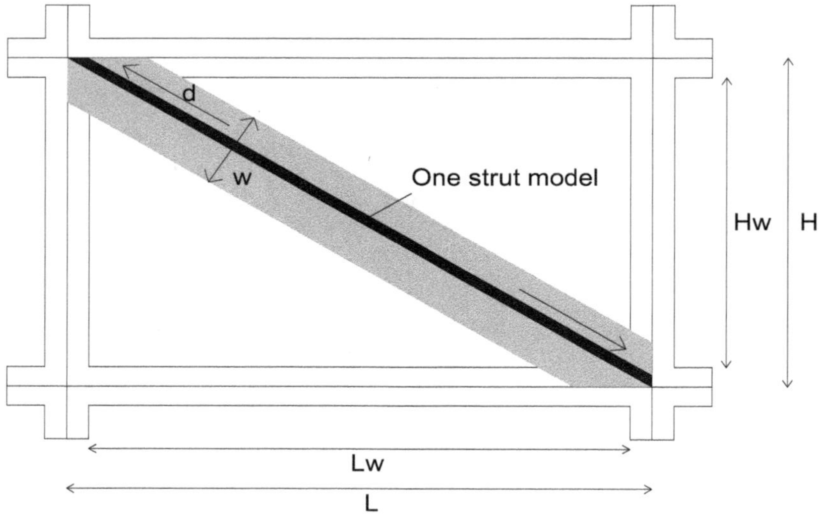

Figure 11. Infill masonry walls numerical modelling strategy: equivalent one strut model (w—strut width; L_W—panel length; H_W—panel height; H—inter-storey height; L—storey length).

Table 2. R_2 reduction factor.

Hw/t	Moderated Damages	Extensive Damages
≤ 21	0.7	0.4
> 21	0—Repairing strategies are needed	

Two different numerical models were built with infill masonry walls: one without consideration of the damages and one considering the infill's damages. The quantification of the level of damage will be explained in Section 4, since the calibration was performed considering the natural frequencies.

4. Modal Analyses

4.1. Introduction

Throughout the present section, the modal analyses results will be presented and discussed. Starting from the model without infill masonry walls (Model 1) then the results from the model considering the infill masonry walls (Model 2) and, finally the results of the model considering the damaged masonry infill walls (Model 3). All the numerical model results will be compared with the experimental results; thus, the impact of the infill masonry walls presence in the structure vibration modes will be discussed. The comparison and discussion of the modal analyses results from the three numerical models will be presented in terms of natural frequencies and vibration modes of the structures.

4.2. Model Without Infill Masonry Walls—Model 1

From the modal analysis of the model without infill masonry walls it was collected the first four vibration modes and the corresponding natural frequencies and are summarized in Table 3.

Table 3. Modal analysis results: model without infill masonry walls (Model 1).

Schematic Layout	Vibration Mode Deformed Shape	Natural frequency [Hz]
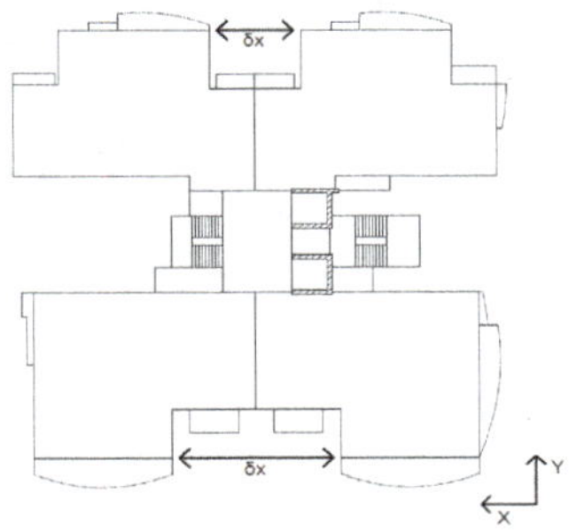	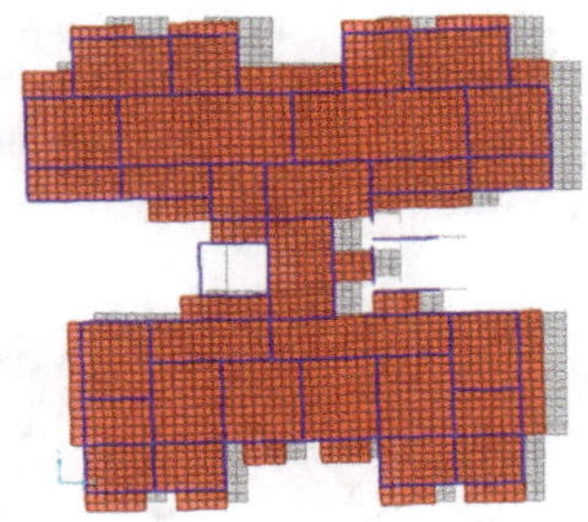	$f_1 = 0.52$
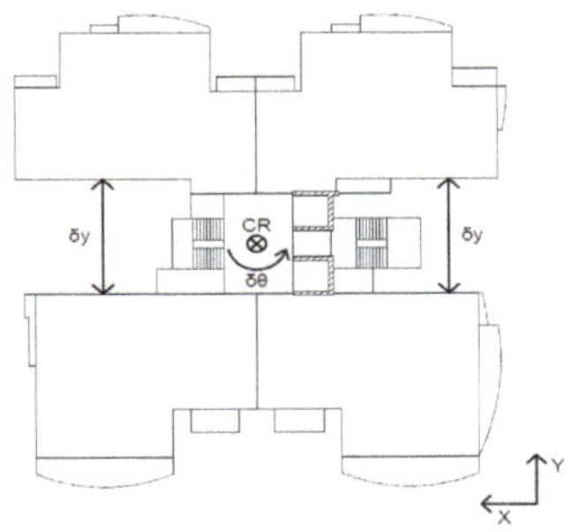	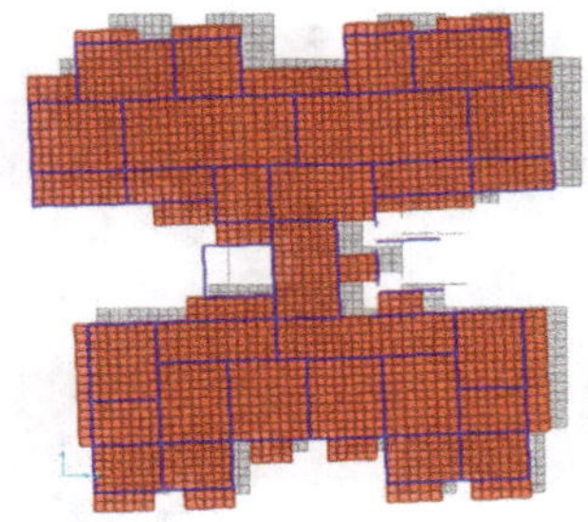	$f_2 = 0.57$
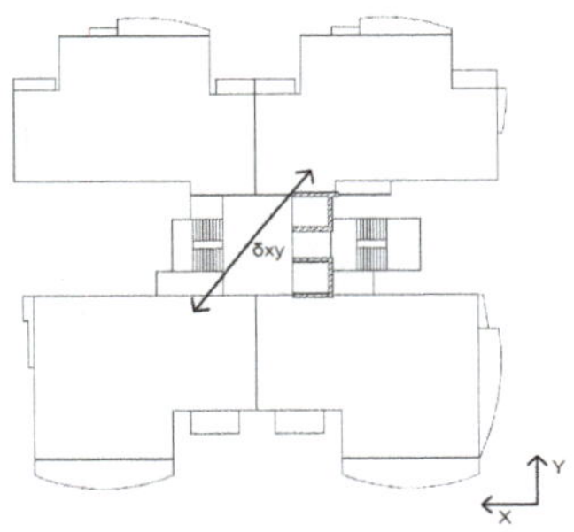	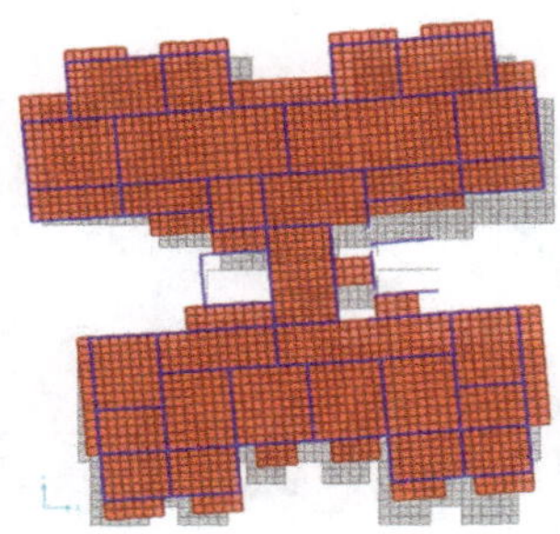	$f_3 = 2.14$
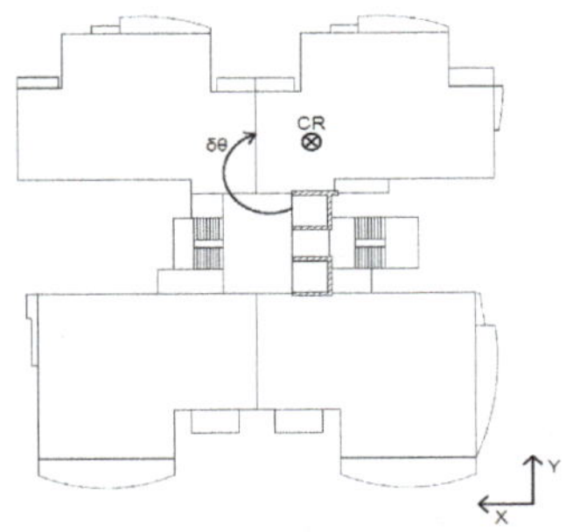	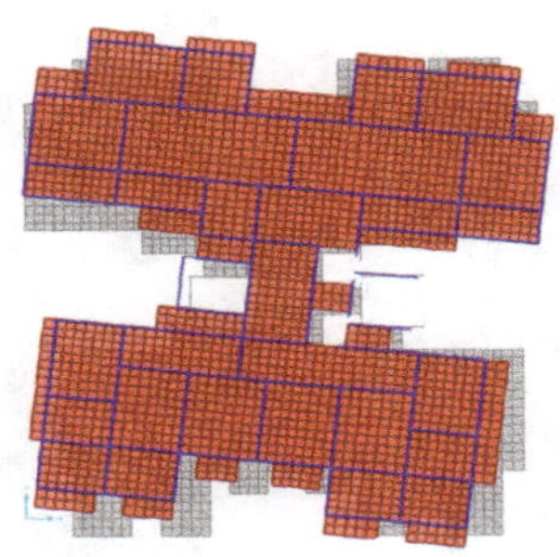	$f_4 = 2.37$

4.3. Model with Infill Masonry Walls (Not Damaged)—Model 2

From the modal analysis of the model with infill masonry walls (not damaged), first four vibration modes and the corresponding natural frequencies were collected and are summarized in Table 4.

From the comparison between the numerical (Model 1 and Model 2) and the experimental frequencies, it can be observed that, as expected, that the frequencies of the model without infill masonry walls (Model 1) are quite lower than the experimental ones. The first and second frequencies are 15% and 25% lower, respectively. This is justified by the absence of infill masonry walls that results in the reduction of the global structure lateral stiffness. Regarding the model considering infill walls (Model 2), it is observed that the numerical frequencies are higher than the experimental ones. This can be accepted, since the experimental results are related to a structure with the panels damaged. The first frequency is 28% and the second one is 15%, respectively. A brief summary of the modal analysis results of the Model 1 and Model 2 is presented in Table 5.

Table 4. Modal analyses results: model with undamaged infill masonry walls.

Schematic Layout	Vibration Mode Deformed Shape	Natural Frequency [Hz]
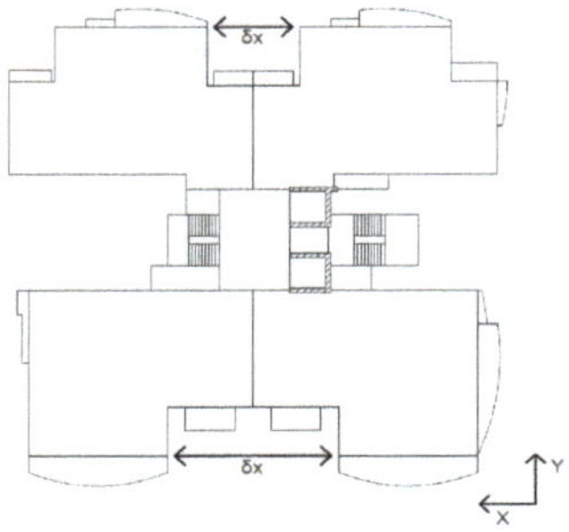	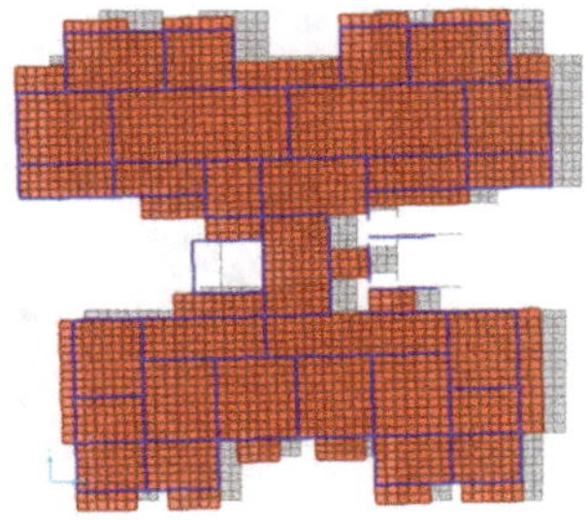	$f_1 = 0.78$
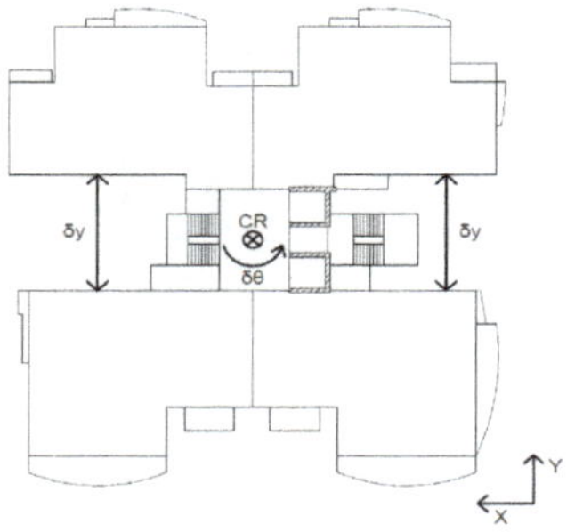	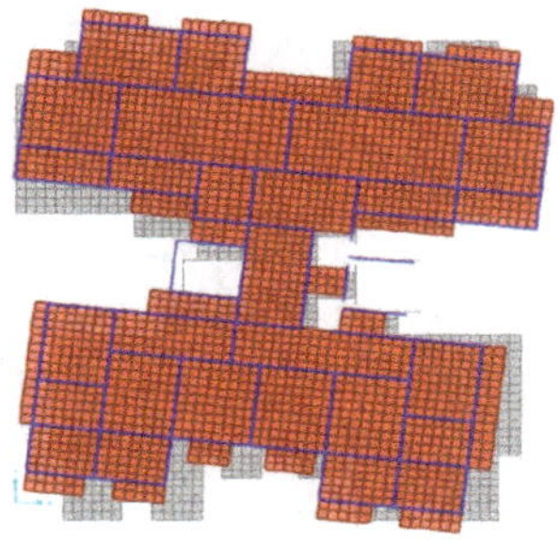	$f_2 = 0.86$
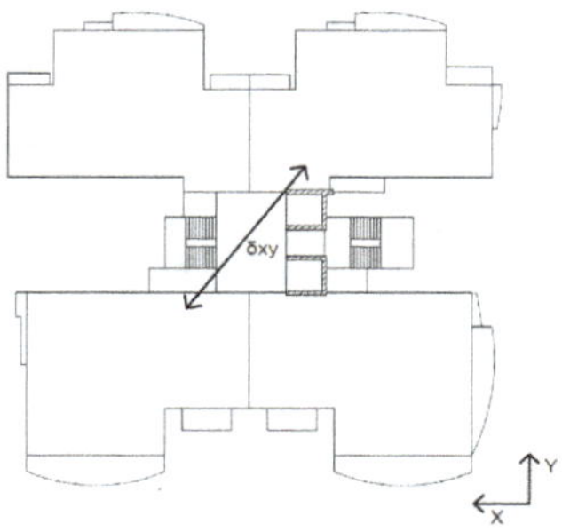	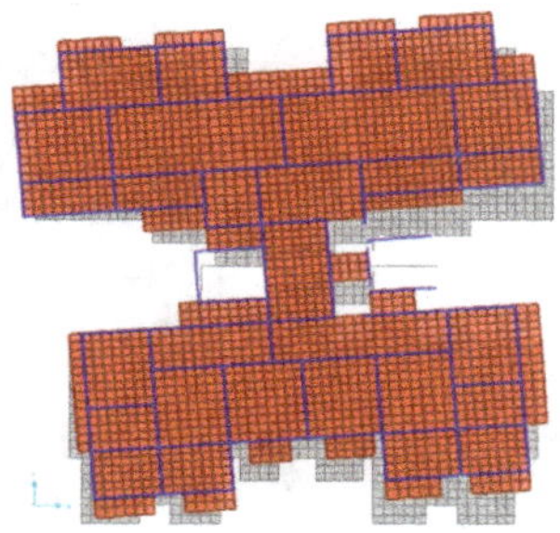	$f_3 = 2.52$
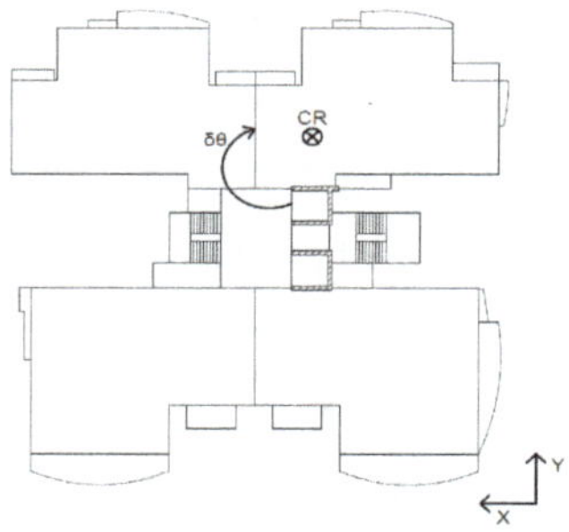	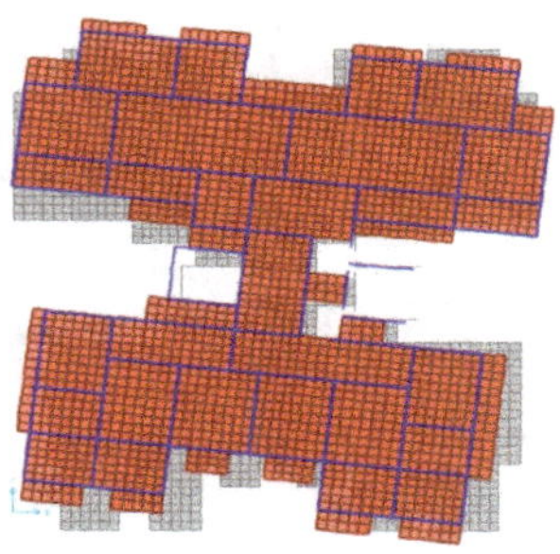	$f_4 = 2.86$

Table 5. Modal analysis results: model with undamaged infill masonry walls.

Type of Results	Natural Frequencies	
	1st Vibration Mode	2nd Vibration Mode
Experimental	0.61	0.75
Model 1	0.52	0.57
Model 2	0.78	0.86

4.4. Model with Infill Masonry Walls (Damaged)—Model 3

The assignment of different levels of damage to the infill masonry walls was based in the Al-Chaar [21] proposal, and a reduction factor R_2 variation between 0.7 and 0.4 was considered according to the state of damage. Due to the high number of panels, it is very difficult to assume a reduction factor for each panel. Thus, three different levels of damages were assumed based on the survey damage assessment (Figure 12).

Table 6. Modal analysis results: model with infill masonry walls (damaged)—Model 3.

Schematic Layout	Vibration Mode Deformed Shape	Natural Frequency [Hz]
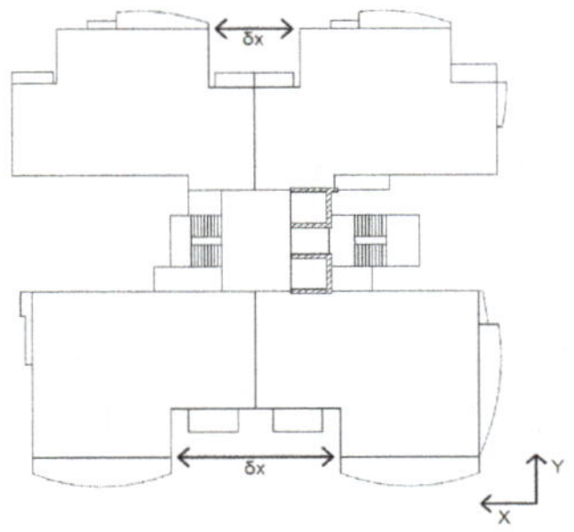	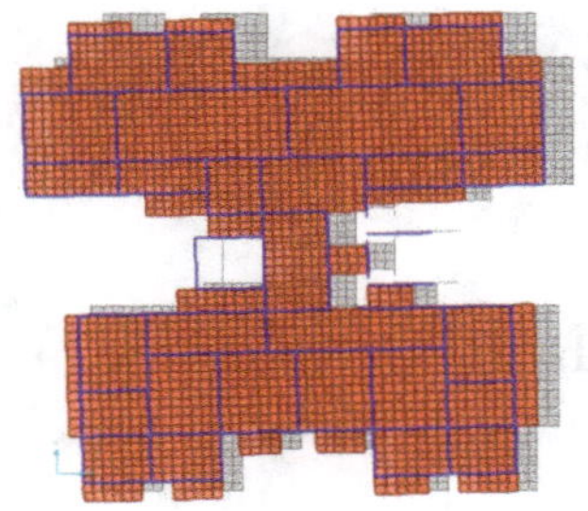	$f_1 = 0.65$
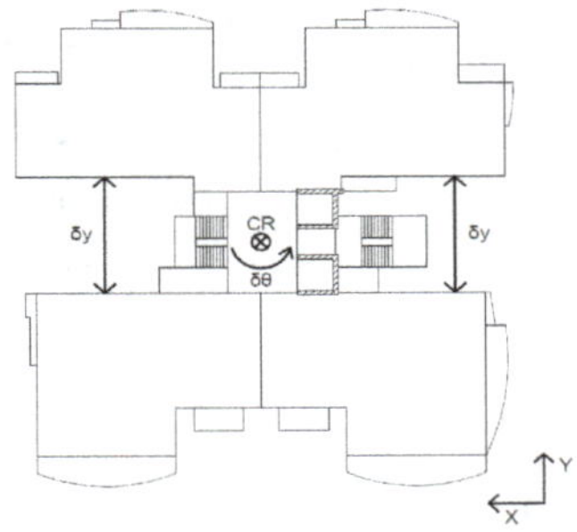	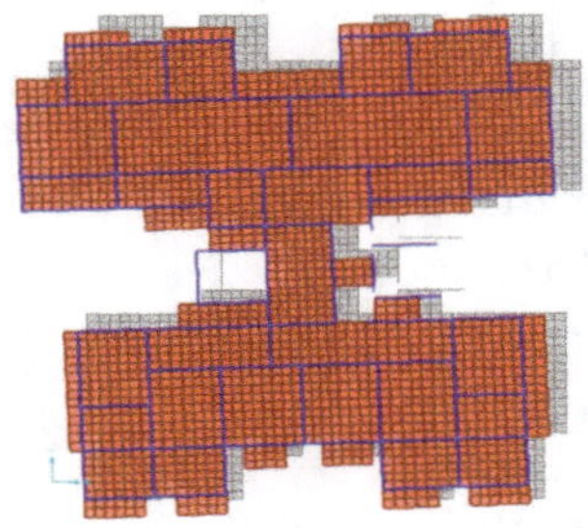	$f_2 = 0.77$
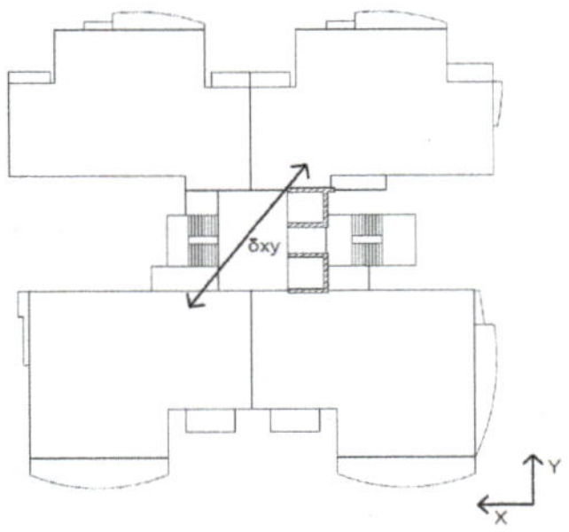	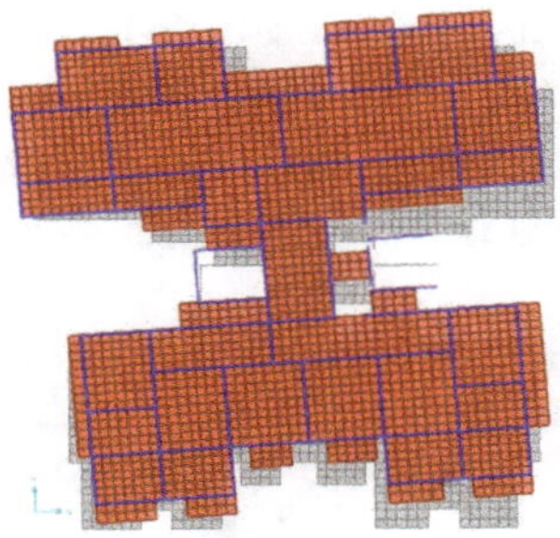	$f_3 = 2.26$
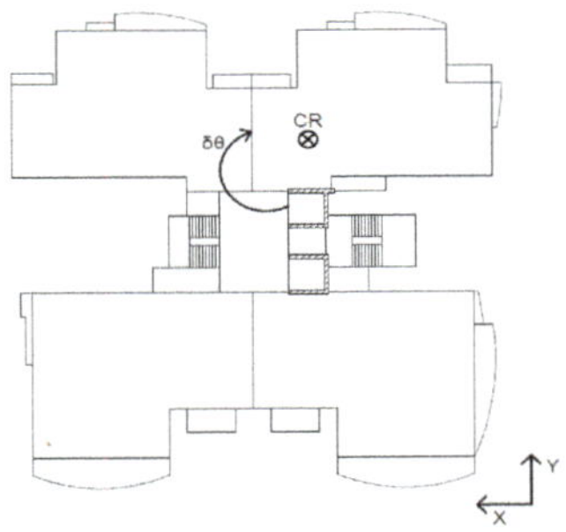	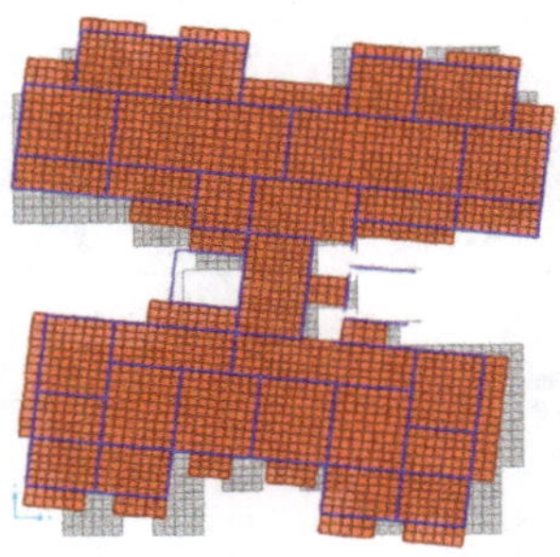	$f_4 = 2.56$

For the first level of damage, a reduction factor of 0.3 was assumed between the 1st and 6th storeys. For the second level of damage, between the 6th and 10th storeys, a reduction factor of 0.5 was assumed. Finally, for the third level of damage, a coefficient of 0.7 was assumed. From the modal analysis of the model with infill masonry walls (damaged), the first four vibration modes and the corresponding natural frequencies were collected and are summarized in Table 6.

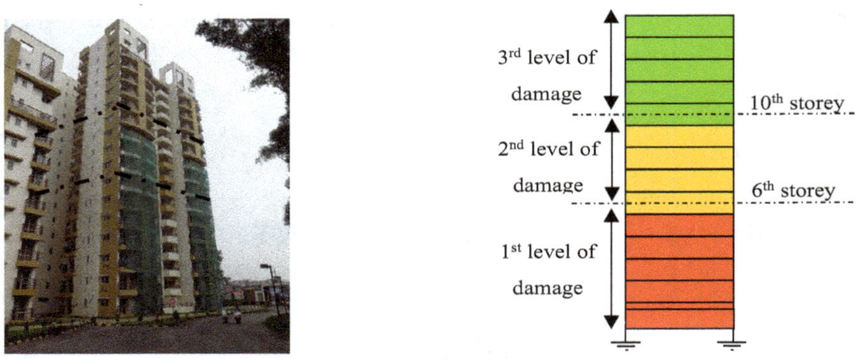

Figure 12. Schematic layout of the infill masonry wall level of damages assumed.

4.5. Global Comparison

From the global comparison of the results from the modal analyses, it can be noticed that the presence of the infill masonry walls increases the global lateral stiffness and thus the natural frequencies. However, the present study shows that the impact of the infill panel damages results in the reduction of the natural frequencies. At this point, the experimental results are important to discern which modelling strategy is more appropriate to simulate this structure. Table 7 summarizes the results from the modal analysis obtained by the three numerical modes and in the ambient vibration test. By the analysis of the first natural frequency (translational mode along direction X), the highest and the lowest value were achieved by Models 2 and 1 respectively. Model 2 is 28% higher than the experimental one and Model 1 is 15% lower. Concerning the comparison between the Model 3 and the experimental one, a small difference was obtained, namely 5% higher.

The second vibration mode, characterized by a translational mode along direction Y, similar results were found; the highest result was obtained by Model 2 with a frequency equal to 0.86 Hz (15% higher than the experimental one) and the lowest result was obtained by Model 1 (24% lower than the experimental one). Once again, Model 3 reached the result with higher accuracy, namely 0.77 Hz (2.6% higher than the experimental one). Similar observations can be drawn for the third vibration mode.

Table 7. Modal analyses results: global comparison.

Model	Vibration Modes		
	f_1	f_2	f_3
Model 1	0.52	0.57	2.14
Model 2	0.78	0.86	2.52
Model 3	0.65	0.77	2.26
experimental	0.61	0.75	2.39

5. Linear Elastic Dynamic Analysis

5.1. Introduction

With the aim to assess the effect of the presence of infill walls in the structural response of the building, linear elastic dynamic analyses were carried out. Since this specific study is not related to the seismic vulnerability assessment of the structure, it was intended to perform only one dynamic linear elastic analysis with only one accelerogram for all the three numerical models. The selected accelerogram was the Gorkha earthquake, which is plotted in Figure 13. From the analysis of the spectral acceleration, it can be observed that the peak spectral acceleration occurs for natural periods between 0.24 s to 1.1 s. Looking for the natural periods of the numerical models, it seems that Model 2 (with undamaged infills) is the one closest to this range of natural periods (T = 1.25 s). This indicates that Model 2 will be the one subjected to higher seismic loading demands in this analysis.

For this analysis a stiffness reduction of the RC elements was considered according to the Greek code recommendations [23], which indicate reduction factors in structural analyses and assessment, and not the design of new structures. A reduction factor of 0.2 was considered for internal columns, 0.4 for external columns, 0.5 for cracked stiffness cores, 0.3 for noncracked stiffness cores and 0.6 for beams.

The analysis results will be analysed and discussed in terms of inter-storey displacements profiles, inter-storey drift ratios, inter-storey shear and inter-storey seismic loading. Four vertical alignments were analysed with the aim of accurately assessing the structural response (Figure 14).

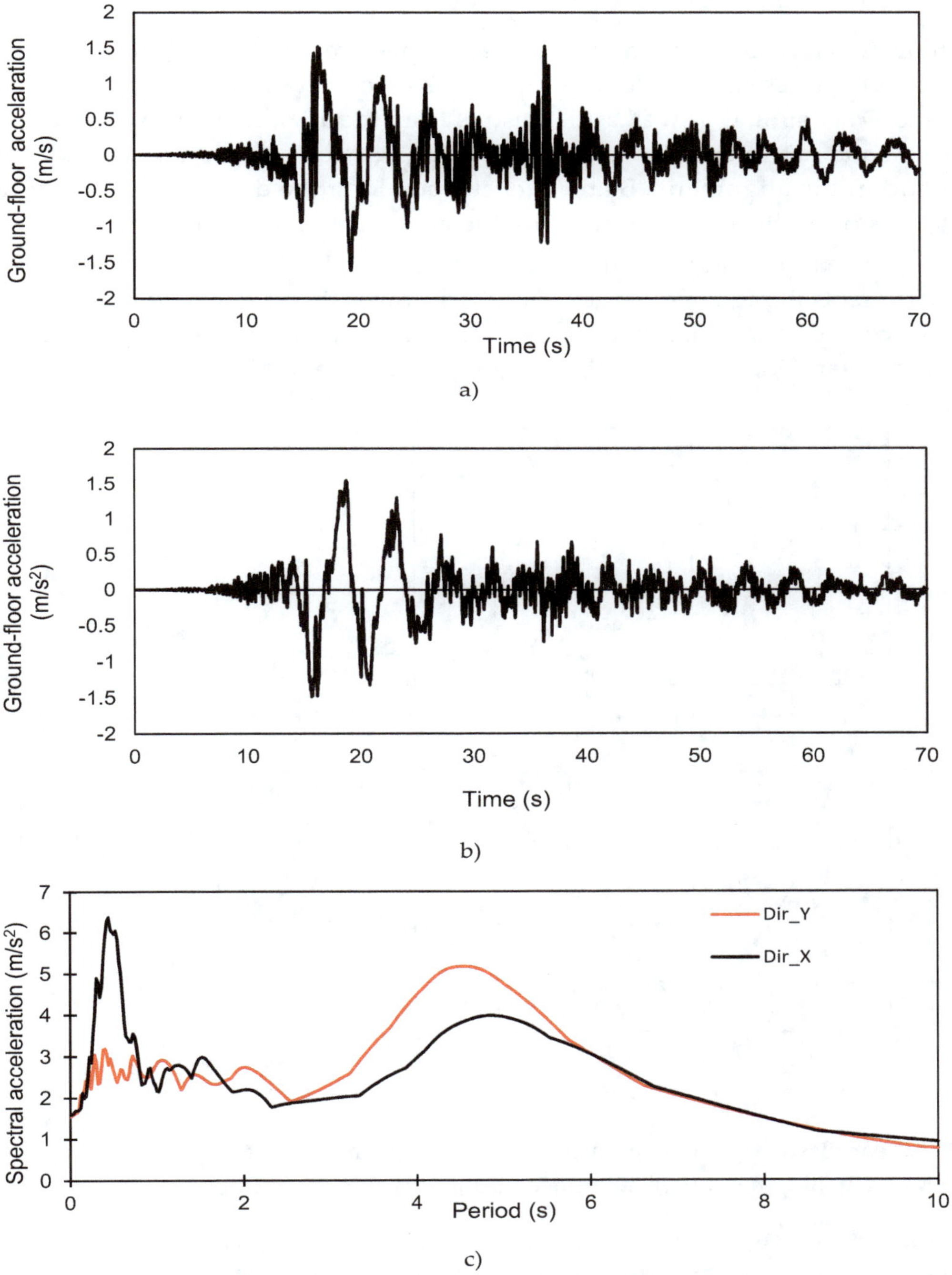

Figure 13. Linear elastic dynamic analysis: (**a**) accelerogram (direction X); (**b**) accelerogram (direction Y); and (**c**) spectral acceleration.

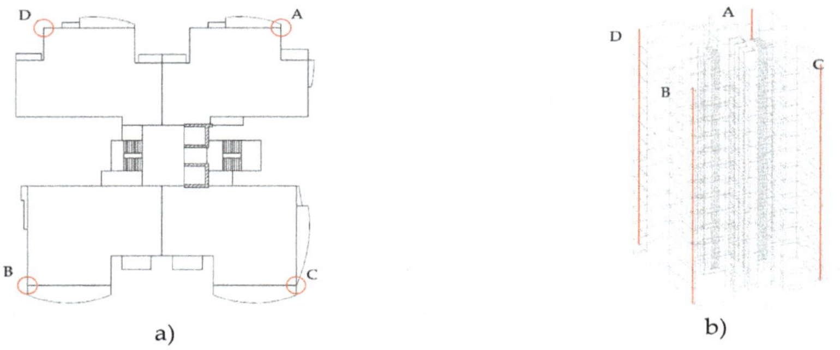

Figure 14. Linear elastic dynamic analysis: vertical alignments under study (**a**) plan view; (**b**) 3D view.

5.2. Inter-storey Displacements And Drift Ratio Profiles

Figure 15 presents the maximum inter-storey displacements of Model 3 along the four vertical alignments, from which it is easy to identify similar behaviour. It can be observed that the results from alignments A and D are similar as well as the results from the alignments B and C. It is possible to find that the maximum and minimum displacements in the alignments A and D occurred along direction Y, on the other hand in the alignments B and C occurred along the direction X. This difference can be justified by the torsional effect characterized by the geometry of this structure.

Based on this maximum and minimum inter-storey displacements, it can be also concluded that the A and D alignments along the direction X are 30% lower than the results obtained by the B and C alignments. The same does not occur for the inter-storey displacements along direction Y, where the results are quite similar. Once again this can be justified by the torsion effect.

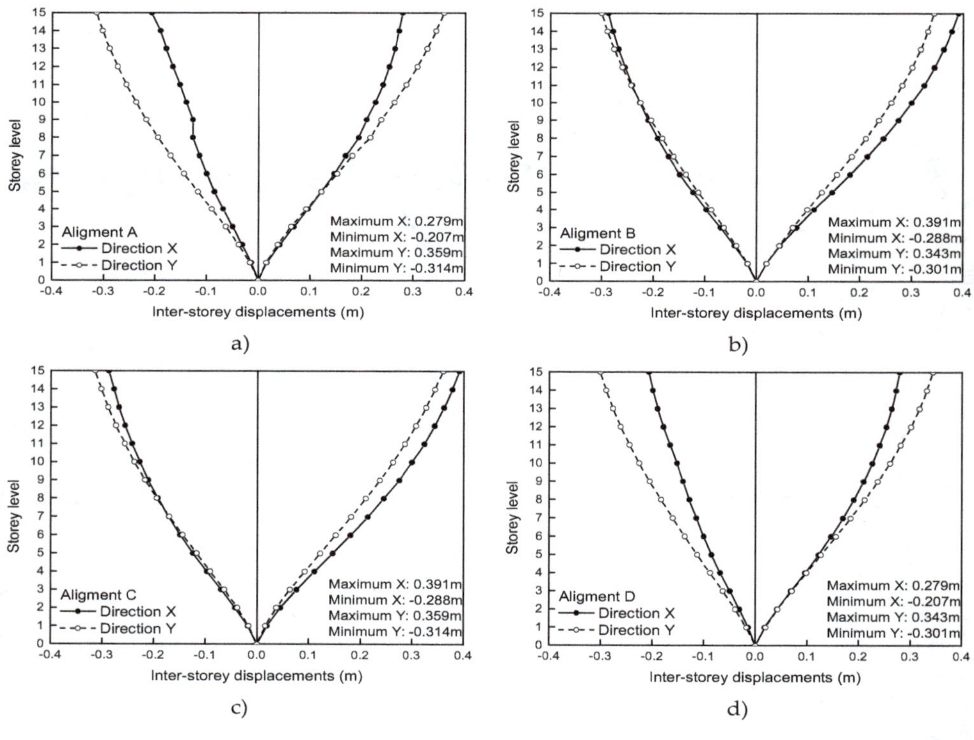

Figure 15. Linear elastic dynamic analysis result: maximum inter-storey displacement of Model 3 (**a**) alignment A; (**b**) alignment B; (**c**) alignment C and; (**d**) alignment D.

We also studied the level of the expected infills damage based on the relationship between the maximum inter-storey drift (ISD) ratio and the corresponding level of expected damage proposed by different authors. Based on the strut model, Magenes and Pampanin [24] proposed an empirical damage evaluation of the infill panels that corresponded to certain limit state, depending on the axial deformation. FEMA-306 [25] and FEMA-307 [26] documents indicates also reference values of ISD ratio. The drift limit proposed for brick masonry is 1.5%, and the drift limit for the beginning of the diagonal cracking which is 0.25% can also be found in these documents, as can be observed in Figure 16.

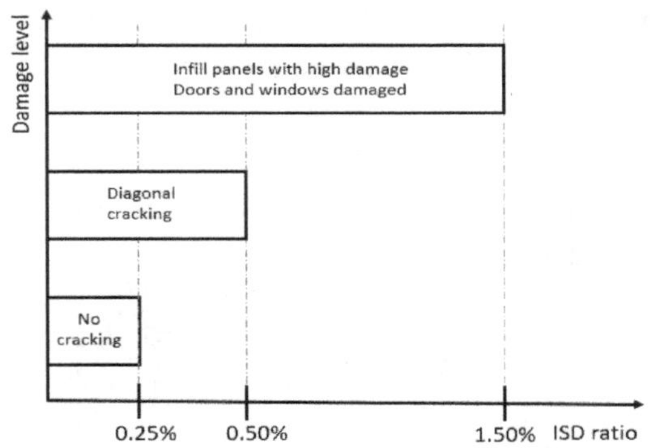

Figure 16. Relationship between maximum ISD ratio and infill masonry damage level proposed by FEMA-306.

From the analysis of the maximum inter-storey drift (ISD) ratio profiles (Figure 17), along the alignment A, the following observations can be drawn, namely:

(i) In all the numerical models, in both directions, the higher ISD ratio values occurred in the intermediate storeys (between storeys 3 and 7);
(ii) Globally, it can be observed that the higher ISD ratios occurred in the direction Y, with exception of Model 3 where it is observed the opposite. This can be justified by the irregularity of the damages observed in the infill masonry walls, which reduced the global lateral stiffness and strength and increased the lateral deformation;
(iii) From the comparison between all the numerical models, it can be observed that Model 1 reached the highest ISD ratio values along the direction X, by achieving a maximum value of 0.85%. Model 2 reached the lowest one with a maximum ISD ratio of 0.43%. Concerning the response of Model 3, the impact of the infill panel damages increased the ISD ratio to a maximum value of 0.72%, which is 15% lower than the value obtained by Model 1 (Figure 16);
(iv) Concerning the direction Y (Figure 16), different results were found since Model 2 and Model 3 reached similar response by reaching a maximum ISD ratio around 0.57%. On the other hand, Model 1 reached again the maximum ISD with a value of 0.83%.
(v) Regarding the drift limits, it is possible to observe that along the damages are concentrated in the storeys above the 8th; most of them had diagonal cracking as observed in situ and reported in Section 2.3. The numerical results are quite similar—from the ISD ratio envelope of Model 2 and the storeys that exceeded the drift limits, as well as in comparison with the observed post-earthquake damages.

A similar trend was observed in the alignment C, however higher ISD ratios were achieved. Starting from the results of Model 1 (Figure 18a), the maximum ISD reached in the direction X is 26% higher than the one in direction Y. The storeys that obtained high level of ISD were the storeys between 2 and 8.

Concerning Model 2 (Figure 18b), lower differences were noticed in both directions. In fact, a small variation of around 5% is observed with the highest one along direction Y. In this case, it is not visible a special concentration of the drift demand along the building height.

Finally, from the analysis of the Model 3 response, the higher level of ISD ratios along the direction X is again visible, which reached a maximum value of 77% higher than the one reached in direction Y. The higher levels of drift occurred along the storeys 2 until 9, with storey 4 having the maximum ISD.

From the global analysis of the ISD ratios, a similar trend is once again visible, namely in the direction X the Model 1 and 2 achieved the highest and the lowest values (Figure 18b), respectively. Model 3 reached a maximum drift of 0.73%, which is 38% lower than the one reached by Model 1.

Concerning direction Y, once again Model 1 reached the maximum ISD ratio followed by the Models 2 and 3, respectively. The maximum value was around 0.92%, which is 39% and 42% higher than the Models 2 and 3, respectively as observed in the alignment A results.

Finally, regarding the drift limits exceedance it is possible to draw the same observations performed for the alignment A, which basically shows that the storeys between the 2nd and 8th storeys were most affected by the earthquake motion.

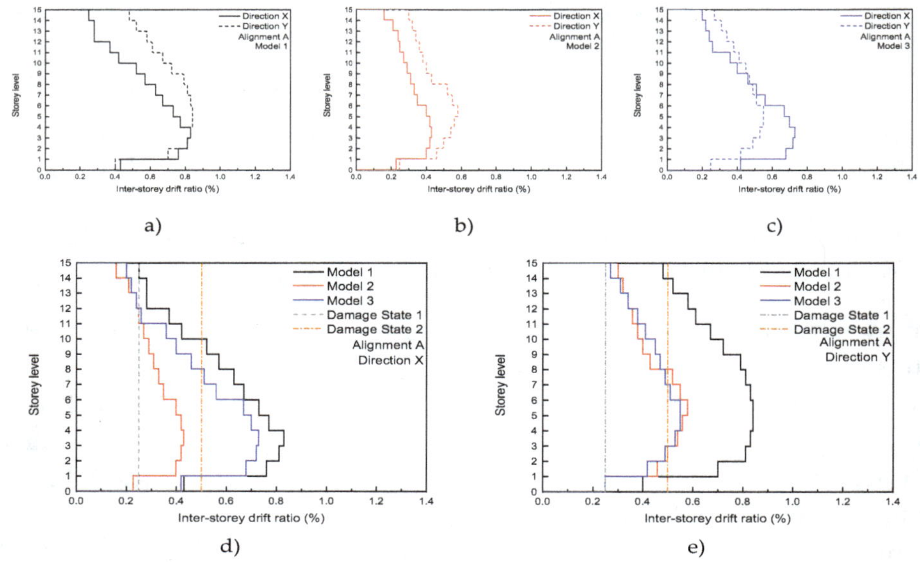

Figure 17. Linear elastic dynamic analysis result: maximum inter-storey drift ration (alignment A) (**a**) Model 1; (**b**) Model 2; (**c**) Model 3; (**d**) global comparison (Direction X); and (**e**) global comparison (direction Y).

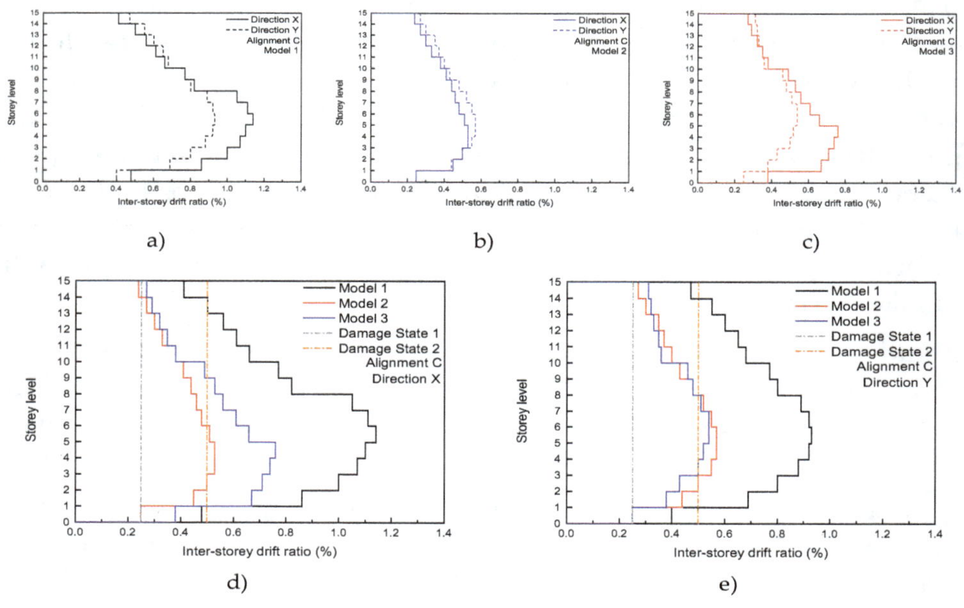

Figure 18. Linear elastic dynamic analysis result: maximum inter-storey drift ration (alignment C) (**a**) Model 1; (**b**) Model 2; (**c**) Model 3; (**d**) global comparison (Direction X); and (**e**) global comparison (direction Y).

5.3. Inter-storey Shear and Seismic Loading Envelopes

Concerning the analysis of the inter-storey shear, the maximum shear reached in each storey was extracted from the results, which are plotted in Figure 19. From the results it is possible to observe that the higher values were reached along the direction X (except Model 3). From the plots, it is visible that in Model 1 reached lower maximum storey shear variation along the storey height (Figure 19a). Otherwise, Model 3 was the one where a larger difference was noticed (Figure 19c). Model 2 appears to be an intermediate situation, such in direction X or Y (Figure 19b). From the global comparison along direction X (Figure 19d), it is possible to observe that Model 3 achieved results about 15% higher than Model 1 and 8% more than Model 2. The presence of the infill walls resulted in the increment of the storey shear, which could lead to significant impact on the frame columns and/or beam-column joints. Concerning direction Y (Figure 19e), it can be observed that the maximum values were reached by Model 2, followed by the Model 1 and then Model 3. Model 1 results were about 23% higher than that of Model 3, which is very similar to the model without infill walls.

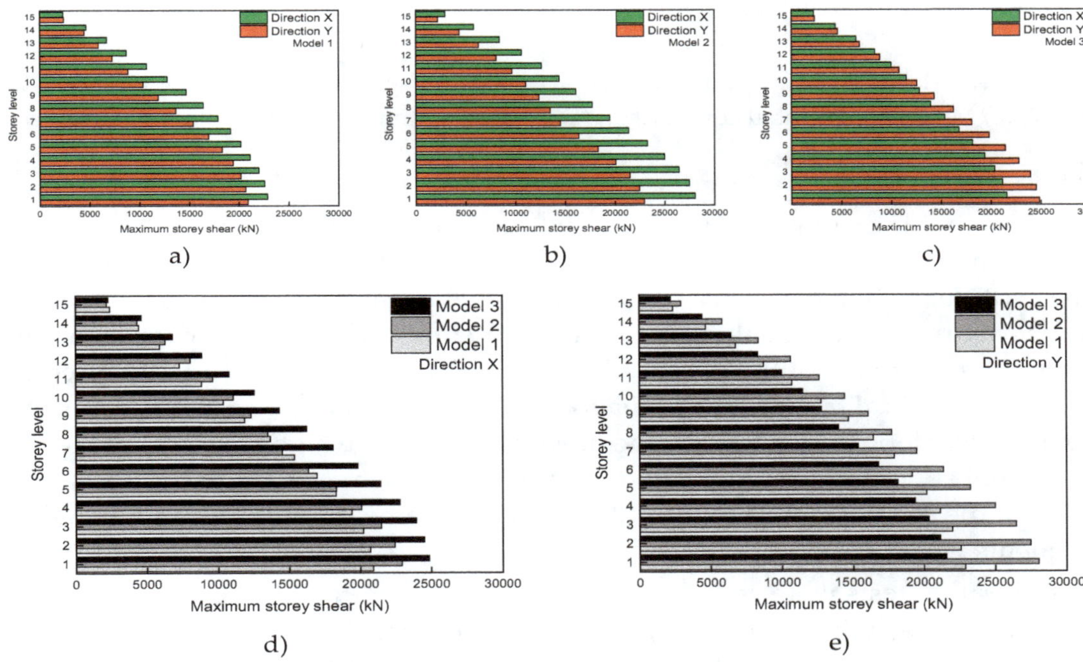

Figure 19. Linear elastic dynamic analysis result: maximum storey shear (**a**) Model 1; (**b**) Model 2; (**c**) Model 3; (**d**) global comparison (Direction X); and (**e**) global comparison (direction Y).

Finally, concerning the maximum seismic loading (Figure 20), it is visible that once again the higher values can be found along direction X in Model 1 and in Model 2. In other hand, Model 3 reached the higher maximum seismic loading along direction Y. By analysing the envelopes, it is possible to observe that in Model 1, the maximum seismic loading occurred for the storeys 8, 7 and 6 and the maximum ones in the 14th and 15th storeys (Figure 20a). Higher differences among storeys are visible in Model 2 (Figure 20b), namely the maximum values were reached along Storeys 5, 6 and 4 and finally in 14 and 15. Finally, in Model 3 (Figure 20c), the maximum seismic loadings were reached in the 8th, 7th and 9th storeys and finally in the top storeys (15 and 14).

From the global comparison between the numerical models it seems that along the direction X from the bottom until the 6th storey that Model 2 reached higher seismic loadings. From the 6th storey until 14th, Model 3 exceeded Model 2's seismic loadings. On the 15th storey, Model 1 achieved the highest value (Figure 20d). Finally, along direction Y, it seems that Model 2 always reached the highest seismic loadings when compared with the remaining ones. From the plot (Figure 20e), it is visible that the maximum seismic loadings were about 20% and 10% higher than Model 1 and 3 respectively.

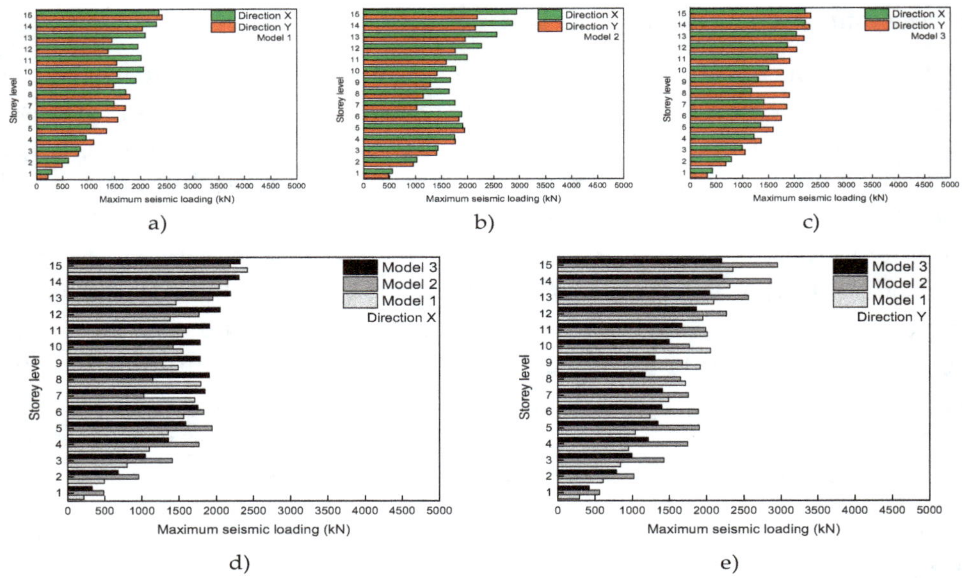

Figure 20. Linear elastic dynamic analysis result: maximum seismic loading envelope (**a**) Model 1; (**b**) Model 2; (**c**) Model 3; (**d**) global comparison (Direction X); and (**e**) global comparison (direction Y).

6. Conclusions

The main aim of this manuscript was to study the impact of the infill masonry walls presence in the seismic performance of a 15-storey high-rise building located in Nepal. The building was subjected to the Gorkha earthquake sequence in 2015 and was visited by an international team that performed a complete damage survey assessment report (herein presented). From the damage assessment, it was concluded that the major part of the observed damages found were related to the presence of infill masonry walls, namely local failure such as diagonal cracking, shear sliding cracks and detachment of the panel from the envelope frame. It was concluded that due to the high flexibility of the RC structure, the infill walls were subjected to significant deformations, which resulted in the observed extensive damages. Due to the seismic design of the RC structural elements, significant damages within the structural elements were not observed.

Ambient vibration tests were carried out to collect the vibration modes and the corresponding natural frequencies of the building under study. From the results, it was visible that the first and second vibration modes are characterized by slight torsion, which is due to the building geometry. The results obtained were used to calibrate the numerical model built in the software SAP2000. Additionally, two different numerical models were also built, considering different modelling strategies related to the infill masonry walls (without infill walls and with undamaged infill walls).

From the modal analysis, it was possible to observe that the infills presence increased the frequencies about 30%. Any modification of the of the vibration modes due to the infill's presence was not visible. From the comparison between the models with undamaged and damaged infill panels, it was found that neglecting the panel damage could result in differences between 10–20%.

Linear elastic analyses were carried out to assess the impact of the infill panels in the expected dynamic response of the structure. From the results, it was observed that the infill panels presence increased significantly the storey shear, and the maximum base shear about 20%. This important increasing of shear loadings due to the infills' presence is very important, since in the case of structures designed only to support gravity loads, it should be analysed carefully if the building' foundations and vertical elements are capable to support those shear loadings. From the results, it can be also concluded that the presence of the infill walls reduced the ISD ratio, however when compared with the drift limits proposed by FEMA-306, it was observed that a large part of the infill panel's damages occurred above the 8th storey. It was also observed that the presence of infill masonry walls contributed to the increasing of the torsion effect. From this study, it is evident that the infill masonry walls played an important role in the seismic performance of the structure, highlighting the need to consider these elements during the design of new structures and/or the structural safety assessment of existing structures.

Author Contributions: Conceptualization, A.F. and N.V.-P.; methodology, A.F. and N.V.-P.; software, A.F., validation, A.F. and N.V.-P; formal analysis, A.F.; investigation, H.V. and A.A.; resources, H.V. and A.A.; writing—original draft preparation, A.F.; writing—review and editing, A.F. and N.V.-P; visualization, A.F.; supervision, N.V.-P., H.V. and A.A. and P.D.; project administration, H.V. (PDF) Seismic Performance of High-Rise Condominium Building during the 2015 Gorkha Earthquake Sequence. Available from: https://www.researchgate.net/publication/330673852_Seismic_Performance_of_HighRise_Condominium_Building_during_the_2015_Gorkha_Earthquake_Sequence [accessed Feb 02 2019].

Acknowledgments: This work was financially supported by: UID/ECI/04708/2019—CONSTRUCT—Instituto de I&D em Estruturas e Construções funded by national funds through the FCT/MCTES (PIDDAC) and by national funds through FCT—Fundação para a Ciência e a Tecnologia, namely through the research project P0CI-01-0145-FEDER-016898—ASPASSI—Safety Evaluation and Retrofitting of Infill masonry enclosure Walls for Seismic demands. The authors also acknowledge the constructive comments and suggestions given by the anonymous reviewers that improved the quality of the manuscript.

References

1. Hermanns, L.; Fraile, A.; Alarcón, E.; Álvarez, R. Performance of buildings with masonry infill walls during the 2011 Lorca earthquake. *Bull. Earthq. Eng.* **2014**, *12*, 1977–1997. [CrossRef]
2. De Luca, F.; Verderame, G.; Gómez-Martinez, F.; Pérez-García, A. The structural role played by masonry infills on RC buildings performances after the 2011 Lorca, Spain, earthquake. *Bull. Earthq. Eng.* **2014**, *12*, 1999–2006. [CrossRef]
3. Romão, X.; Costa, A.A.; Paupério, E.; Rodrigues, H.; Vicente, R.; Varum, H.; Costa, A. Field observations and interpretation of the structural performance of constructions after the 11 May 2011 Lorca earthquake. *Eng. Fail. Anal.* **2013**, *34*, 670–692. [CrossRef]
4. Furtado, A.; Rodrigues, H.; Varum, H.; Costa, A. Evaluation of different strengthening techniques' efficiency for a soft storey building AU-Furtado, André. *Eur. J. Environ. Civ. Eng.* **2017**, *21*, 371–388. [CrossRef]
5. Brando, G.; Rapone, D.; Spacone, E.; O'Banion, M.S.; Olsen, M.J.; Barbosa, A.R.; Faggella, M.; Gigliotti, R.; Liberatore, D.; Russo, S.; et al. Damage Reconnaissance of Unreinforced Masonry Bearing Wall Buildings After the 2015 Gorkha, Nepal, Earthquake. *Earthq. Spectra* **2017**, *33*, S243–S273. [CrossRef]
6. Gautam, D.; Rodrigues, H.; Bhetwal, K.K.; Neupane, P.; Sanada, Y. Common structural and construction deficiencies of Nepalese buildings. *Innov. Infrastruct. Solut.* **2016**, *1*, 1. [CrossRef]
7. Barbosa, A.R.; Fahnestock, L.A.; Fick, D.R.; Gautam, D.; Soti, R.; Wood, R.; Moaveni, B.; Stavridis, A.; Olsen, M.J.; Rodrigues, H. Performance of Medium-to-High Rise Reinforced Concrete Frame Buildings with Masonry Infill in the 2015 Gorkha, Nepal, Earthquake. *Earthq. Spectra* **2017**, *33*, S197–S218. [CrossRef]
8. Varum, H.; Furtado, A.; Rodrigues, H.; Oliveira, J.; Vila-Pouca, N.; Arêde, A. Seismic performance of the infill masonry walls and ambient vibration tests after the Ghorka 2015, Nepal earthquake. *Bull. Earthq. Eng.* **2017**, *15*, 1185–1212. [CrossRef]
9. Carvalheira, A. Estudo da Influência Das Paredes de Alvenaria no Comportamento Sísmico de Um edificio no Nepal Com 15 Pisos Elevados. Master's Thesis, FEUP–Faculdade de Engenharia da Universidade do Porto, Porto, Portugal, 22 July 2016.
10. Russo, S. Integrated assessment of monumental structures through ambient vibrations and ND tests: The case of Rialto Bridge. *J. Cult. Herit.* **2016**, *19*, 402–414. [CrossRef]
11. ARTeMIS Extractor Pro. *Pro Structural Vibration Solutions*; ARTeMIS Extractor: Aalborg, Denmark, 2009.
12. Cunha, Á.; Caetano, E. Experimental modal analysis of civil engineering structures. *Sound Vib.* **2006**, *40*, 12–20.
13. SAP2000. *Integrated Finite Element Analysis and Design of Structures Basic Analysis Reference Manual*; Computer and Structures, Inc.: Berkeley, CA, USA, 2009.
14. Dal Cin, A.; Russo, S. Annex and rigid diaphragm effects on the failure analysis and earthquake damages of historic churches. *Eng. Fail. Anal.* **2016**, *59*, 122–139. [CrossRef]
15. Furtado, A.; Rodrigues, H.; Arêde, A.; Varum, H.; Grubišić, M.; Šipoš, T.K. Prediction of the earthquake response of a three-storey infilled RC structure. *Eng. Struct.* **2018**, *171*, 214–235. [CrossRef]

16. Furtado, A.; Rodrigues, H.; Varum, H.; Arêde, A. Mainshock-aftershock damage assessment of infilled RC structures. *Eng. Struct.* **2018**, *175*, 645–660. [CrossRef]
17. Asteris, P.G.; Cavaleri, L.; Di Trapani, F.; Sarhosis, V. A macro-modelling approach for the analysis of infilled frame structures considering the effects of openings and vertical loads. *Struct. Infrastruct. Eng.* **2015**, *12*, 551–566. [CrossRef]
18. Ricci, P.; Domineco, M.; Verderame, G. Empirical-based out-of-plane URM infill wall model accounting for the interaction with in-plane demand. *Earthq. Eng.Struct. Dyn.* **2017**, *47*, 802–827. [CrossRef]
19. Crisafulli, F.; Carr, A. Proposed macro-model for the analysis of infilled frame structures. *Bull. N. Z. Soc Earthq. Eng.* **2007**, *40*, 69–77.
20. Asteris, P.; Cotsovos, D.; Chrysostomou, C.; Mohebkhah, A.; Al-Chaar, G. Mathematical micromodelling of infilled frames: State of the art. *Eng. Struct.* **2013**, *56*, 1905–1921. [CrossRef]
21. Al-Chaar, G.; Issa, M.; Sweeney, S. Behavior of masonry-infilled nonductile reinforced concrete frames. *J. Struct. Eng.* **2002**, *128*, 1055–1063. [CrossRef]
22. Paulay, T.; Priestley, M.J.N. *Seismic Design of RC and Masonry Buildings*; John Wiley: Hoboken, NJ, USA, 1992; ISBN 0-471-54915-0.
23. EAK. *Greek Code for Seismic Resistant Structures, arthquake Resistant Planning and Protection*; Ministry of Environment Planning and Public Works: Athens, Greece, 2000.
24. Magenes, G.; Pampanin, S. Seismic Response of Gravity-Load Design Frames with Masonry Infills. In Proceedings of the 13th World Conference on Earthquake Engineering, Vancouver, BC, Canada, 1–6 August 2004.
25. FEMA306. *Evaluation of Earthquake Damaged Concrete and Masonry Wall Buildings: Basic Procedures Manual*; FEMA-306–Applied Technology Council: Washington, DC, USA, 1998.
26. FEMA307. *Evaluation of Earthquake Damaged Concrete and Masonry Wall Buildings—Technical Resources*; FEMA: Washington, DC, USA, 1998.

9

Evaluation of the Performance of Unreinforced Stone Masonry Greek "Basilica" Churches when Subjected to Seismic Forces and Foundation Settlement

George C. Manos *, Lambros Kotoulas and Evangelos Kozikopoulos

Lab. Strength of Materials and Structures, Aristotle University, 541 24 Thessaloniki, Greece; kotoulaslambros@gmail.com (L.K.); vago_kozi@outlook.com.gr (E.K.)
* Correspondence: gcmanos@civil.auth.gr

Abstract: Unreinforced stone masonry made of low strength mortar has been used for centuries in forming old type stone masonry churches of the "Basilica" typology. The seismic performance of such stone masonry structures damaged during recent strong seismic activity in Greece, combined with long term effects from foundation settlement, is presented and discussed. A simplified numerical process is presented for evaluating the performance of such damaged stone masonry structures, making use of linear and non-linear numerical tools and assumed limit-state failure criteria. In order to obtain a quantification of the in-plane sliding shear failure criterion, a number of stone masonry wallets were built with weak mortar and were tested in the laboratory. Through the comparison of the obtained numerical predictions with the observed structural behaviour for selected cases of stone masonry "Basilica" churches, the validity of the applied simplified numerical process is demonstrated. It is shown that reasonable approximation of the observed performance of such structures can be obtained when the assumed failure criteria are realistic.

Keywords: stone masonry; weak mortar; foundation settlement; seismic actions; observed performance; linear and non-linear numerical tools; stone masonry wallets; shear-sliding tests

---

1. Introduction

During the last fifty years, various parts of Greece have been subjected to a number of damaging earthquakes ranging from Ms = 5.2 to Ms = 7.2 on the Richter scale. Some of these events, not necessarily the most intense, occurred near urban areas [1]. One of the most demanding tasks for counteracting the consequences of all these seismic events was the effort to ensure the structural integrity of old masonry structures that had sustained considerable damage. In this framework, it was essential to study their structural system and to investigate the most significant causes of structural damage.

From such an investigation two main contributing factors have come to light. The first factor is the pre-existing state of stress and strain either from previous earthquake events and/or from long term permanent foundation settlement. The effect of foundation deformability is significant for structures that are currently designed and constructed with contemporary materials and construction techniques [2,3]. The effects of interaction between old masonry structures and deformable layers of supporting soil are far more significant. Some of the most celebrated cases are masonry towers that are inclined, like the tower of Pisa in Italy, due to soil deformability; in some cases such inclination led to total collapse. An in-depth presentation of the causes of soil settlement and its effect on old masonry structures as well as remedial measures is given by Croci [4] in the chapter with the relevant subtitle. Stone masonry bridges are another structural type that suffer from foundation settlement [5]. In this case, foundation deformability, which results from long term river flow or short term turbulent river flow from flooding, also leads to the collapse of such stone masonry structures [6,7]. The worst case scenario for the various masonry structural elements is the accumulation of stress and strain from such long term effects and the absence of any appropriate counter-measures. The second factor is a strong seismic ground motion and the earthquake forces generated by it [1,8–14].

Such seismic events result in stress demands that many times exceed the capacities of unreinforced masonry structural elements and their connections; this is more likely to occur for masonry structures which have already accumulated a considerable pre-existing state of stress and strain, as described in References [4,10,12]. The combination of these two contributing factors can lead unreinforced masonry structural members and their connection to a critical state that is accompanied by significant structural damage and partial collapse.

The combination of these two actions must be seen in a relatively wide time window when one studies their effect on Cultural Heritage Structures. M. Cerone et al. [15], by investigating the influence of the soil together with earthquake forces on the Colosseum in Rome (Italy), concluded that the registered collapses are due to the combination of soil movement together with the earthquake activity over the centuries and the continued lack of maintenance. The most detrimental state of stress is that resulting from uneven foundation settlement [4].

Heavy structural damage develops as a result of such actions because the resistance of unreinforced weak-mortar masonry to tensile or shear stress demands is rather low and is accompanied by a brittle type behaviour. In many cases, this resistance is in effect even lower than assumed due to poor maintenance conditions.

Initially, a summary of observed case studies is presented that ascertains the previously stated rationale. Next, a simplified numerical process is presented which can assist the evaluation of the performance of such stone masonry structural systems in the framework of either explaining the observed structural damage or predicting it in advance in order to prohibit its development with certain retrofitting counter-measures [16]. Summary results from specific tests are also presented towards verifying assumed in-plane sliding shear strength values.

From a variety of stone masonry structures a particular typology is selected to be investigated here. This is because its use is widespread and as such its performance has been studied by the authors for quite some time. This structural system is utilized in many Greek Christian churches, with a number of variations in plan and height [10,12], belonging to the so called "Basilica" typology which is one of the oldest structural forms.

This "Basilica" structural system is of rectangular shape, formed by relatively thick peripheral masonry walls; a semi-cylindrical apse is usually part of the East wall, whereas the interior is divided into a number of naves by longitudinal colonnades of various dimensions and shapes, as shown in Figure 1a. The roofing system develops mainly in the longitudinal direction and usually rises for the central nave at a higher level than that of the side-naves, being seen as an elevated extension of the interior colonnades.

The roofing system that covers the side naves is partially supported on the peripheral walls and is usually lower than the roofing of the central nave (Figure 1b). In some instances this structural type takes the simplest form of one nave with no internal separations.

(a) (b)

Figure 1. (**a**) The Basilica structural system with the interior colonnade of the central nave; (**b**) The Basilica structural system with the peripheral longitudinal and transverse walls.

The longitudinal and transverse walls of such structures are usually long and thick with very large in-plane stiffness. These walls are interconnected to form the main part of the total 3-D structural system, which mainly resists the horizontal earthquake forces [10,12]. Moreover, any change of shape imposed by soil-foundation settlement, because of the large in-plane wall stiffness, leads to considerable stressing of these planar structural elements and their interconnections [4]. In some cases, these planar masonry walls as well as the internal colonnades support a simple or complex relatively stiff masonry vaulting that is in turn protected by a wooden roof (Figure 2a). The soil-foundation settlement and the resulting state of stress-strain that develops on these planar masonry walls spreads also to such masonry vaulting. Therefore, the resulting structural damage develops in either the planar walls or/and the vaulting. Such relatively heavy masonry vaulting also generates large inertia forces in case of a strong earthquake ground motion, which in turn result in large tensile and shear stress demands that can be detrimental for the structural performance of the masonry planar walls and/or the masonry vaulting. Figure 2b depicts a typical damage pattern with the collapse of the central masonry dome [14].

(a) (b)

Figure 2. (**a**). Masonry vaulting; (**b**). Church of delle Anime Sante, L'Aquila earthquake 2009.

2. Summary of Observed Case Studies

2.1. Foundation Settlement

In what follows, two distinct cases of foundation settlement are presented. In the first case the uneven deformability of the soil caused considerable damage to the "Basilica" church of the Assumption of the Virgin Mary at Dilofo-Voio-Kozani, built in 1844 A.D. Its basic dimensions are 21.5 m long (together with the apse of the East wall) 11 m wide and a maximum height to the top of the wooden roof of 10 m.

A stone masonry vaulting superstructure (Figure 2a) rests on the peripheral walls as well as on internal five-column twin colonnades. The largest portion of the South-West plan of this church is founded on relatively hard soil (weathered flysch layers) whereas the North-Eastern part, from the North-East corner up to the East part of the North wall, is founded on silty clays that were deposited over the years on top of the flysch layers prior to building the church. Soft soil layers were also added to compensate for the natural slope at this location in order to form a horizontal plane to construct the foundation of the church. A number of boreholes were made close to the East (Figure 3a) and North (Figure 3b) sides; they revealed that at the North-East corner the silty clay layers extended to a depth of 7 m.

(a) (b) (c)

Figure 3. (**a**) Heavy damage of the East wall. South-East view; (**b**) Heavy damage of the North wall. North-West view; (**c**) Heavy damage of the vaulting superstructure.

The uneven settlement of the foundation of the stone masonry walls as well as of the internal columns caused considerable structural damage that progressed to the partial collapse of the whole North-East part. An external view of this church from the South-East is shown in Figure 3a. Extensive damage developed due to considerable settlement at the South-East corner. A very wide crack started off at the top of the East peripheral wall near the apse and propagated towards the bottom of this wall near the South-East corner. Furthermore, a wide crack propagates through the North peripheral wall from top to bottom as is shown in Figure 3b. It must be noted that the thickness of these masonry walls varies from 750 mm to 800 mm. From this extensive peripheral wall damage the vaulting system that is supported by these peripheral walls also suffered extensive severe cracking that eventually led to its partial collapse. The partially collapsed central dome is shown in Figure 3c.

Another case of foundation settlement of a Post-Byzantine three-nave Basilica is that of the church of Profitis Ilias at Siatista–Kozani, built in 1701 A.D. on the top of a hill. This time the damage, caused by the long term foundation settlement combined with the strong earthquake motion of May 1995, is less spectacular than that shown in Figure 3a–c. This church is a three-nave "Basilica" with a wooden roof without a stone masonry vaulting system. The horizontal dimensions of this church are 23.25 m in length and 16.60 m in width. The top of the roof lies at 7.1 m from the floor level of the interior of this church. The naves are formed by four-column twin colonnades (Figure 4a) built with stone masonry. The structural damage is in the form of inclination of the South longitudinal wall outwards, shown in Figure 4b together with its temporary wooden shoring. This is accompanied by extensive cracking at the connections of this wall with the East and West exterior masonry walls as well as with the mid-transverse wall. Cracking is also evident at the arches of the internal colonnades.

(a) (b)

Figure 4. The "Basilica" church of Profitis Ilias at Siatista – Kozani (**a**) Longitudinal cross-section; (**b**) Wooden shoring of South longitudinal wall.

2.2. Damage due to Strong Earthquake Ground Motion

Figures 5 and 6 depict the in-plane typical damage patterns of the longitudinal and transverse masonry walls for a number of stone masonry "Basilica" churches. Figures 7 and 8 depict out-of-plane typical damage patterns of the longitudinal and transverse walls for two cases of stone masonry "Basilica" churches. Initially, a simplified assumption is that the various planar and vaulting masonry structural elements are well interconnected at their intersections as well as with the wooden roof and the foundation. The simplified numerical evaluation process presented in Section 3 is based on this assumption.

However, in many cases, like those depicted in Figures 7 and 8, failure of the connection between the wooden roof and the masonry wall results in loss of support for parts of the masonry walls leading to out-of-plane partial collapse. Therefore, it is realistic to consider the effect of limit stage conditions at the interconnection of the various masonry structural elements either at the corners of longitudinal and transverse planar walls or at locations where planar walls are joined with the wooden roof or the soil-foundation interface. This is studied in Section 5 utilizing the strength values listed in Table 2.

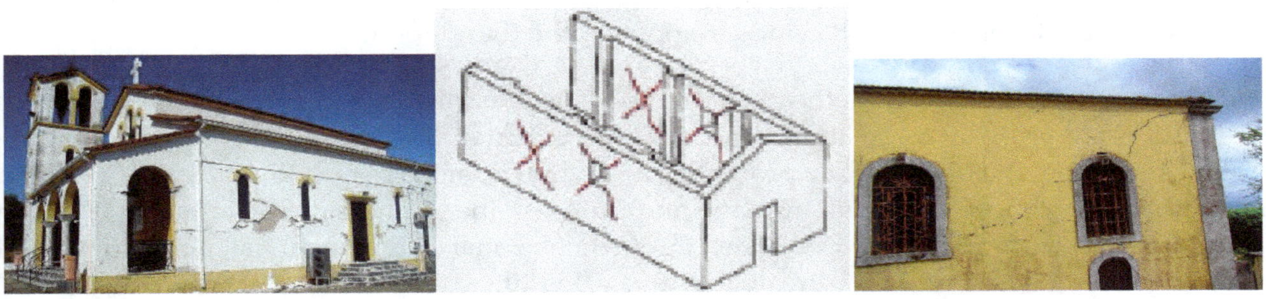

Patras-Greece Earthquake 2008 Kefalonia-Greece Earthquake 2014

Figure 5. Typical damage patterns of longitudinal walls.

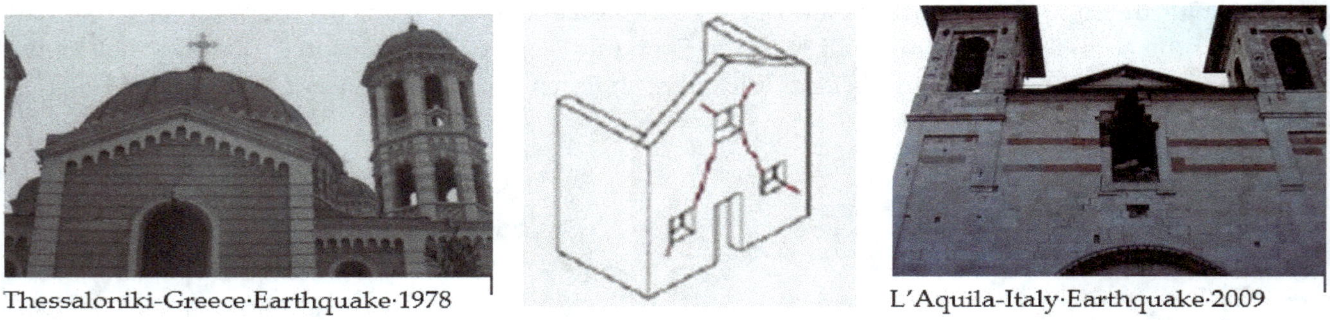

Figure 6. Typical in-plane damage of transverse wall.

Figure 7. Typical out-of-plane damage of longitudinal wall. Kozani-Greece Earthquake 1995.

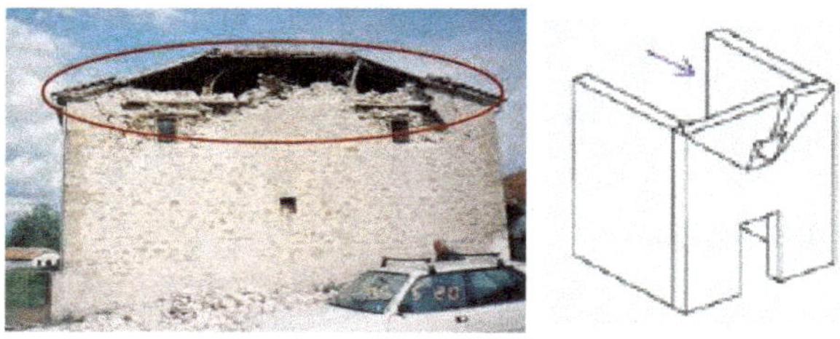

Figure 8. Typical out-of-plane damage of longitudinal wall. Kozani-Greece Earthquake 1995.

3. Simplified Numerical Evaluation Process Assuming Non-Failing Masonry Wall Inter-Connections

In evaluating the dynamic and earthquake response of such masonry structures it is initially assumed that the various planar and vaulting masonry structural elements are well interconnected at their intersections as well as with the wooden roof and the foundation.

At this *"first stage"* evaluation, the performance of each particular structural element is assessed individually, assuming that these interconnections are withstanding the imposed demands without any form of damage. Assumed elastic properties are adopted for each individual masonry structural element in order to approximate its in-plane and out-of-plane stiffness characteristics. The actual main architectural features are used to form a three dimensional (3-D) numerical model of the whole structural system, as shown in Figures 2a and 9. This linear elastic numerical model is further simplified by utilizing shell elements for numerically simulating each masonry structural element thus approximating the in-plane and flexural (out-of-plane) stiffness without having to actually portray the masonry structural thickness in this numerical approximation.

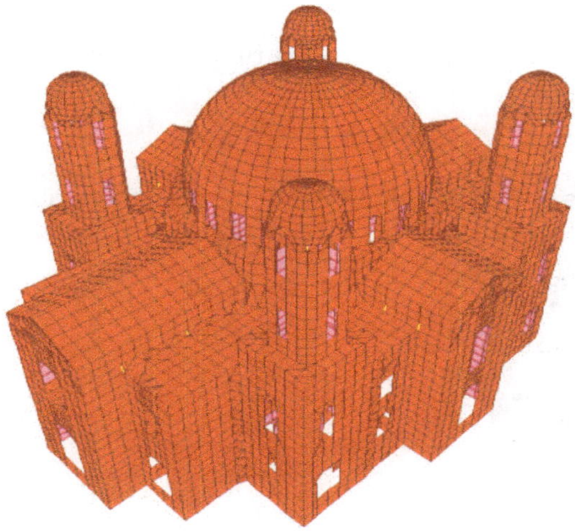

Figure 9. Formation of the 3-D numerical approximation.

Towards this objective, appropriate software packages are utilized in order to form this 3-D numerical model [17,18]. It is important at this stage to carefully check all the intersections in order to ensure that there is compatibility in the finite element representation of the 3-D actual structural system, despite the simplification introduced by the use of shell finite elements. The deformability of the foundation is also approximated in two different ways. First, deformable supports are placed under the foundation; these supports have elastic properties equivalent to the deformability properties of the underlying soil layers. Alternatively, layers of deformable soil are used and placed under the foundation of this numerical simulation. Manos and Kozikiopoulos [19] utilized in-situ measurements from a bell tower in Kefalonia island in order to approximate the stiffness characteristics of the underlying soil layers. Moreover, a 3-D finite element approximation of flexible soil layers was utilized to obtain the stiffness properties of equivalent link elements used to replace the soil layer in a 3-D finite element representation of the structure together with its flexible foundation. This 3-D numerical approximation is subjected to a variety of load combinations that include the gravitational forces as well as snow or earthquake loads, described by relevant design provisions. The outcome of such a numerical study is deformation and stress demands S_{Ed} for each masonry structural element. These stress demand values (S_{Ed}) are next utilized together with corresponding capacity values S_{Rd} obtained on the basis of assumed strength values for the stone masonry for the studied churches. A set of such assumed strength values are listed in Table 2. In order to utilize current provisions for the design of masonry structural elements the numerically obtained deformation and stress demands for each masonry structural element is uncoupled into its in-plane and out-of-plane part. One of the main difficulties in assessing the capacity values for old stone masonry construction is the lack of experimentally verified strength values. In order to partially overcome this difficulty a number of specimens (Table 1 column 1) were built employing irregular stones with a cubic compressive strength of 60 MPa and low strength mortar with a mean cubic compressive strength equal to 0.85 MPa. The mortar joints were relatively thick (approximately 25 mm). These specimens were approximately 370 mm by 270 mm in plan and 270 mm height. Each specimen was placed in a testing rig hosting a vertical jack with a load cell and a flat sliding bearing resting at the top surface of each specimen; each specimen had its bottom part securely fixed as shown in Figure 10a,b. In addition, a horizontal actuator was securely attached at the top part of each specimen in order to apply a horizontal load in a gradually increasing manner, keeping at the same time the vertical load constant at a predetermined level. The aim of this experimental setup was to force each specimen to fail in an almost horizontal sliding mode at an equivalent mortar joint located between its top and bottom part, as is shown in Figure 10b. The final objective of this experimental sequence was to be able to quantify the shear strength against the sliding mode of failure (f_{vk}) through the parameters included in a "Mohr-Coulomb" shear strength criterion as is expressed by Equation (1). The shear strength of the stone masonry when the normal stress is zero is denoted by f_{vko}. The compressive axial stress acting on the bed joint is denoted by σ_n and μ is an assumed value for the static coefficient of friction.

$$f_{vk} = f_{vko} + \mu\,\sigma_n \quad (1)$$

Table 1. Comparison between measured and predicted shear strength values.

Code Name of Tested Specimen	Measured Value f_{vk} (Mpa)	Applied Level of Normal Stress σ_n (Mpa)	Predicted f_{vk} (Mpa)	Ratio Measured f_{vk}/Predicted f_{vk}
(1)	(2)	(3)	(4)	(5)
Sample 1	0.396	0.53	0.359	1.103
Sample 2	0.41	0.61	0.395	1.038
Sample 3	0.305	0.46	0.327	0.933
Sample 4α	0.20	0.30	0.255	0.784
Sample 4β	0.375	0.54	0.363	1.033

(a) (b)

Figure 10. (**a**). Short stone masonry specimen subjected to sliding shear; (**b**) Sliding mode of failure of short stone masonry specimen.

Table 1 (column 2) lists the measured shear strength values together with the corresponding values of the compressive stress normal to the equivalent bed-joint (σ_n) applied during testing (Table 1 column 3). Employing formula 1 with values f_{vko} = 0.12 MPa and μ = 0.45 the predicted sliding shear strength values are found, listed in Table 1 column 4. Reasonably good agreement is obtained between measured and predicted sliding shear strength, as is indicated by the relevant ratio of measured over predicted sliding shear strength with values listed in column 5 of the same Table.

The in-plane shear capacity of masonry structural elements based on such a "Mohr-Coulomb" failure envelope, as defined through Equation (1) with a normal stress (σ_n) acting simultaneously, is also employed by Euro-Code 6 [20]. In this case, the value of the static friction coefficient is assumed to be equal to 0.4. The strength values listed in Table 2 are based on the Euro-Code 6 shear strength envelope, assuming values of f_{vko} = 0.16 MPa and of μ = 0.4 (Table 2, column 1). Similarly, low strength values were assumed for the tensile strength normal (f_{xk1}) and parallel (f_{xk2}) to an equivalent horizontal joint (Table 2, column 2). The shear capacity defined in this way corresponds to the mechanism resisting the sliding mode of failure. Tomazevic [21] proposed a procedure, developed by Turnsek and Cacovic [22], towards estimating the shear capacity corresponding to the mechanism resisting the diagonal tension mode of failure for a masonry structural element having a height (h) and a length (l). In this case the shear strength ($\tau_{\max}$) is given by the following relationship.

$$\tau_{\max} = f_{xk1}\sqrt{(f_{xk1} + \sigma_n)}/b \text{ where : } b = h/l \quad (2)$$

Table 2. Assumed Mechanical Characteristics of the Stone Masonry in N/mm^2 (MPa).

Shear Strength f_{vko} for Zero (σ_n) Normal Stress (MPa)	Tensile Strength Normal (f_{xk1})/Parallel (f_{xk2}) to Bed-Joint (MPa)	Compressive Strength f_k (MPa)	Young's Modulus E (MPa)	Poisson's Ratio
(1)	(2)	(3)	(4)	(5)
0.16	0.15/0.60	3.50	1000	0.2

The quality of the used stones may vary, therefore the stone compressive values, listed below, are indicative. It is frequently reported that the variability of the quality of mortar as well as that of the techniques used in the stone masonry construction are far more important in influencing the strength values that are most significant in defining the earthquake capacity of the various structural elements. A number of destructive and non-destructive techniques have been employed in the past in the framework of investigation procedures for the diagnosis of historic masonries [23]. However, most of these procedures are of a qualitative nature. The determination of the most significant strength values for particular masonry construction is quite a demanding task to be performed in-situ, which is fulfilled only in limited cases. As an alternative, one can use relevant information from controlled laboratory experiments, like those performed by Vintzileou [24]. A number of stone masonry wallets were built [24] with low strength lime mortar (mortar compressive strength 0.80 MPa, stone compressive strength 50 MPa), including timber ties connected within these masonry wallets in various ways. The resulting compressive strength of the wallet without timber ties was found equal to 0.47 MPa. In the framework of ongoing research, stone masonry wallets of similar dimension were built at Aristotle University, having a cross section 500 mm × 600 mm and a height of 830 mm also with low strength mortar (mortar cubic compressive strength 1.10 MPa, stone cubic compressive strength 60 MPa); this resulted in wallet compressive strength values equal to 1.5 MPa. In this experimental sequence a number of wallets, which included timber ties, were also tested. These wallets including the wooden ties, when tested in compression, resulted in a moderate increase (10% to 20%) of the initially measured compressive strength.

It was observed during the stone masonry wallet compressive tests conducted at Aristotle University that a moderate increase in the mortar compressive strength with the addition of pozzolan in the mortar (mortar compressive strength 1.28 MPa) resulted in a threefold increase of the compressive strength of the corresponding stone masonry wallet (approximately 5.0 MPa). Therefore, the compressive strength value of 3.50 MPa adopted here, listed in column 3 of Table 2, seems to be a reasonable assumption. In any case, in all the examined structures the compressive limit-state scenario is very remote. In columns 4 and 5 of Table 2 the adopted values of Young's modulus and Poisson's ratio are also listed. Apart from the wooden inserts that are included in stone masonry walls the influence of spandrels that bridge the door and window openings must also be briefly discussed. Past research demonstrated that spandrels can have a significant influence on the capacity of masonry walls [25–28]. Spandrels are constructed in a variety of forms employing as basic materials in old stone masonry construction: masonry, wooden or even iron parts. Due to this variety of materials and construction techniques and the lack of measured strength values for the structures studied in the present work this influence, although important, is not investigated in any detail. In order to estimate the performance of each masonry structural element the following inequalities are employed:

$$R_i = S_{Rdi}/S_{Edi} > 1 \tag{3}$$

$$R_i = S_{Rdi}/S_{Edi} < 1 \tag{4}$$

S_{Edi} represents the demand posed for each masonry structural element as it results from the simplified numerical simulation; S_{Rdi} is the corresponding capacity value which is obtained on the basis of assumed strength values for the stone masonry (Table 2). Note that no safety coefficients are used in estimating the capacity values at this stage of the evaluation process. Inequality 3 signifies safe structural performance. Inequality 4 denotes that the predicted structural performance exceeds a certain limit state thus signifying the development of structural damage corresponding to the specific limit state that is exceeded. These corresponding capacity over demand ratio ($\boldsymbol{R_i}$) values are used in this simplified numerical evaluation process; a ratio ($\boldsymbol{R_i}$) value smaller than 1 indicates that a distinct limit state has been reached leading to the corresponding failure mode. The following five common structural damage scenarios are stated corresponding to five distinct relevant limit-states through the relevant ratio values ($\boldsymbol{R_i = S_{Rdi}/S_{Edi}}$). Scenario (**a₁**) addresses the in-plane shear limit state which corresponds to a sliding failure mode through the value of the ratio ($\boldsymbol{R_{\tau sli}}$); scenario (**a₂**) addresses the in-plane shear limit state corresponding to a diagonal tension failure mode ($\boldsymbol{R_{\tau dia}}$). Scenario (**b**) corresponds to a compressive mode of failure ($\boldsymbol{R_{\varsigma}}$) whereas scenario (**c**) corresponds to the in-plane tensile limit state ($\boldsymbol{R_{\sigma}}$). Finally, scenario (**d**) corresponds to the out-of-plane tensile limit state ($\boldsymbol{R_M}$). Both scenario (**c**) and scenario (**d**) use the $\boldsymbol{f_{xk1}}$ strength value, listed in Table 2 column 2.

(**$\mathbf{a_1}$**) $\boldsymbol{R_{\tau sli}}$ = shear strength/shear stress demand. $\boldsymbol{R_{\tau sli}} < 1$ signifies in-plane sliding shear mode of failure

(**$\mathbf{a_2}$**) $\boldsymbol{R_{\tau dia}}$ = shear strength/shear stress demand. $\boldsymbol{R_{\tau dia}} < 1$ signifies in-plane diagonal tension mode of failure.

(**b**) $\boldsymbol{R_{\varsigma}}$ = compressive strength/compression stress demand. $\boldsymbol{R_{\varsigma}} < 1$ signifies in-plane compression mode of failure.

(**c**) $\boldsymbol{R_{\sigma}}$ = tensile strength/tensile stress demand. $\boldsymbol{R_{\sigma}} < 1$ signifies tensile mode of failure normal to bed joint (in-plane)

(**d**) $\boldsymbol{R_M}$ = tensile strength/tensile stress demand from out-of-plane flexure. $\boldsymbol{R_M} < 1$ signifies out-of-plane tensile mode of failure normal to bed joint at the extreme fibre.

All masonry parts of the studied structures were examined in terms of in-plane and out-of-plane stress demands posed by the applied load combinations against the corresponding capacities, as these capacities were obtained by applying the "Mohr-Coulomb" criterion of Equation (1) or the stone masonry compressive and tensile strength limits listed in Table 2. Ratio values smaller than one ($\boldsymbol{R_{\tau sli}}$, $\boldsymbol{R_{\tau dia}}$, $\boldsymbol{R_{\varsigma}}$, $\boldsymbol{R_{\sigma}}$, $\boldsymbol{R_M} < 1$) predict the corresponding limit state condition. As can be seen, this methodology is based on combining numerical stress demands resulting from elastic analyses with limit-state strength values. An alternative approach is to incorporate these limit-state strength values in a non-linear push-over type of analysis [29]. As was shown in this study by Manos et al. [29] the above linear-elastic approach is a reasonable approximation of the actual behaviour and of predicting regions of structural damage, being both less complex and time consuming than the corresponding non-linear approach. Manos et al. [30,31] developed a relevant expert system for assessing the various resisting capacities of vertical masonry structural elements.

4. Results from Distinct Case Studies and Discussion

4.1. Numerical Simulation of the "Basilica" Church of Assumption of the Virgin Mary at Dilofo-Voio-Kozani

Summary results of the numerical simulation of the "Basilica" church of the Assumption of Virgin Mary at Dilofo-Voio-Kozani (Section 2, Figure 3a–c), damaged by foundation settlement are presented here. The assumptions presented in Section 3 were followed in forming this numerical model. Two different cases were simulated. In the first case the soil-foundation interface was assumed to be non-deformable (Figure 11a,b) whereas in the second case a deformable soil layer was placed under the foundation exactly at the same location as was identified from the in-situ examination (Figure 12a,b).

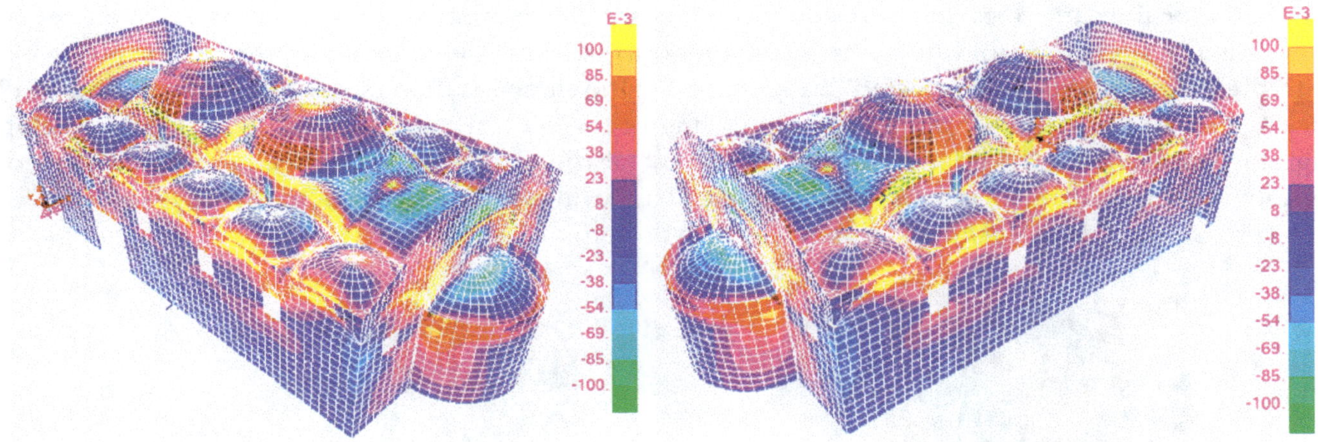

(**a**) σ_{max} = 0.28 MPa (**b**) σ_{max} = 0.28 MPa

Figure 11. Non-deformable foundation (**a**) distribution of maximum tensile stresses South-East view (**b**) distribution of maximum tensile stresses North-East view. The scale at the far right of each plott indicates that the colours used to represent stress values in this numerical representation of the structure are ranging from 0.1 MPa to −0.1 MPa. Yellow colour indicates stress values excedding 0.1MPa.

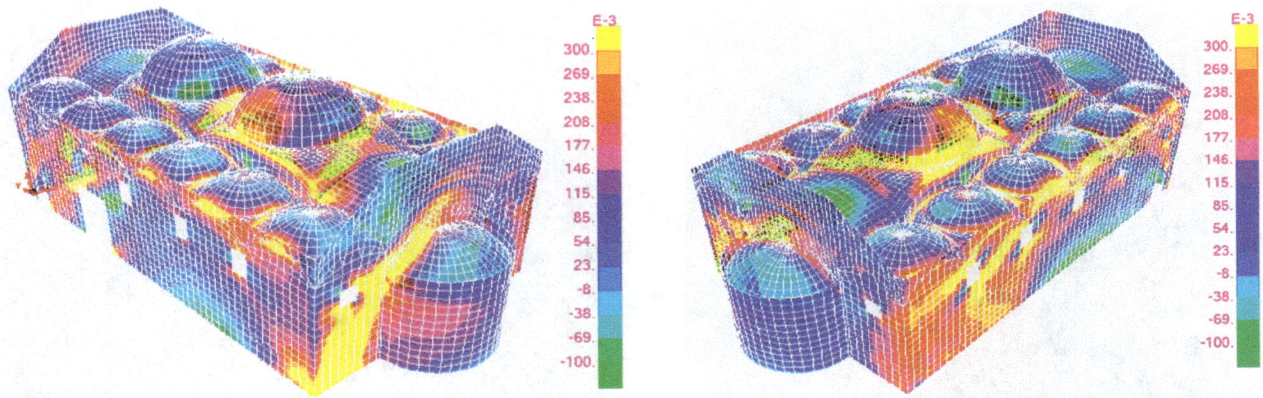

(**a**) σ_{max} = 0.93 MPa (**b**) σ_{max} = 0.99 MPa

Figure 12. Deformable foundation (**a**) distribution of maximum tensile stresses South-East view (**b**) distribution of maximum tensile stresses North-East view. The scale at the far right of each plott indicates that the colours used to represent stress values in this numerical representation of the structure are ranging from 0.1 MPa to −0.1 MPa. Yellow colour indicates stress values excedding 0.3 MPa.

The in-plane maximum tensile stress distribution is shown in these figures. Figures 11a and 12a represent a South-East view of the maximum tensile stress response that can be compared to the actual structural damage shown in Figure 3a whereas Figures 11b and 12b represent a North-East view of the tensile stress response which can be similarly compared to the damage shown in Figure 3b,c. The maximum in-plane tensile stress demands appear at the regions supporting the central vertical domes and have maximum values 0.28 MPa in the case of non-deformable foundation and 0.99 MPa in the case of the deformable foundation, respectively. As can be seen, this maximum in-plane tensile stress value in the case of deformable foundation exceeds by far the tensile strength value adopted in this study and listed in column 2 of Table 2 (0.15 MPa). Therefore, in this latter case the simplified numerical process indicates that scenario c (exceeding the tensile strength) occurs in these locations, thus predicting the development of the relevant structural damage, which correlates reasonably well with observed performance.

In more detail, a comparison of the tensile stress distribution pattern at the South-East corner of the transverse wall shown in Figure 12a correlates well with the observed damage depicted in

Figure 3a for the same location (see also top of Figure 13). Similarly, the maximum in-plane tensile stress distribution at the middle of the North longitudinal wall, depicted in Figure 12b correlates well with the observed damage depicted in Figure 3b for the same location (see also middle of Figure 13). Finally, the area surrounding the supporting ring of the central dome (Figure 12a,b) also develops high in-plane tensile stress demands that exceed by far the assumed strength values which leads to a reasonably good correlation with the extensive structural damage which actually developed in this part (Figure 3c) of this "Basilica" church (see also bottom of Figure 13).

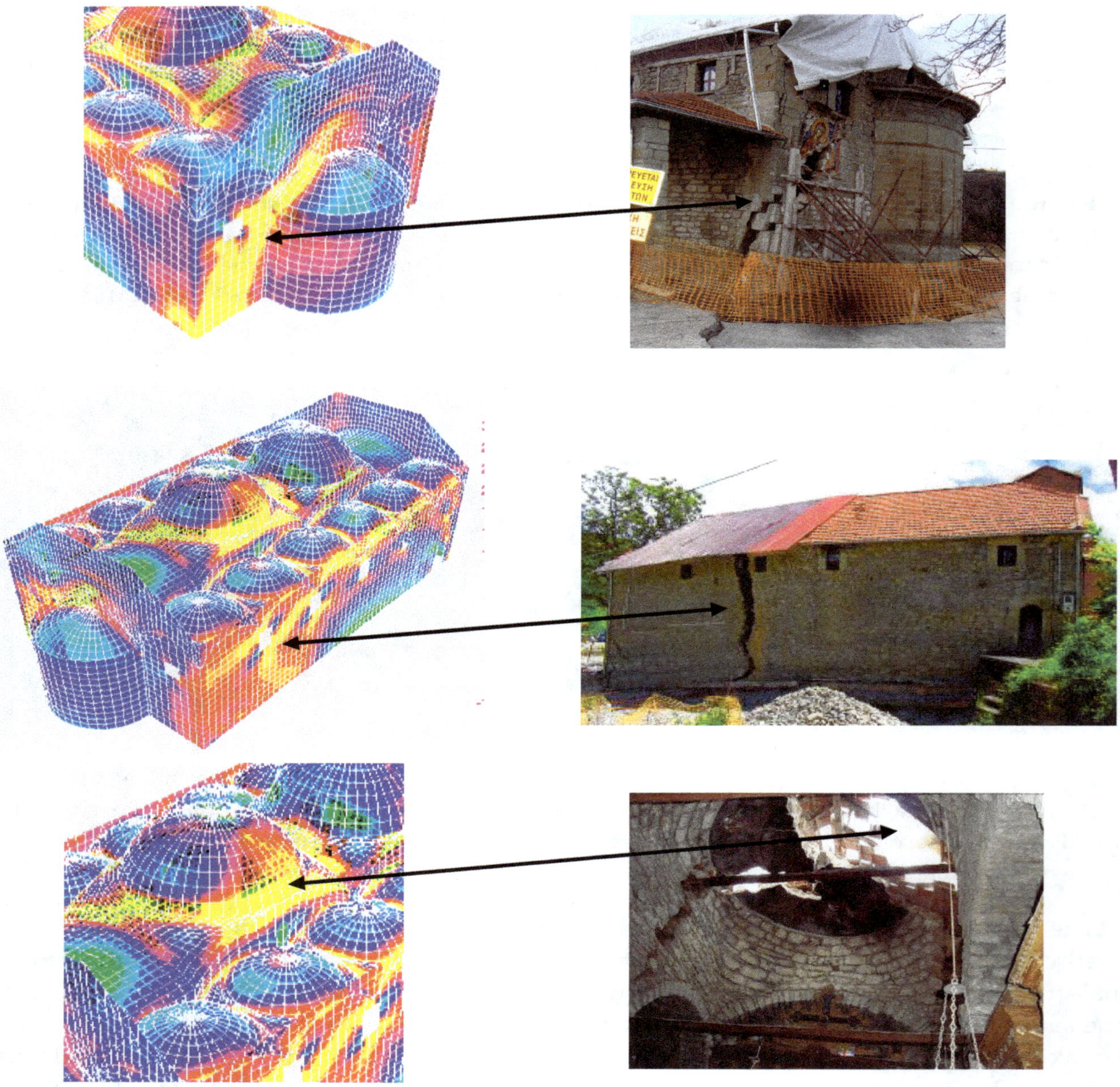

Figure 13. Correlation between predicted (deformable foundation) and observed damage.

4.2. Numerical Simulation of the Dynamic and Earthquake Response of "Basilica" Churches

The dynamic and earthquake response of the 17th century "Basilica" church of St. Marina in Soullaroi, which is located in the island of Kefalonia and was heavily damaged during the 2014 earthquake sequence [11,12], is studied here. A linear elastic numerical simulation was formed as previously described. Thick shell finite elements [17] were employed to numerically simulate all the stone masonry structural elements having a thickness of 750 mm and an assumed Young's Modulus of

1000 MPa (Table 3); 7160 finite elements were employed in total for this numerical simulation with dimensions approximately 300 mm × 300 mm. The foundation deformability was introduced with linear links [17] having axial stiffness equal to either 109 KN/mm, representing relatively hard soil conditions or 24.5 KN/mm, representing a moderately deformable foundation. Moreover, in all cases the vertical walls were connected at the corners with two-node 3-D links in an effort to control the rigidity of these connections as well as to approximate the structural behaviour when the examined structures exhibited heavy damage in these locations (see Section 5). The eigen-periods for the N-S transverse direction translational response are equal to 0.177 s and 0.20 s for the hard and medium soil deformability, respectively. Similarly, the eigen-periods for the E-W longitudinal direction translational response are equal to 0.10 s and 0.12 s for the hard and medium soil deformability, respectively (Figure 14). More information on the dynamic properties of this numerical simulation is given by Manos et al. [12]. The numerical dynamic analyses included the superposition of the gravitational forces with the seismic forces specified on the basis of the ground acceleration that was recorded at two stations during the 3rd February strongest aftershock GEER-EERI-ATC [11], Papaioannou [32]. More information on the location of these recordings in relation to the studied "Basilica" church is given in Manos et al. [12]. The corresponding response curves for the Chavriata recording which is more demanding than the Lixouri ground motion, are shown in Figure 15a,b for the North-South and East-West directions, respectively.

Table 3. Base shear Values (KN) based on the Chavriata response spectra (which is the most demanding).

Studied Church/Soil Conditions	N-S (Transverse x-x)	E-W (Longitudinal y-y)
St. Marina Soullaroi/Hard Soil	5229	8397
St. Marina Soullaroi/Soft Soil	5803	8828

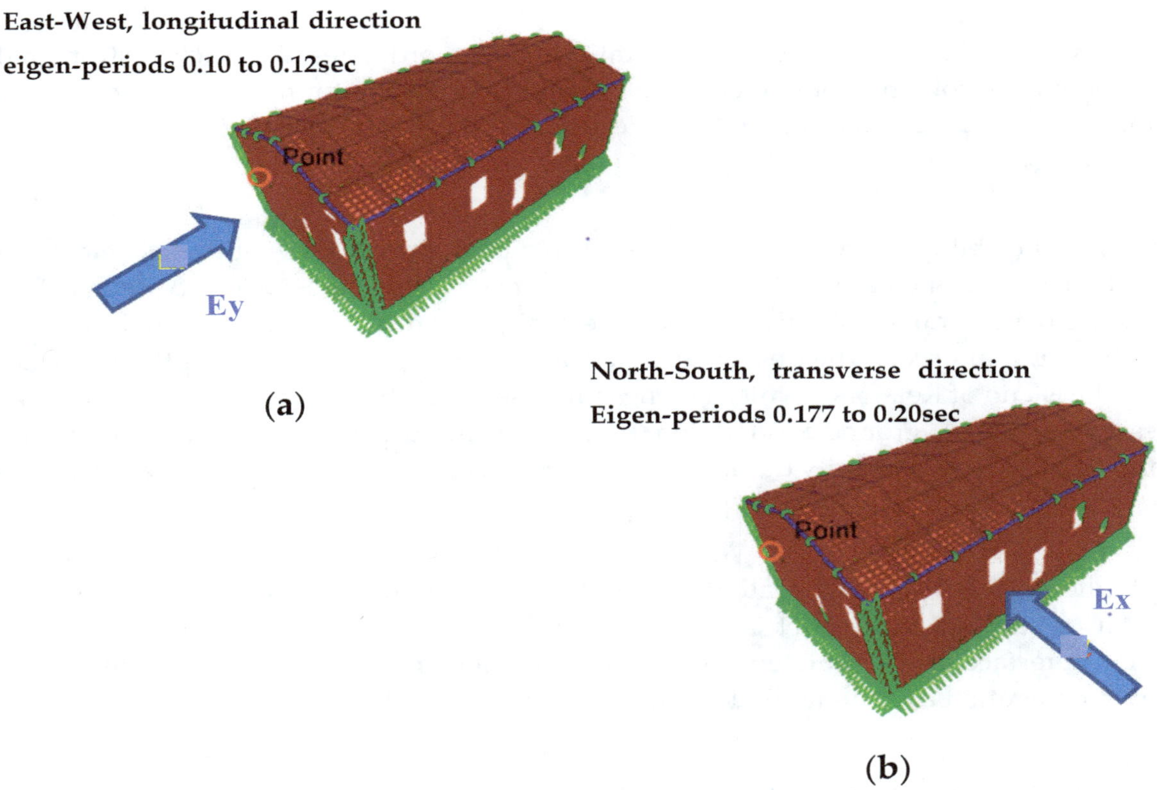

Figure 14. The numerical simulation of the "Basilica" church of St. Marina in Soullaroi in Kefalonia. (**a**) Longitudinal direction; (**b**) Transverse direction.

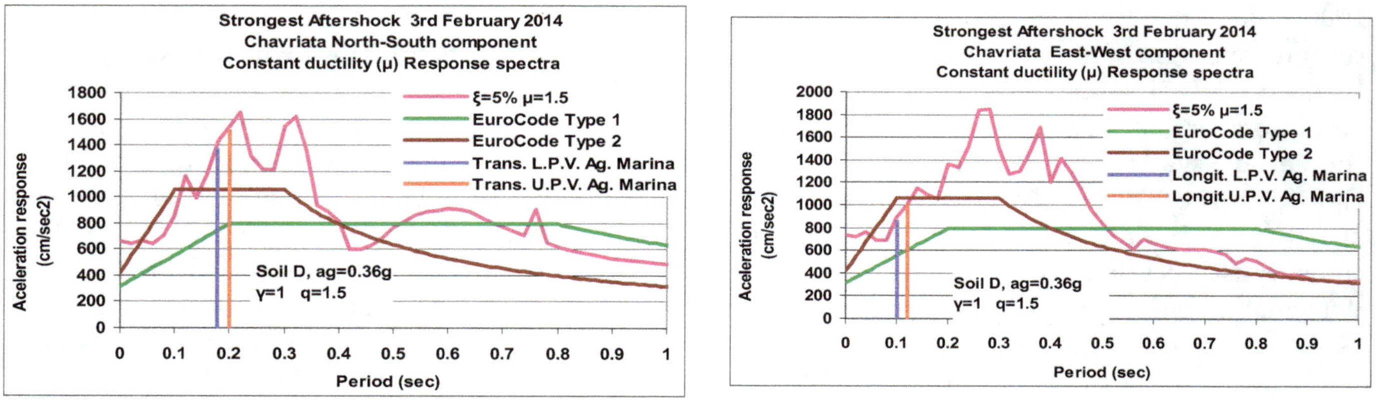

(a) (b)

Figure 15. Constant ductility response spectral curves of the Chavriata record, Kefalonia 2014 earthquake together with the relevant Euro-Code 8 design spectral curves: (**a**) N-S component (**b**) E-W component.

In each one of these figures the relevant constant ductility spectral curves are depicted for a ductility ratio equal to 1.5, which is assumed to be valid for unreinforced masonry structures. Moreover, the Type-1 and Type-2 design spectral curves specified according to Euro-Code 8 [33–35] for design ground acceleration equal to 0.36 g (g is the acceleration of gravity), importance factor equal to $\gamma = 1$, response modification coefficient $q = 1.5$ (for unreinforced masonry) and soil category D. Finally, in each one of these figures the eigen-period values of the main translational eigen-modes in the North-South and East-West direction, as specified above, are also indicated.

As can be seen from Figure 14a,b the acceleration spectral ordinates of the actual strong motion in the North-South direction are more demanding (transverse direction for the church, approximately 1.5 g) than the corresponding values in the East-West direction (longitudinal direction for the church, approximately 1.0 g).

Moreover, the flexibility of the foundation results in a moderate increase of the seismic demands resulting from the similar increase of the corresponding spectral ordinates, as can be seen in Table 3. Such large values of spectral acceleration result in very high seismic load values, listed in Table 3 in terms of base shear, that exceed the design seismic loads resulting from applying the Euro-Code 8 provisions for the island of Kefalonia, which belongs to the seismic zone of Greece [35] with the highest expected design earthquake ground acceleration (0.36 g). It is no surprising that this old unreinforced stone masonry structure was heavily damaged as was the case for a number of similar structures in nearby locations Manos et al. [12].

In what follows, the simplified evaluation process described in Section 3 was followed for the case of St. Marina in Soullaroi church at the island of Kefalonia, Greece, built in 1686 A.D. The soil-foundation interface was simulated with linear link elements having axial stiffness equal to 24.5 KN/mm, representing a moderately deformable foundation [12,19].

Figure 16 depicts summary results of this evaluation process in terms of R_τ, R_σ, R_M ratio values. These ratio values for either the sliding shear ($R_{\tau sli}$,) or the diagonal tension ($R_{\tau dia}$) are indicated with the common ratio value R_τ. Ratio R_ς values, indicating in-plane compressive limit-state, are not shown as this mode of failure is not reached anywhere in this church.

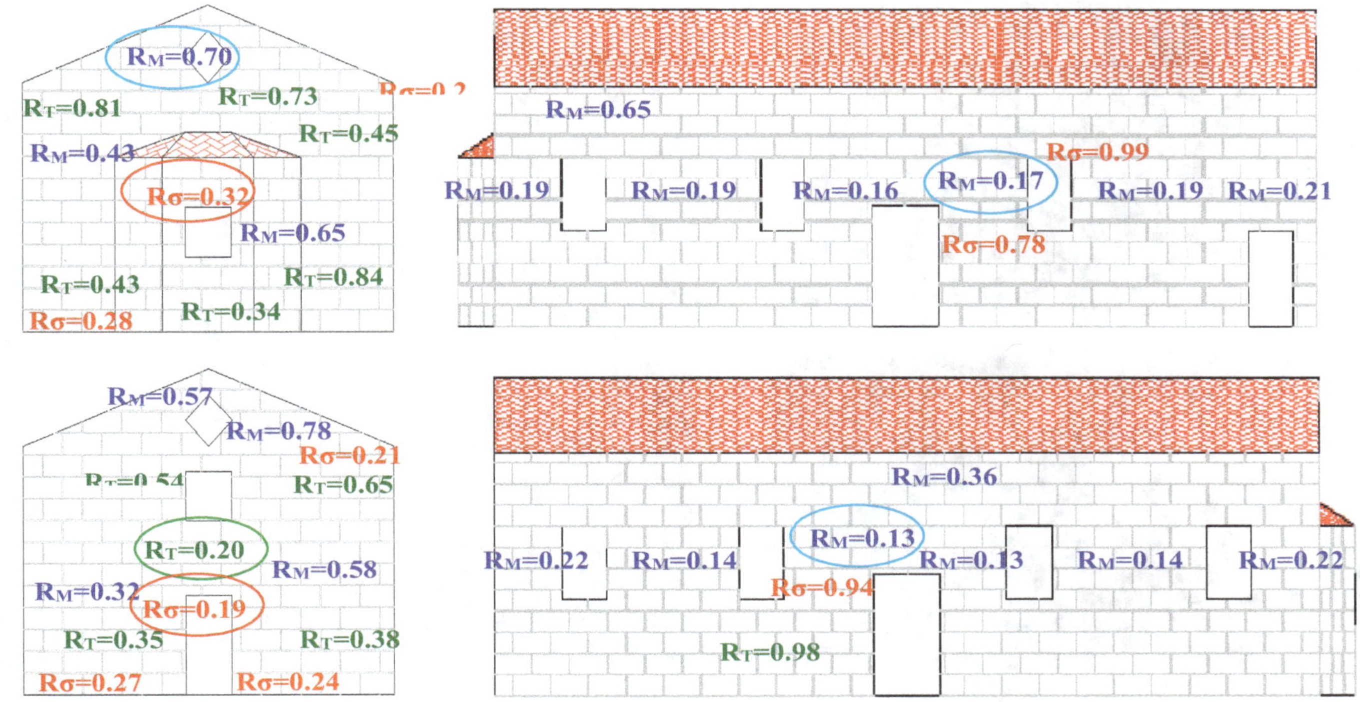

Figure 16. Summary results of the evaluation process in terms of $R\tau$, $R\sigma$, R_M ratio values, for moderately deformable foundation (axial stiffness value of supporting links equal to 24.5 KN/mm).

The following observations can be made on the basis of the R_τ, R_σ, R_M ratio values. It can be seen that all these ratio values are smaller than one (R_τ, R_σ, $R_M < 1$) in numerous locations, indicating that the corresponding limit state has been reached at all these locations of the structure that the relevant ratio value is linked with. For clarity the whole structure is deconstructed part by part in these figures in four walls that form its 3-D stone masonry shell. The East transverse wall is placed at the top left corner of Figure 16 whereas the West transverse wall at the bottom left corner. The South longitudinal wall is placed at the top right corner of Figure 16 whereas the North longitudinal wall at the bottom right corner.

The results of the evaluation include all possible combinations of the gravitational forces with the seismic actions indicated in Figure 14. As can be seen in Figure 15 the R_M ratio values are well below one ($R_M < 1$) mainly for both North and South longitudinal walls indicating that these walls for this church reached a widespread out-of-plane flexural limit state.

This is confirmed by the observed damage, which was also very widespread. In order to avoid partial or total collapse, temporary scaffolding was installed very shortly after the severity of the sustained damage was realized. Furthermore, this evaluation process indicates that the East and West transverse walls reached at their bottom part in-plane tensile and shear limit states with the corresponding ratio values well below one (R_σ <1, R_τ < 1) whereas at the top part of these walls the development of the out-of-plane flexure limit state prevails (R_M < 1).

Again, reasonably good correlation can be seen between predicted limit states, based on the simplified evaluation results of Figure 16, with the observed damage shown in Figure 17. As already discussed in Sections 2 and 3, this simplified numerical evaluation process assumes non-failing inter-connections of masonry elements. Therefore, one could neither predict the severity of the structural damage which developed near the inter-connections between the transverse and the longitudinal walls (Figure 18) nor account for the effect of this type of damage to the state of stress of each individual wall. An effort to study this effect is presented in the following Section 5.

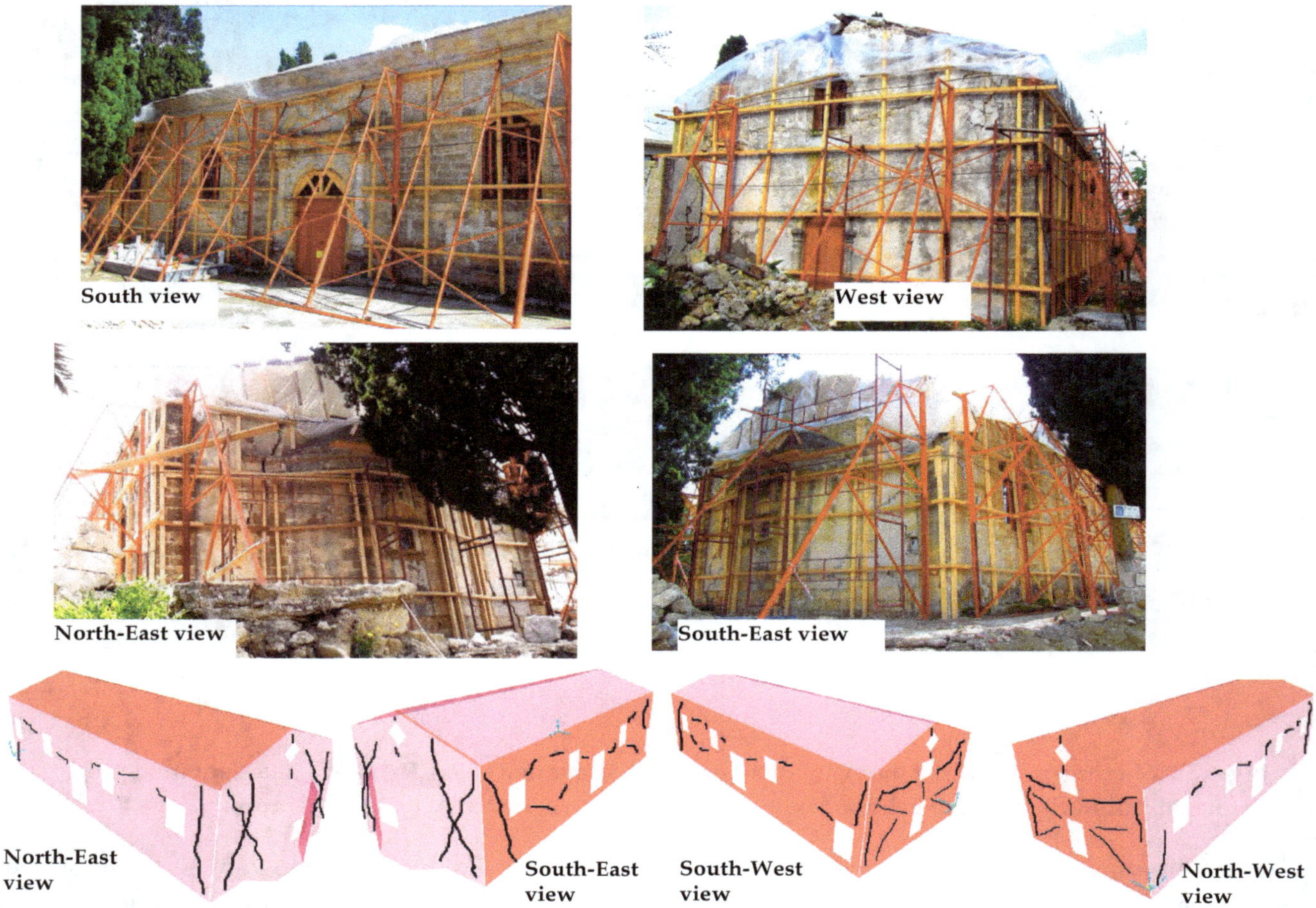

Figure 17. Photographs and sketches of the most severe structural damage.

Figure 18. Damage at the N-W corner.

5. Numerical Simulations of the Seismic Performance Including Non-Linear Response Mechanisms

In this section numerical simulations which include specific non-linear response mechanisms will be presented and discussed. The non-linear mechanisms which are introduced in the numerical simulation of a stone masonry structure of the "Basilica" typology are the following:

(a1) First, the two-node 3-D links at the soil-foundation interface, introduced to account for the soil-foundation deformability, are provided with a tension cut-off limit so they can sustain only compression in their axial direction and no tension (Figure 19a).

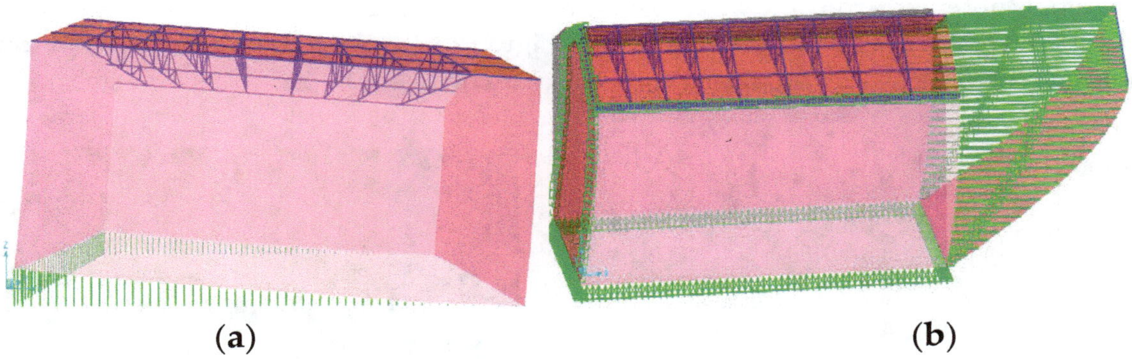

Figure 19. (**a**) Non-linear links to simulate uplifting of the structure at the soil-foundation interface; (**b**) Non-linear links to simulate detachment between the roof and the masonry walls or between vertical walls at the corners.

(b1) Similarly, two-node 3-D non-linear link elements are also utilized in connecting the vertical walls at their corners with a tension cut-off limit in such a way as to transfer compression between the intersecting walls at the corner but limited transfer of tension (Figure 19b).

(c1) A similar connection is used between the wooden elements of the roof and the tympana of the masonry walls at the East and West sides or at the top of the North or South longitudinal masonry walls.

The value of each of these tension cut-off limits was based on the assumed in-plane shear and tensile strength values listed in Table 2 and on the relevant contact surface corresponding to each non-linear link. It must be recognized that this quantification process entails a considerable degree of uncertainty that must be investigated further. Thus, through the proper use of these non-linear 3-D two-node link elements, these non-linear mechanisms are introduced in an effort to simulate numerically the capability of the uplifting of the structure at the soil-foundation interface, the possibility of detachment between the wooden elements of the roof and the masonry walls or between the transverse from the longitudinal walls at the connecting corners (Figures 17 and 18). This alternative approach adopted here employs such realistic multiple selective failure mode scenarios in either the longitudinal or the transverse direction finding for each case a base shear capacity value rather than following the spectrum method. This push-over type of non-linear analysis was applied to the 3-D numerical simulation of a typical stone masonry church with overall dimensions that are representative of numerous other churches in Kefalonia island. These non-linear numerical simulations use a step-by-step push-over type of analysis Manos et al. [14] with the first step being the application of all the permanent vertical loads (100% D). During subsequent time steps the horizontal seismic loads are gradually introduced at each shell element (corresponding to its mass) either in the North-South (y-y, Ey Figure 14a) or East-West (x-x, Ex Figure 14b) directions. This is done gradually by increasing the level of the applied seismic forces by a small amount at each subsequent step. The seismic horizontal displacement attained at each step is checked at the top point of the tympanum of the East wall for the East-West push-over analyses or at the top of the middle of the North wall (where the roof is connected) for the push-over analyses in the North-South direction. The analysis is stopped when a target horizontal displacement is reached.

5.1. Foundation Uplift

The base shear versus horizontal displacement response obtained from these push-over numerical analyses is depicted in Figure 20a,b for the East-West (D + Ey) and North-South (D + Ex) directions, respectively. The difference between the linear response (plotted with the straight lines) and the corresponding non-linear response, resulting from these push-over analyses, becomes evident in these figures. As can be seen in the uplifting of the foundation occurs when the base shear reaches approximately 20,000 KN (Figure 20a) in the East-West and 11,000 KN (Figure 20b) in the North-South direction, respectively. When these base shear values are compared with the corresponding values listed in Table 3, obtained from the linear dynamic spectral analyses that employed the recorded ground motion response spectra (Chavriata record, Figure 15a,b), it can be concluded that foundation uplift cannot develop for this particular "Basilicas" church.

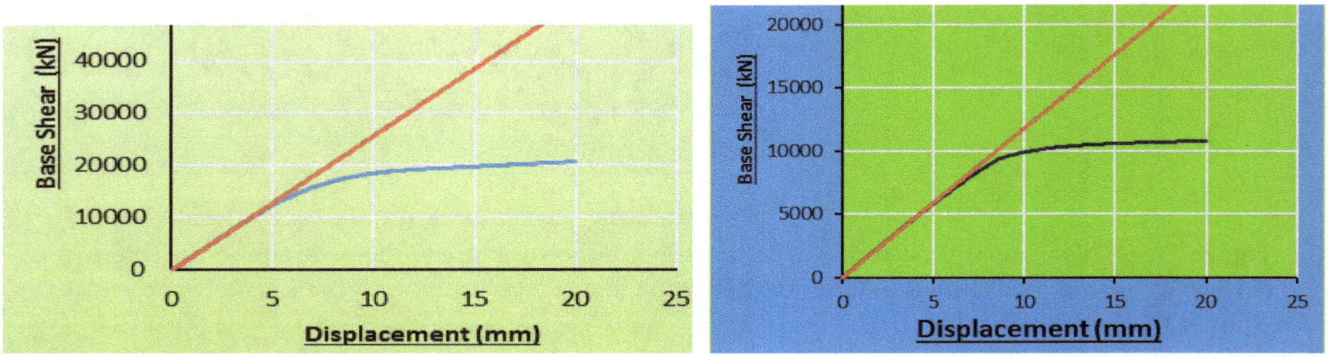

(a) (b)

Figure 20. (**a**) Push-over in the East-West direction (Ey), foundation uplift; (**b**) Push-over in the North-South direction (Ex), foundation uplift.

5.2. Non-Linear Response Simulating the Detachment of the Masonry Walls from the Roof Level as well as at Their Corner Interconnection

Next, the potential of the walls to be detached at the corners where these vertical walls are interconnected as well as at the roof level is examined. This is examined for load combinations D + Ey or D + Ex, whereby the seismic forces Ey (East-West) or Ex (North-South) are applied in a push-over type of non-linear analyses. The obtained base shear versus horizontal displacement response from these push-over numerical analyses is depicted in Figure 21a,b for the East-West (D + Ey) and North-South (D + Ex) directions, respectively. Again, the difference between the linear response (plotted with straight lines) and the corresponding non-linear response becomes evident in these figures. As can be seen in Figure 21a, a 10 mm relative detachment displacement of the West wall occurs for a base shear value approximately 7000 KN in the East-West direction. Similarly, a 10 mm relative detachment displacement of the North wall in the North-South direction (Figure 21b) occurs for a base shear value approximately 5000 KN. In both cases, these values are smaller than the corresponding base shear values listed in Table 3, which were obtained from the dynamic spectral linear analyses that employed the Chavriata ground motion. Therefore, it can be concluded that for these force levels the detachment of these walls at the corners is predicted by combining the results of both numerical simulations. Such predicted wall detachment agrees with the observed performance (Figures 17 and 18).

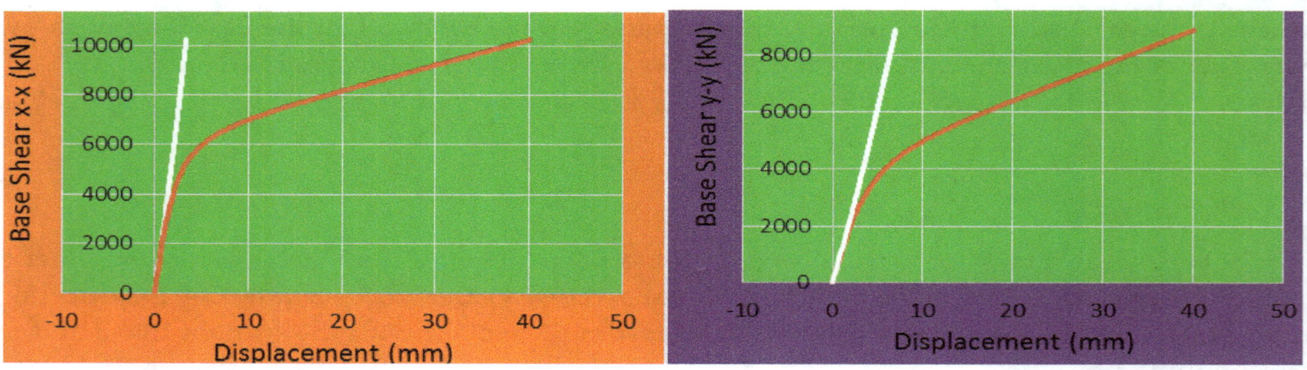

(a) (b)

Figure 21. (**a**) Push-over in the East-West direction (Ey)—detachment of the West wall; (**b**) Push-over in the North-South direction (Ex)—detachment of the North wall.

6. Observed Structural Performance for a Long Period Range

In the previous sections, typical damage sustained by stone masonry "Greek Basilica" churches was presented together with the underlying causes due to natural hazards. Some of these churches span a period of over 800 years [1,9,12,13,29,30] with the majority of them being approximately 300 years old. One significant factor is the variability of the severity of the natural hazards that are linked with the structural response of the foundation and/or the superstructure. Apart from this variability of the severity of the actions it must also be stressed that the structural performance at any given time depends on the structural maintenance for each individual church. The heavy damage that was observed for numerous Christian churches during the L'Aquila earthquake [14] created new stimulus for research on the effectiveness of various structural maintenance techniques applied in the past to old masonry structures. Despite this controversy, it must be agreed that neglecting structural maintenance for such old masonry structures is in the long term synonymous to severe structural damage. An effort was made here to demonstrate that the most critical combination is a strong earthquake event subjecting an old masonry structure, which is already in a state of pre-existing state of stress and deformation from uneven foundation settlement, to considerable earthquake forces. Greece is divided into three seismic zones with Kefalonia island being in the most intense seismic zone [35]. All the churches in Argostoli, the capital of this island, are relatively new, built after the destruction of the old churches during the catastrophic seismic sequence of 1953 (Figure 22). During the seismic sequence of 2014 these new churches met the earthquake demands very successfully; however, at the Western region of this island surrounding the town of Lixouri, the 2nd largest city in the island, the seismic ground motion was more severe, as shown by the acceleration recordings of the ground motion [11,12]. Numerous churches in this region developed heavy structural damage reaching collapse, like the church of St. Marina Soullaroi presented in Sections 4 and 5, although they did not suffer greatly during the 1953 earthquake sequence. The structural maintenance measures that were applied to these churches after the 1953 event proved to be insufficient for the demands generated by the 2014 earthquake sequence [11,12]. Similar observations can be made for the old masonry churches in the Kozani prefecture and the structural damage they developed during the 1995 earthquake sequence [9]. This region was seismically "quiet" for many centuries before this damaging 1995 earthquake sequence. Consequently, it was believed that earthquake strong motion was not an actual hazard for either old or new structures in this region. The most spectacular damage to an old stone masonry church was that of Taxiarchis in the outskirts of Eani (12th century AD). Despite its small size, the walls of this church were totally destroyed, indicating the severity of the ground motion as well as the lack of maintenance (Figure 23). This monument remains in this damaged condition till today. Many other churches in this region with less heavy structural damage have experienced successful structural rehabilitation.

Figure 22. Images of total destruction to old masonry construction in Argostoli, Kefalonia island during the 1953 earthquake sequence.

Before the 1995 earthquake sequence

Damaged by the 1995 earthquake

Figure 23. The church of Taxiarchis in Eani, Kozani prefecture.

A heavy task after numerous strong earthquake events during the last 50 years in many regions of Greece has been the structural rehabilitation of old masonry churches. This is also the case for masonry churches which sustained heavy structural damage in the city of Kalamata during the 1986 strong earthquake sequence. Numerous churches in this city as well as in the surrounding area were damaged. Figures 24 and 25 depict two such cases located in the centre of Kalamata. The heavy structural damage of the peripheral walls in these two churches was accompanied by the collapse of the central dome and the partial collapse of the bell towers; similar damage patterns were observed in many churches during the 2009 L'Aquila earthquake [14]. These figures also show the condition of these two churches today. It must be underlined that due to this structural rehabilitation effort for numerous cultural heritage structures in Greece over the last fifty years very valuable scientific and technical knowledge has been gained in this field.

Damaged by the 1986 earthquake sequence

Restored after the 1986 earthquake sequence

Figure 24. The church of Ipapanti in the centre of Kalamata.

Damaged by the 1986 earthquake sequence   Restored after the 1986 earthquake sequence

Figure 25. The church of Holy Apostles in the centre of Kalamata.

Certain comments are also due relevant to the numerical tools presented in Sections 3–5. It was stressed, when outlining the numerical process in Section 3 and the corresponding numerical predictions of the structural performance in Sections 4 and 5, that both numerical approaches are based on simplified assumptions. In both approaches, that is the linear numerical simulation which assumes non-failing masonry wall inter-connections or the non-linear numerical simulation which assumes flexible masonry wall inter-connections or foundation uplift, one of the major obstacles is to be able to quantify all these numerical assumptions with realistic limit-state values. This difficulty is due to the immense variability of old stone masonry in terms of materials and construction techniques and the subsequent lack of relevant in-situ or laboratory measurements. Recordings of the dynamic response of a particular structure from in-situ man-made excitations can be utilized in order to validate a given numerical simulation [6,7]. During the past decade, numerous researchers have proposed the application of complex numerical simulations for predicting the performance of old stone masonry structures like then ones investigated here. However, these complex numerical simulations are faced with the already mentioned obstacles; that is the immense variability of old stone masonry in terms of materials and construction techniques and the subsequent lack of relevant in-situ or laboratory measurements. Therefore, it is essential not to rely on the complexity of the numerical modelling; instead to follow a number of limit state scenarios, like those presented in Sections 4 and 5, ensuring that these are based on a certain degree of realism for every particular case being studied [36].

7. Conclusions

1. A systematic study of the performance of damaged stone masonry structures representing "Basilica" Christian churches substantiates two fundamental causes; the long-term permanent uneven foundation settlement combined with seismic forces generated from relatively strong earthquake ground motions.

2. A simplified evaluation process is presented based on a dynamic linear elastic numerical simulation for obtaining the imposed demands on the various structural elements. Next, towards predicting the structural performance use is made of strength over demand ratio values ($\boldsymbol{R_\tau}$, $\boldsymbol{R_\varsigma}$, $\boldsymbol{R_\sigma}$, $\boldsymbol{R_M}$). These are derived using the numerically predicted demands as well as capacities based on assumed strength properties of the masonry for distinct failure modes which correspond to in-plane shear, compression or tension as well as out-of-plane flexure limit states.

3. Limited experimental results are also presented in an effort to verify up to a point the validity of the assumed strength values for the masonry at hand.

4. This simplified evaluation process, is applied to selected cases of "Basilica" Christian churches. It is demonstrated that reasonably good agreement can be obtained when compared to observed behaviour either for a case of foundation settlement or a case of strong earthquake excitation. This fact must be considered together with the simplified process basic limitation which is the assumption that the interconnections of masonry walls between themselves or with the roof remain in-tact.

5. In order to deal with this limitation, push-over step-by-step non-linear numerical analyses were next performed. These non-linear analyses include the potential of foundation uplift as well as the potential of the masonry walls to be detached from their interconnections either at their corners or at their connection with the roof.

6. It is shown that both these type of analyses, that is the simplified dynamic linear elastic analysis and the "push-over" non-linear step-by-step non-linear analysis can be used in a combined way in order to achieve a more realistic prediction of the expected performance of such stone masonry structures.

7. The necessity to obtain a more comprehensive set of measured strength properties for such type of masonry construction must be also underlined. This is necessary in order to increase the confidence on the validity of simple or complex numerical approximations.

8. Complex numerical simulations in predicting the performance of old masonry structures should not be considered a' priori with a high degree of confidence. They are also faced with the immense variability of old stone masonry in terms of materials and construction techniques and the subsequent lack of relevant in-situ or laboratory measurements. Therefore, it is essential not to rely on the complexity of the numerical modelling; instead to follow a number of limit state scenario, like those presented in Sections 4 and 5, ensuring that these are based on a certain degree of realism for every particular case being studied.

Author Contributions: G.C.M., L.K. and E.K. had equal contribution. In addition, G.C.M. wrote the manuscript.

References

1. Manos, G.C. Consequences on the urban environment in Greece related to the recent intense earthquake activity. *Int. J. Civ. Eng. Architect.* **2011**, *5*, 1065–1090.
2. Zhang, N.; Tan, K.-H. Effects of support settlement on continuous deep beams and STM modelling. *Eng. Struct.* **2010**, *32*, 361–372. [CrossRef]
3. Naser, M.Z.; Hawileth, R.A. Predicting the response of continuous RC deep beams under varying levels of differential settlement. *Front. Struct. Civ. Eng.* **2018**. [CrossRef]
4. Croci, G. *The Conservation and Structural Restoration of Architectural Heritage*; Computational Mechanics Publication: Southampton, UK; Boston, MA, USA, 1998; ISBN 1-85312-4826.
5. Jones, C.J.F.P. Defects originating in the ground. In *The Maintenance of Brick and Stone Masonry Structures*; Sowden, A.M., Ed.; 1990; Chapter 9; ISBN 0419-14930-9. Available online: https://www.amazon.com/Maintenance-Brick-Stone-Masonry-Structures/dp/0419149309 (accessed on 29 April 2019).
6. Manos, G.C.; Nick Simos, N.; Kozikopoulos, E. The Structural Performance of Stone-Masonry Bridges. In *Structural Bridge Engineering*; InTechOpen: London, UK, 2016; Chapter 4; ISBN 978-953-51-2689-8. Print ISBN 978-953-51-2688-1. [CrossRef]
7. Simos, N.; Manos, G.; Kozikopoulos, E. Near- and far-field earthquake damage study of the Konitsa stone arch. *Eng. Struct.* **2018**, *177*, 256–267. [CrossRef]
8. Sergio, L. Damage assessment of churches after L'Aquila earthquake (2009). *Bull. Earthq. Eng.* **2011**. [CrossRef]
9. Manos, G.C.; Soulis, V.J.; Karamitsios, N. The Performance of Post-Byzantine churches during the Kozani-1995 Earthquake—Numerical Investigation of their Dynamic and Earthquake Behaviour. In Proceedings of the 15WCEE, Lisbon, Portugal, 24–28 September 2012.
10. Manos, G.C.; Soulis, V.; Felekidou, O.; Matsou, V. A Numerical Investigation of the Dynamic and Earthquake Behaviour of Byzantine and Post-Byzantine Basilicas. In Proceedings of the 3rd International Workshop on Conservation of Heritage Structures Using FRM and SHM, Ottawa, ON, Canada, 11–13 August 2010.
11. GEER-EERI-ATC. *Cephalonia GREECE Earthquake Reconnaissance January 26th/February 2nd 2014*. Report No. GEER-034, Report date 06-06-2014. Available online: http://www.geerassociation.org/index.php/component/geer_reports/?view=geerreports&id=32 (accessed on 29 April 2019). [CrossRef]
12. Manos, G.C.; Kozikopoulos, E. The Dynamic and Earthquake Response of Basilica Churches in Kefalonia-Greece including Soil-Foundation Deformability and Wall Detachment. In Proceedings of the CompDyn 2015, Crete, Greece, 25–27 May 2015.

13. Manos, G.C. The Seismic Behaviour of Stone Masonry Greek Orthodox Churches. *J. Architect. Eng.* **2016**, *1*, 44–53. [CrossRef]
14. Modena, C.; Casarin, F.; da Porto, F.; Munari, M. L'Aquila 6th April 2009 Earthquake: Emergency and Post-emergency Activities on Cultural Heritage Buildings. In *Earthquake Engineering in Europe, Geotechnical, Geological, and Earthquake Engineering*; Garevski, M., Ansal, A., Eds.; Springer Science+Business Media B.V.: Berlin, Germany, 2010; Volume 17, Available online: https://link.springer.com/book/10.1007/978-90-481-9544-2 (accessed on 29 April 2019). [CrossRef]
15. Cerone, M.; Viscovic, A.; Carriero, A.; Sabbadini, F.; Capparela, L. The Soil Stiffness Influence and the Earthquake Effects on the Colosseum in Roma. In Proceedings of the 2nd Internationl Conference on Studies in Ancient Structures, Istanbul, Turkey, 9–13 July 2001; pp. 421–426.
16. Manos, G.C.; Karamitsios, N. Numerical simulation of the dynamic and earthquake behavior of Greek post-Byzantine churches with and without base isolation. In *Earthquake Engineering Retrofitting of Heritage Structures, Design and Evaluation of Strengthening Techniques*; Syngellakis, S., Ed.; Wessex Institute of Technology: Southampton, UK, 2013; pp. 171–186. ISBN 978-1-84564-754-4. eISBN 978-1-84564-755-1.
17. *SAP2000, Integrated Software for Structural Analysis and Design*; Computers and Structures Inc. Available online: https://www.csiamerica.com/products/sap2000 (accessed on 29 April 2019).
18. Abaqus Unified FEA-SIMULIA™ by Dassault Systèmes. Available online: https://www.3ds.com/products-services/simulia/products/abaqus/ (accessed on 29 April 2019).
19. Manos, G.C.; Kozikopoulos, E. In-situ measured dynamic response of the bell tower of Agios Gerasimos in Lixouri-Kefalonia, Greece and its utilization in the numerical predictions of its earthquake response. In Proceedings of the CompDyn 2015, Crete, Greece, 25–27 May 2015.
20. European Committee for Standardization. *Euro-Code 6: Design of Masonry Structures, Part 1-1: General Rules for Building. Rules for Reinforced and Un-Reinforced Masonry*; EN 1996-1-1; European Committee for Standardization: Brussels, Belgium, 2005.
21. Tomaževič, M. Shear resistance of masonry walls and Eurocode 6: shear versus ten-sile strength of masonry. *Mater. Struct.* **2009**, *42*, 889–907. [CrossRef]
22. Turnsek, V.; Cacovic, F. Some experimental results on the strength of brick masonry walls. In Proceedings of the 2nd International Brick-Masonry Conference, Stoke-on-Trent, UK, 12–15 April 1971; pp. 149–156.
23. Binda, L.; Saisi, A.; Tiraboschi, C. Investigation procedures for the diagnosis of historic masonries. *Contruct. Build. Mater.* **2000**, *14*, 199–233. [CrossRef]
24. Vintzileou, E. Effect of Timber Ties on the Behavior of Historic Masonry. *J. Struct. Eng.* **2008**, *134*. [CrossRef]
25. Cattari, S.; Lagomarsino, S. A strength criterion for the flexural behaviour of spandrels in un-reinforced masonry walls. In Proceedings of the 14th WCEE, Beijing, China, 12 October 2008.
26. Beyer, K.; Mangalathu, S. Review of strength models for masonry spandrels. *Bull. Earthq. Eng.* **2013**, *11*, 521–542. [CrossRef]
27. Betti, M.; Galano, L.; Vignoli, A. Seismic response of masonry plane walls: A numerical study on spandrel strength. *AIP Conf. Proc.* **2008**, *1020*, 787–794.
28. Beyer, K.; Mangalathu, S. Numerical study on the peak strength of masonry spandrels with arches. *J. Earthq. Eng.* **2014**, *18*, 169–186. [CrossRef]
29. Manos, G.C.; Soulis, V.J.; Diagouma, A. Numerical Investigation of the behaviour of the church of Agia Triada, Drakotrypa, Greece. *Adv. Eng. Softw.* **2007**, *39*, 284–300. [CrossRef]
30. Manos, G.C.; Kotoulas, L.; Felekidou, O.; Vaccaro, S.; Kozikopoulos, E. Earthquake damage to Christian Basilica Churches—Application of an expert system for the preliminary in-plane design of stone masonry piers. In Proceedings of the International Conference on STREMAH 2015, A Coruna, Spain, 13–15 July 2015; WIT Press: Southampton, UK, 2015.
31. Manos, G.C.; Kotoulas, L. Unreinforced stone masonry under in-plane state of stress from gravitational and seismic actions. Measured and predicted behavior. In Proceedings of the CompDyn2017, Rhodes Island, Greece, 15–17 June 2017.
32. Papaioannou, C. *Strong Ground Motion of the 3rd February 2014, (M = 6.0) Kefalonia Earthquake*. Institute of Earthquake Engineering and Engineering Seismology Report. 2014. Available online: http://www.itsak.gr/news/news/79 (accessed on 29 April 2019).

33. European Committee for Standardization. *Euro-Code 8: Design of Structures for Earthquake Resistance—Part 1: General Rules, Seismic Actions and Rules for Buildings*; Final Draft prEN 1998-1; European Committee for Standardization: Brussels, Belgium, 2003.
34. Manos, G.C. *International Handbook of Earthquake Engineering: Codes, Programs and Examples*; Seismic Code of Greece; Paz, M., Ed.; Chap-Man and Hall: London, UK, 1994; Chapter 17; ISBN 0-412-98211-0.
35. *Provisions of Greek Seismic Code with Revisions of Seismic Zonation*; Government Gazette: Athens, Greece, 2003.
36. Limoge Schraen, C.; Giry, C.; Desprez, C.; Ragueneau, F. Tools for a large-scale seismic assessment method of masonry cultural heritage. In *Structural Studies, Repair and Maintenance of Cultural Heritage, STREMAH XIV*; WIT Press: Southampton, UK, 2015; ISBN 978-1-84564-968-5.

10

Seismic Evaluation and Strengthening of an Existing Masonry Building in Sarajevo, B&H

Naida Ademović [1,*], Daniel V. Oliveira [2] and Paulo B. Lourenço [2]

1 Faculty of Civil Engineering in Sarajevo, University of Sarajevo, Sarajevo 71 000, Bosnia and Herzegovina

2 ISISE, Institute of Science and Innovation for Bio-Sustainability (IB-S), Dept. of Civil Engineering, University of Minho, Guimarães 4800-058, Portugal; danvco@civil.uminho.pt (D.V.O.); pbl@civil.uminho.pt (P.B.L.)

* Correspondence: naidadem@yahoo.com

Abstract: A significant number of old unreinforced load-bearing masonry (URM) buildings exist in many countries worldwide, but especially in Europe. In particular, Bosnia and Herzegovina has an important stock of masonry buildings constructed from the 1920s until the 1960s without application of any seismic code, due to their nonexistence at that time. With the 1963 Skopje earthquake, this class of buildings were shown to be rather vulnerable to seismic actions, which exhibited serious damage. This article assesses the seismic vulnerability of a typical multi-storey residential unreinforced load-bearing masonry building located in the heart of Sarajevo, which may be exposed to an earthquake of magnitude up to 6 by Richter's scale. The buildings of this kind make up to 6% of the entire housing stock in the urban region of Sarajevo, while in Slovenia this percentage is much higher (around 30%). The analysis of a typical building located in Sarajevo revealed its drawbacks and the need for some kind of strengthening intervention to be implemented. Additionally, many structures of this type are overstressed by one to two additional floors (not the case of the analyzed structure) constructed from 1996 onwards. This was due to the massive population increase in the city center of Sarajevo and further increased the vulnerability of these buildings.

Keywords: unreinforced load-bearing masonry; strengthening intervention; non-linear analysis

---

1. Introduction

Bosnia and Herzegovina lie in the heart of South Eastern Europe, which is marked as one of the European regions with rather complex tectonic formations. Many researchers investigated this region, i.e., Northwestern Balkans, Northwest Bosnia and Herzegovina. One of the features which makes South Eastern Europe interesting for researchers, is the fact that all the earthquakes that struck this region until now were earthquakes with a shallow focus. Uniform hazard spectra for this region were produced by several researchers [1,2] after the study of the influence of deep and shallow geology. In 2018, the seismic hazard map for Bosnia and Herzegovina, defined in terms of peak ground acceleration (PGA) with a return period of 475 years (Figure 1), as defined in Eurocode 8, was constructed, and is now part of the National Annex in BAS EN 1998-1: 2018 [3].

When considering 1944 recorded earthquakes, the focal depth of 64% of them is only 10 km, while earthquakes with a focal depth in the range from 11 to 20 km make up 30% of the sample and around 4.5% of the earthquakes were with a focal depth in the range from 21 to 30 km; which leaves just 1.7% with a focal depth larger than 30 km [4]. The shallow focus is one of the most destructive features of earthquakes in Bosnia and Herzegovina [5,6].

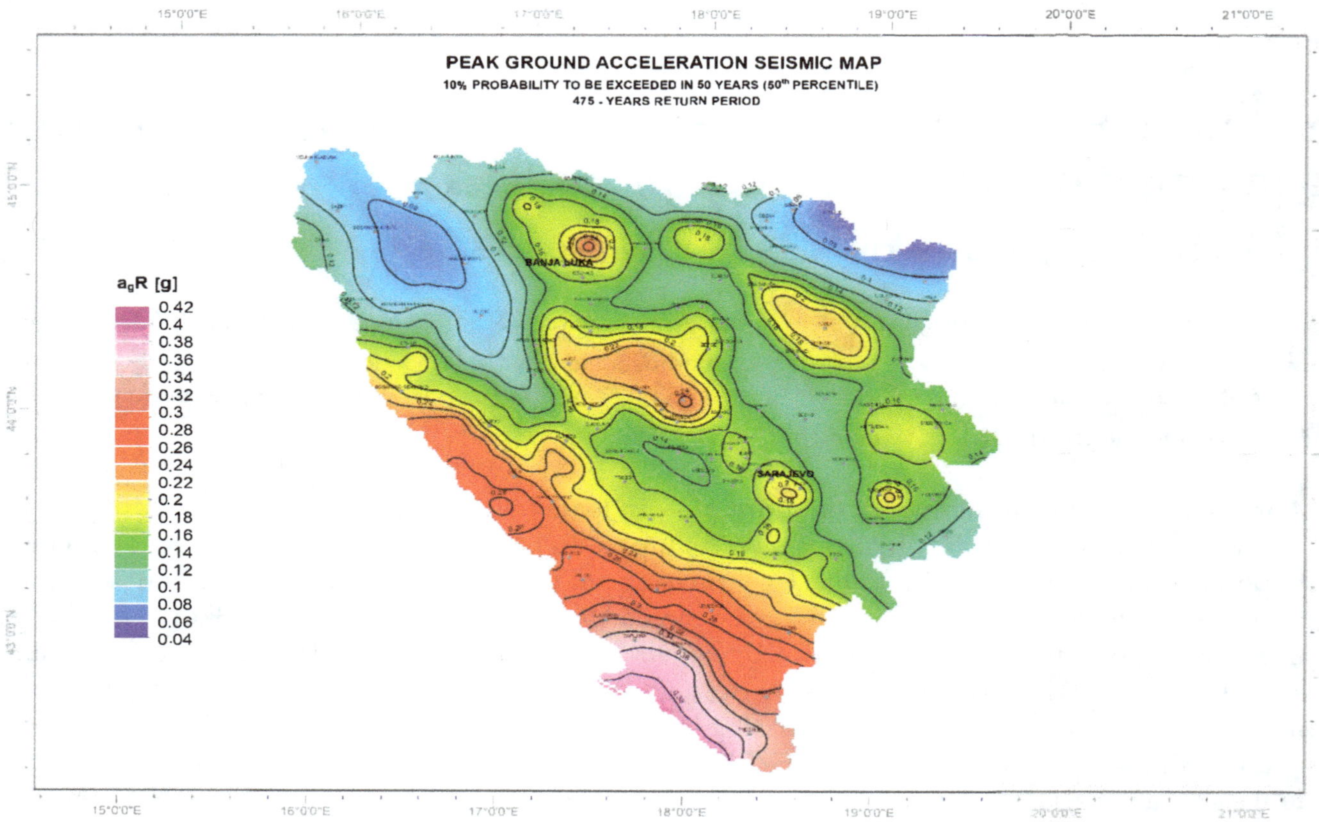

Figure 1. Seismic hazard map for Bosnia and Herzegovina in terms of peak ground acceleration (PGA) (475 years return period) [3].

In this paper, a building structure built in 1957 and typical of the Balkans region is elaborated. These type of buildings can be found in all towns in Slovenia (Figure 2a) and they make up to 30%of the entire housing stock in Slovenia [7]. According to the national census [8], the buildings of this type make up 6% of the housing in the urban part of the Sarajevo city center. This construction type was built from 1920 until 1963 in the entire region of ex-Yugoslavia, especially in the Republic of Macedonia, Montenegro, and Bosnia and Herzegovina, mainly in the urban areas. At that time, no seismic codes existed in this region, so these structures were constructed with no seismic regulations. The 1963 Skopje earthquake showed the drawbacks of this type of structures (Figure 2b), which were severely damaged due to inadequate wall concentration in the predominant earthquake direction, low resistance of the loadbearing system, high height of the unreinforced masonry structure (URM), etc. It is only after this catastrophic event that these type of structures were addressed in the first Temporary Seismic Code produced in 1964 and later upgraded after the 1969 Banja Luka earthquake (which had a focal depth of only 25 km, and a magnitude of 6.4 by Richter's scale).

This was one of the most devastating earthquakes registered in this region. Due to the vast devastation caused by the Banja Luka earthquake, micro-zonation of the urban part of the city was done in 1972. According to that data, the expected average ground acceleration for Banja Luka is 0.18g [9]. Regardless of this fact, the area of Banja Luka in the seismological map of Bosnia and Herzegovina was defined as an area of maximum registered intensity of 9 by the Mercalli–Cancani–Sieberg intensity scale [10] for a return period of 500 years. More recently, Lee et al. [11] conducted the micro-zonation of Banja Luka in the light of performance-based earthquake-resistant design.

(a) (b)

Figure 2. (**a**) Typical building in Slovenia; (**b**) devastation of a typical building after the 1963 Skopje earthquake [4].

According to the new seismic hazard code for Bosnia and Herzegovina, defined in terms of peak ground acceleration (PGA) with a return period of 475 years (Figure 1), Sarajevo can experience earthquakes with a peak ground acceleration of 0.18g [3]. Such conditions make this structure rather vulnerable as it was constructed without the application of any seismic measures.

As this structure is typical for the wider region of the Western Balkans, the main purpose of this research is to investigate the seismic safety of the building if exposed to ground motions compatible with Eurocode 8 [3]. On the basis of obtained numerical results, possible intervention measures for improving the seismic response of the structure are also analyzed and proposed.

2. Description of the Case Study

The analyzed building is located in the urban part of Sarajevo (Figure 3a,b) and it has a rectangular plan, with length L = 38 m and width B = 13 m (Figure 3c). The total height of the structure with the basement is 21 m (Figure 3d). The main load bearing walls are located in the Y direction, while the load bearing walls in the X direction are attenuated by many openings. The structure is composed of a basement made of reinforced concrete walls. Dimensions of the walls differ and the inner walls in the Y direction are 38 cm thick, while in the X direction the thickness of the outer walls is 30 cm and of two inner walls is 25 cm. The walls in both directions on all the floors are made of solid brick masonry (dimensions of bricks 250 × 120 × 65 mm) and a façade part (non-load bearing) made of hollow bricks 125 mm thick. The slabs are made of semi-prefabricated "Herbst" concrete hollow elements, joists, and a concrete slab of 6 cm. The total thickness of the slab at the storey's is 26.5 cm, and the same construction is kept for the roof with thickness increased to 43.5 cm. The structure is an unreinforced masonry building with prefabricated slabs, behaving as rigid diaphragms [5,6].

The vulnerability of the building under study is connected to several issues. One of the essential requirements for adequate seismic response is not satisfied, as there is the lack of load-bearing elements in one of the structure's main directions. The loadbearing walls are mainly placed in the transversal direction (Y), while walls in the longitudinal (X) direction are weakened by a large number of openings (Figure 2). The structure has a basement, ground floor, and five storeys, and it was constructed as an unreinforced masonry building. The walls are made of solid bricks connected with lime mortar providing a low compressive strength of masonry. By taking into account the EMS 98 [12], the structure is associated with vulnerability class C. This structure (unconfined masonry, at most 60 years old with reinforced concrete floors) can have different damages grades. For a seismic intensity VII, the damage grade would be 2 (moderate damage), and as the intensity increases the damage grade increases by one level: Seismic intensity VIII provides substantial to heavy damage; seismic intensity IX provides very heavy damage and seismic intensity X provides destruction. Additionally, according to the regulations regarding technical standards for the construction of buildings in seismic areas from 1991 [10], there is a clear limitation regarding the number of storeys for unreinforced masonry structures in respect to the seismic intensity level, as shown in Table 1.

Table 1. Allowed number of storeys for different masonry structures [10].

Type of Masonry Structure	Seismic Intensity Degree		
	IX	VIII	VII
URM	-	G + 1	G + 2
Confined masonry	G + 2	G + 3	G + 4
Reinforced masonry	G + 7	G + 7	G + 7

(G stands for ground floor).

This type of structure is characterized by two longitudinal façade walls with a large number of openings, while in the transverse direction the exterior walls have only one opening at the height of each floor. Other transverse (inner) walls have door openings of 2.3 m^2–6.9 m^2. The total area of the openings in the basement for the longitudinal direction is 19.8%. In the transverse direction, openings in the outer wall occupy only 8.6% of the wall surface. A significant percentage of the aperture is located in the longitudinal walls in the amount of as much as 46% of the wall surface. Thus, the lateral resistance in the longitudinal direction is significantly lower than the lateral resistance in the transverse direction.

Sarajevo is located in the seismic intensity zone VII, thus limiting the number of storeys to G + 2 (see Table 1), while the analyzed structure is G + 5. It is evident that the limitation according to Table 1 is not respected, as this regulation was enforced only in the 1980s.

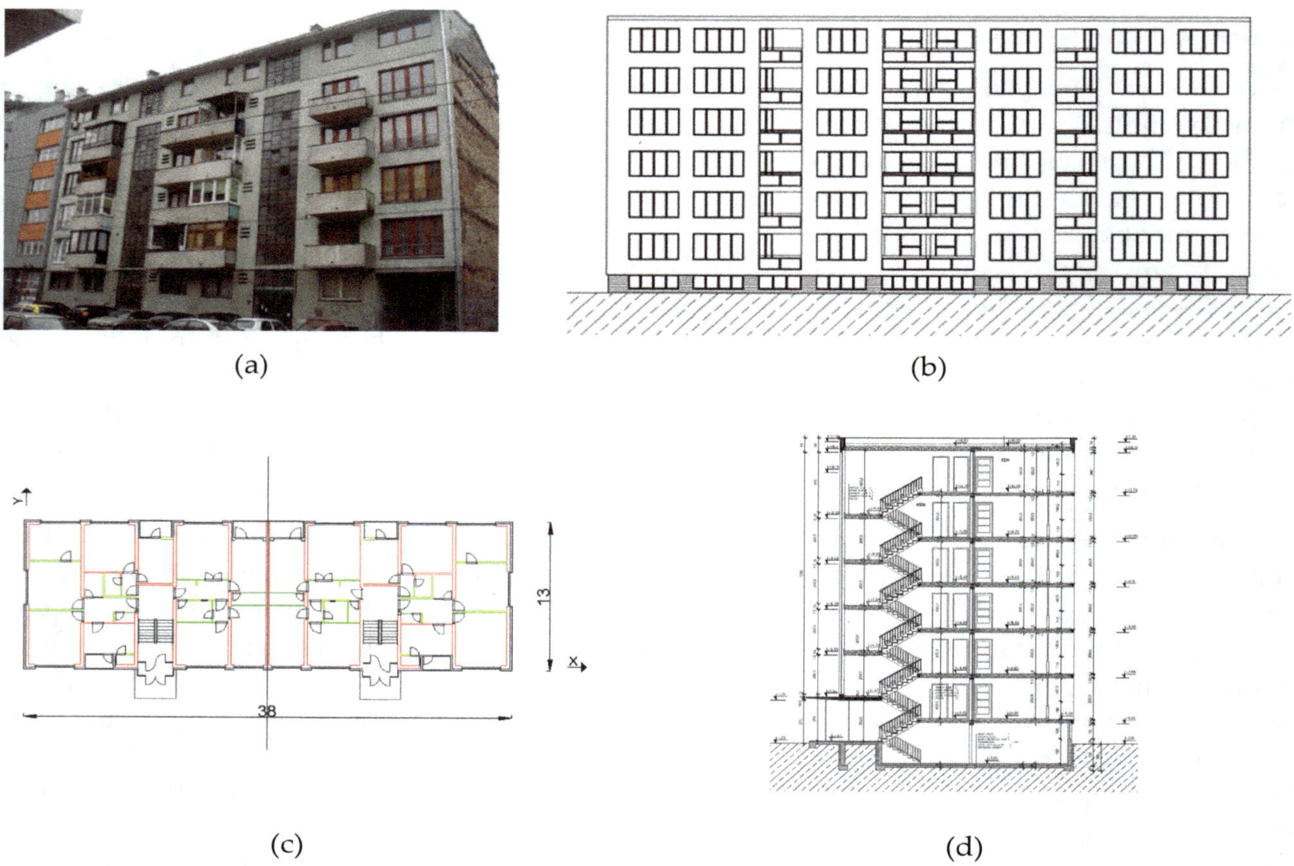

Figure 3. (**a**) Analyzed building located in Sarajevo; (**b**) east façade; (**c**) plan view of the typical floor and labeling of coordinates (X and Y); (**d**) typical cross section.

In order to obtain the necessary information regarding the physical and mechanical characteristics, several on-site and laboratory tests were done. Laser distancemeters and total stations were used for verification of the geometric data and as a result, drawings of the building were produced (Figure 3b–d). Brick units and concrete compressive strengths were determined on samples from the building (Figure 4a). Five series of two bricks were taken out from representative locations in the structure and their compressive strength was tested in accordance with the ex-Yugoslavian standards [13] (Figure 4b). Locations of the extracted brick samples are marked by numbers from 1 to 5 (Figure 4a). The mean value of brick compressive strength was 19.4 MPa, while the minimum value was 13.4 MPa putting it into a class of M15 (15 MPa), which fulfills the requirements for a load-bearing wall [13]. Compressive strength of concrete was determined using six cylindrical samples in accordance with the regulations defined in standard [14]. The mean value of concrete compressive strength was 25.1 MPa, while the minimum value was 22.8 MPa putting it into a class of MB25, which is equivalent to C20/25 in Eurocode 2 [15], and the reinforcement was ø = 14 mm, type of steel GA240/360 (smooth bars with yield stress 240 MPa, and 360 MPa ultimate strength). Based on results from the experimental campaign, the calculated compressive strength of masonry according to Eurocode 6 [16] was 4.1 N/mm^2, which was used as the basis for the calculation of other mechanical characteristics. The slabs were made out of concrete class C20/25 and a reinforcement ø = 8 mm and the same type of steel was used as in the basement. All other characteristics were determined from the characteristic compressive strength of the concrete according to Eurocode 2 [15].

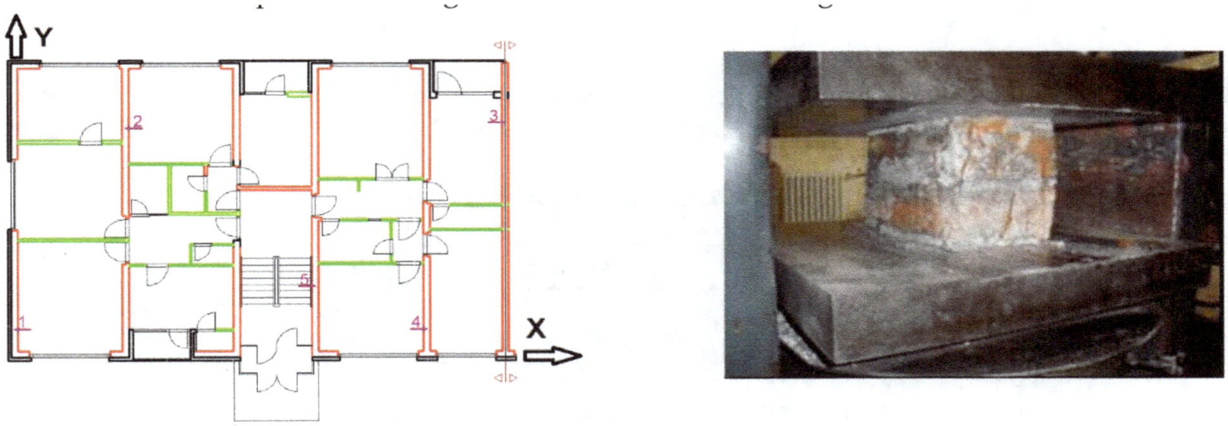

(a) (b)

Figure 4. (**a**) Location of the samples for determination of the compressive strength of bricks; (**b**) tested sample.

The building, depicted in Figure 3a, was modelled using a finite element model (FEM) and the equivalent frame model (EFM), with the application of DIANA [17] (Figure 5a) and 3MURI [18] (Figure 5b), respectively. DIANA software [19–24], as well as 3Muri [24–30], have been widely applied to model masonry structures. The structure was modeled with a curved shell (quadrilateral element CQ40S type) element. In FEM an element size was 0.25 m having a total of 84,523 nodes and 28,522 elements, while in the EFM, the number of 3D nodes was 218, 34 2D nodes, and 506 elements for the entire structure (Figure 5a,b). In FEM only half of the structure was modeled and analyzed as the structure is symmetric, employing in total 15,759 CQ40S elements and 45,443 nodes (Figure 5c). The applied adequate boundary conditions are indicated in Table 2. Material nonlinearity properties of masonry [31] were taken into account with the application of the total strain fixed crack model, as defined in DIANA [17]. The post-cracked shear behavior was introduced by a very low shear retention factor. The physical and material characteristics taken in the analysis are given in Table 3.

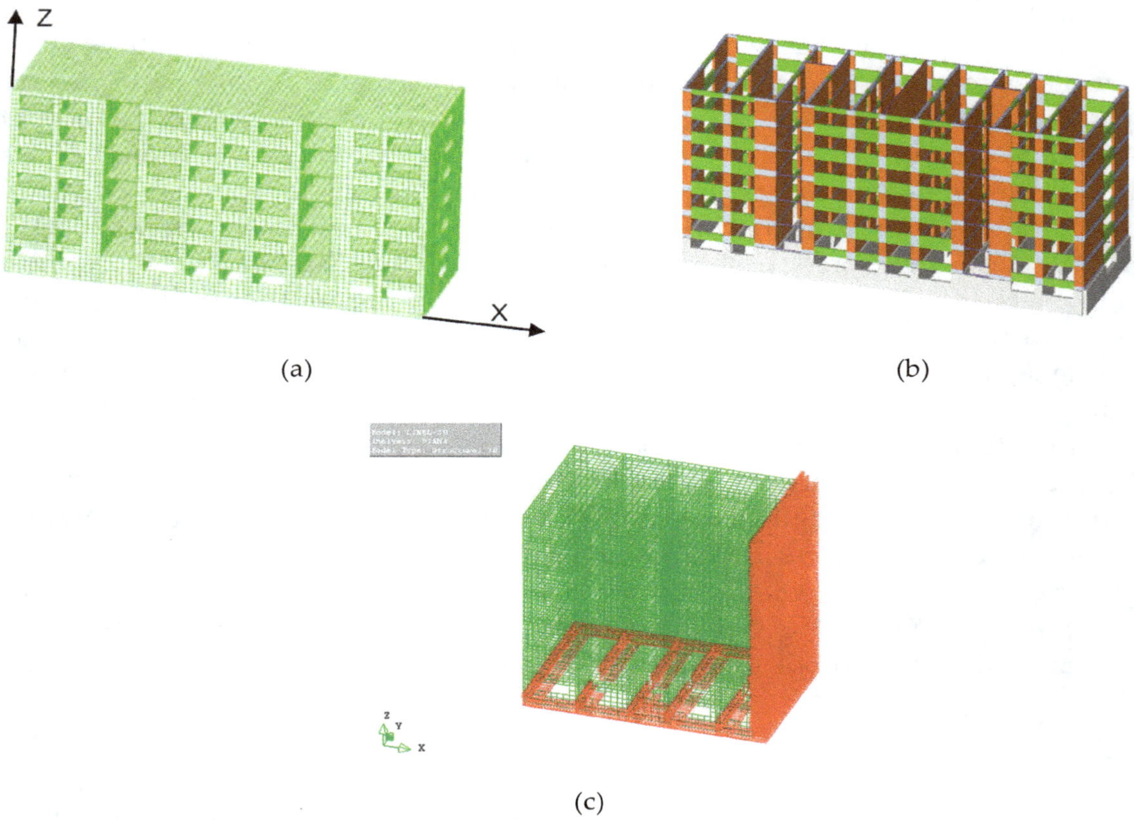

Figure 5. Models used for safety assessment: (**a**) Finite element model (FEM) in DIANA; (**b**) EFM in 3MURI; (**c**) FEM model in DIANA (only half is modelled to keep the computational effort acceptable).

Table 2. Constrain Summary.

Support	Translation	Rotation
Plane of symmetry	X	Y; Z
Base of structure	X; Y; Z	X; Y; Z

Table 3. Physical and material characteristics.

Element	Thickness (m)	Compressive Strength f_c (N/mm^2)	Compressive Fracture Energy G_{fc} (N/mm)	Tensile Strength f_t (N/mm^2)	Tensile Fracture Energy G_t (N/mm)	Modulus of Elasticity E (N/mm^2)	Poisson Ratio υ	Density ϱ (kg/m^3)
				Masonry				
Façade walls	0.375	4.07	6.51	0.20	0.10	4070	0.20	2700*
Inner walls	0.250							1900
Element	**Thickness (m)**	**Compressive Strength f_c (N/mm^2)**		**Tensile Strength f_{ct} (N/mm^2)**		**Modulus of Elasticity E (N/mm^2)**	**Poisson Ratio υ**	**Density ϱ (kg/m^3)**
				Concrete				
Façade walls	0.380	24		2.2		30,000	0.20	2400
Inner walls	0.250							
Floor	0.265							2190
Roof	0.435							2050

* On the basis of experimental test, the value of density of this type of masonry is 1900 kg/m^3, however, in order to take into account the non-bearing façade walls in respect of the mass, the value has been proportionally increased, while keeping the thickness of d = 25 cm enabling the stiffness to remain intact.

Dynamic characteristics of the structure are important for understanding its dynamic behavior under earthquake actions. The natural frequencies and mode shapes of the entire structure were determined by modal analysis. The first eigen frequency in FEM was 2.19 Hz, while in EFM it amounted to 1.96 Hz and the first mode in both cases was the translation in the X direction [5,6,29]. Furthermore, the values of frequencies are consistent with the data provided by Tomaževič [32], indicating: "For higher structures even up to 11-storeys the value are close to 2 Hz even though buildings have been built with different materials". This was used as a verification parameter of the model as no ambient vibration tests were conducted on the structure.

Figure 6a shows the labeling of the walls in the structure in the two directions as well as the numbers of the nodes on the top floor of the structure. The node used to monitor the top displacement is located at the axis of symmetry in the center of the structure (referred to as node 44014) (Figure 6b). For more details regarding the numerical model of the structure and material characteristics the reader is referred to references [5,6].

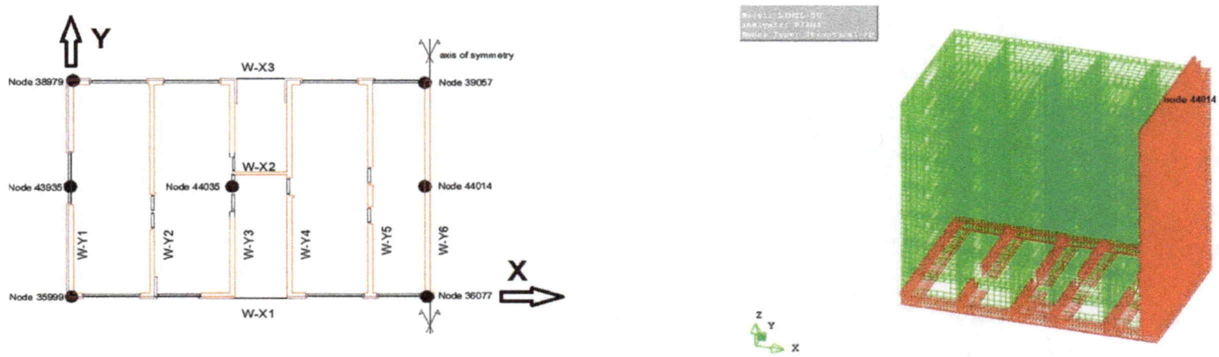

(a) (b)

Figure 6. (**a**) Location of the nodes; wall labeling; (**b**) location of the control node 44014.

3. Results of the Unstrengthened Configuration

The FEM model was used to perform pushover and time history analysis, while the EFM model was only used to conduct pushover analysis. Comparing the pushover curves obtained from the two models, the difference in the maximum load coefficient was in the range of 6.4% to 6.9%, which is considered as an acceptable range [5,6]. Here, the load coefficient is the ratio between the base shear forces due to pushover (y direction) and the sum of all vertical loads (z direction).

Results show that the EFM model is stiffer in comparison to the FEM model. The difference in the stiffness could be attributed to the rigid connections between the spandrel and the pier elements. Spandrels mainly influence the boundary conditions of piers [33]. The time history analysis in FEM was compared with pushover analysis in EFM just in the view of distribution of pattern damage and it was evident that it follows the same sequence [5,6]. The location of the cracks were around the opening and, as the windows are very close to the corner of the walls (75 to 100 cm), this is weakening the connection between the walls causing additional concentration of stresses and crack development. The major damage is concentrated on the ground floor. The two models gave results that were in excellent correlation, however for the sake of computational efficiency, it was decided to continue the further analysis with the EFM model.

In order to generate the response of the structure to various PGA values, the structure was exposed to the short-period 1979 Petrovac strong earthquake motion registered at Montenegro. This acceleration record is commonly used for seismic structural assessment throughout the region [34,35]. In order to take into account the ground motion in Sarajevo, according to the new seismic hazard map, the accelerogram from the Petrovac earthquake (with a PGA of 0.43 g) was scaled down to 0.1 g (Figure 7) and 0.2 g. Seismosignal software [36] was used for scaling and filtering the acceleration record. The code response spectrum used in the EFM model is presented in Figure 8 for 0.1 g.

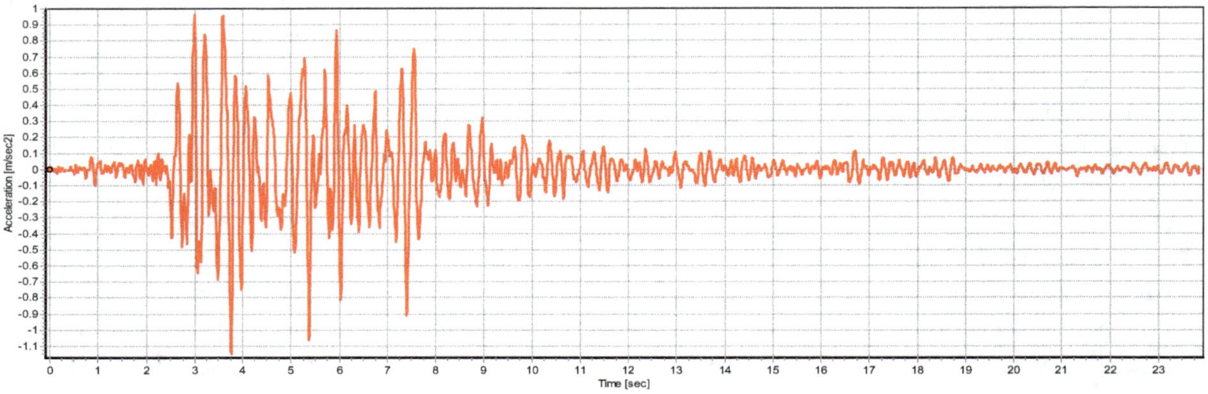

Figure 7. Petrovac earthquake acceleration scaled to 0.1 g [36].

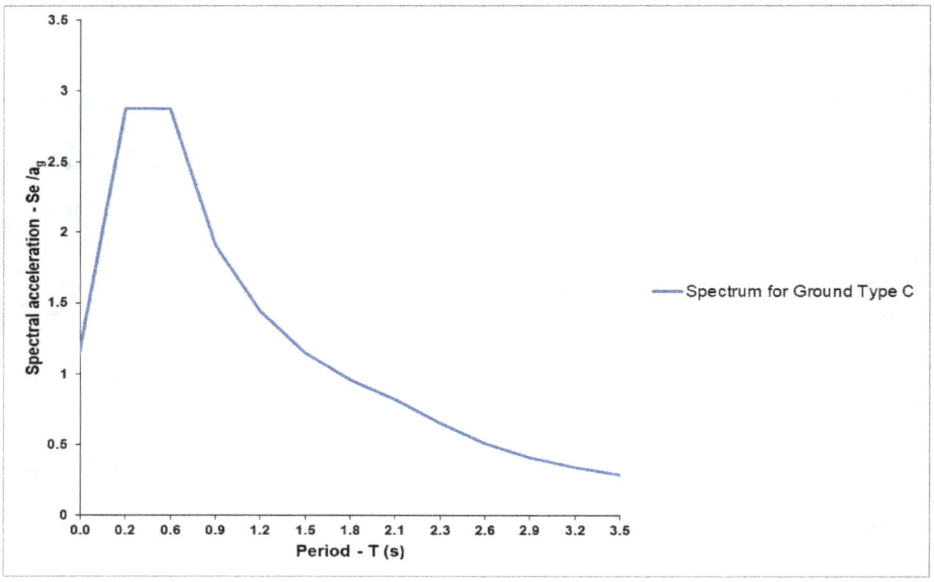

Figure 8. Elastic response spectra for Sarajevo according to Eurocode 8 for 0.1 g.

Due to the earthquake action of 0.1 g, in the Y direction, formation of typical diagonal cracks (Figure 9a) were noticed with concentration of the damage at the ground level (Figure 9b) (connection between the concrete basement and masonry structure) after 24 seconds.

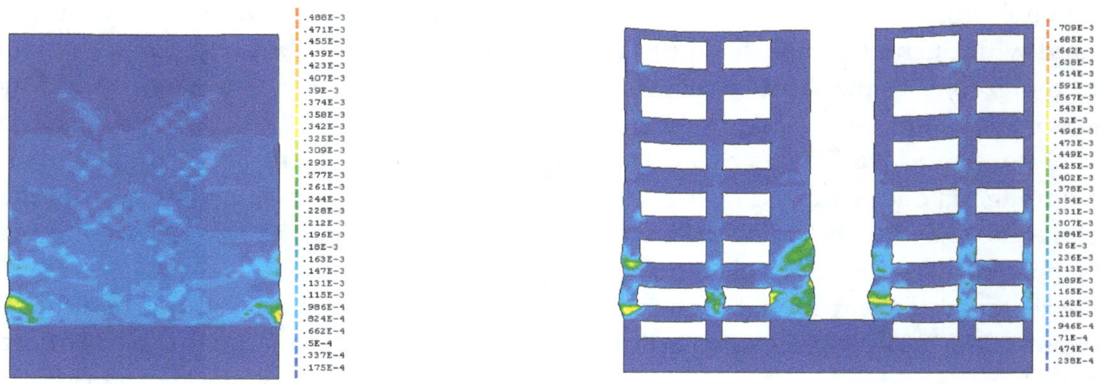

(a) Wall W-Y6

(b) Wall W-X1

Figure 9. (**a**) Typical diagonal X-shaped crack due to shear in the perpendicular wall; (**b**) concentration of damage at the ground floor in the longitudinal wall (0.1g scaled accelerogram).

On the other hand, it was only after 3.89 seconds that the structure showed significant damage, which caused the collapse of the structure, when the structure was exposed to 0.2 g of the same earthquake (Figure 10a,b).

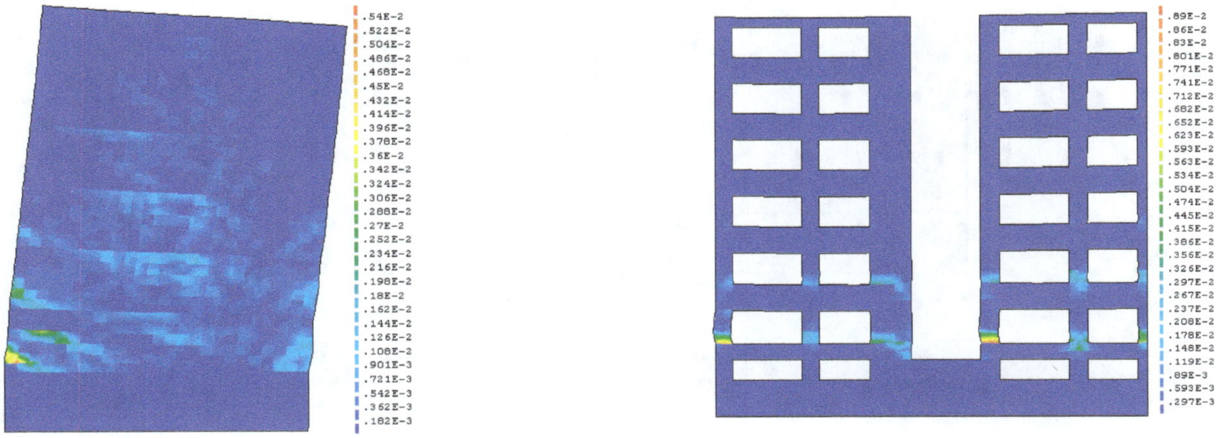

(a) Wall W-Y6 (b) Wall W-X1

Figure 10. (**a**) Damage of the perpendicular wall; (**b**) longitudinal wall due to 0.2 g scaled accelerogram.

From Figure 11 it is evident that, once the structure is exposed to the Petrovac ground motion scaled to 0.2 g, it undergoes very large displacements with a maximum of 29.33 cm after only 3.89 s, which is 2.6 times larger than when the structure was exposed to a 0.1 g scaled accelerogram. It is evident that these types of structures are highly vulnerable to such seismic actions and strengthening measures should be proposed.

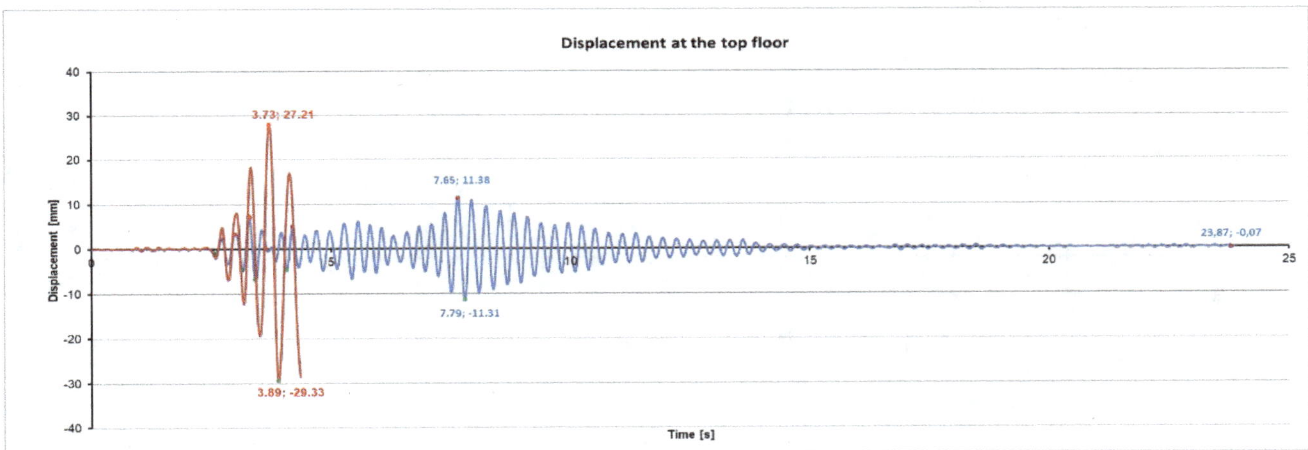

Figure 11. Displacement of the node 44014 at the top floor due to action of 0.1 g (indicated by a blue color) vs 0.2 g scaled accelerogram (indicated by the red color).

The maximum displacement in the case of 0.1 g is rather small and it amounts to 11.38 cm while for the case of higher seismic activity (0.2 g) the displacement increased to 29.33 cm, which is 2.6 times larger. The maximum inter-storey drift per each floor was calculated separately (as it does not happen simultaneously). The maximum value of inter-story drift of each floor was identified, and an envelope of the inter-story drift was created. Further, the maximum inter-storey drift (envelope) moved from 0.8 % to 5.8 % (Figure 12) as the PGA increased from 0.1 g to 0.2 g, making the structure more vulnerable, and showing that severe damage will occur if exposed to an earthquake of higher PGA (0.2 g).

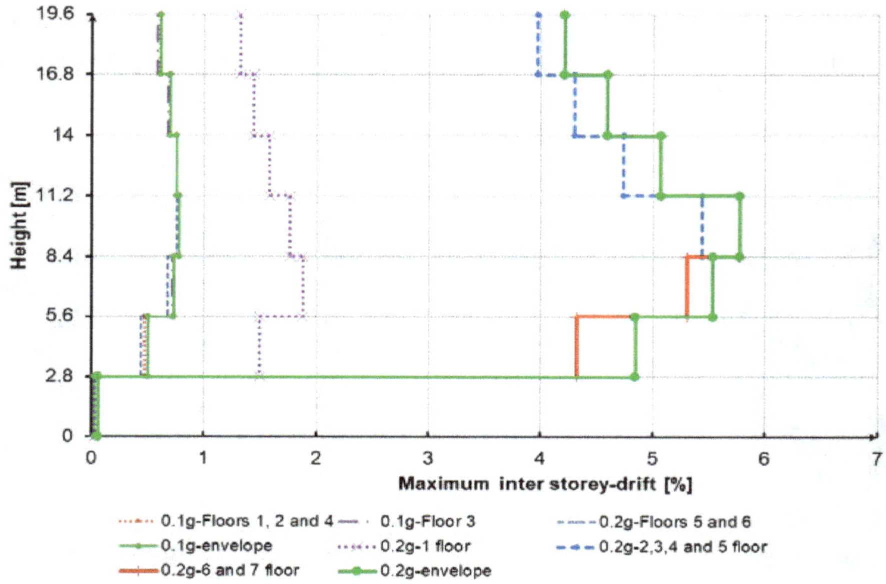

Figure 12. Maximum inter-storey drift for PGA 0.1g and PGA 0.2g scaled accelerograms.

The hysteresis curve originated by the 0.2 g scaled accelerogram was determined for the control node 44014 and the relationship between the seismic coefficient (ratio between base shear and gravity forces) and displacement has been obtained as shown in Figure 13. The maximum seismic coefficient of 0.37 was reached while the maximum displacement was 29.3 mm. From Figure 13, energy dissipation was observed due to flexural and shear phenomena. However, in order to understand the behavior of a structure besides the amount of dissipated energy, it is important to know the equivalent viscous damping ratio and dissipated energy ratio.

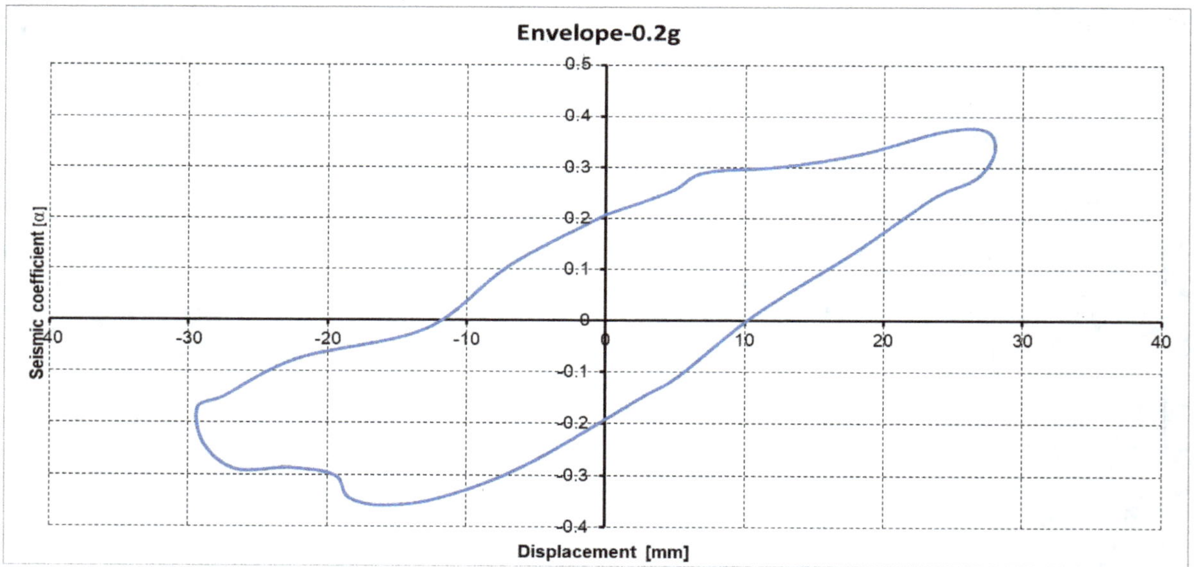

Figure 13. Envelope of the hysteresis curve in the transversal direction.

The structural analysis of the case study was also performed with the EFM (additional information could be found in references [5,6,29]). Here, only the results in the X direction of the façade walls will be shown. Damage was again concentrated in the lower floor of the structure with bending and compression failure at the ground level of the structure (Figure 14). The observed damage shown in Figure 14 (damage in masonry piers) is consistent with the damage observed on the building after the Skopje earthquake (Figure 2b). As results obtained by both FEM and EFM approaches regarding the pushover analysis were in a very good agreement, it was decided not to report the respective results here. In addition, due to the much less computational time required, it was decided to conduct further analysis in the X direction utilizing only the EFM approach. As stated before, one of the major flaws of this type of building is the nonexistence of adequate number of load bearing walls in the longitudinal (X) direction and the large number of openings causing a lower lateral stiffness in this direction with respect to the transversal (Y) direction.

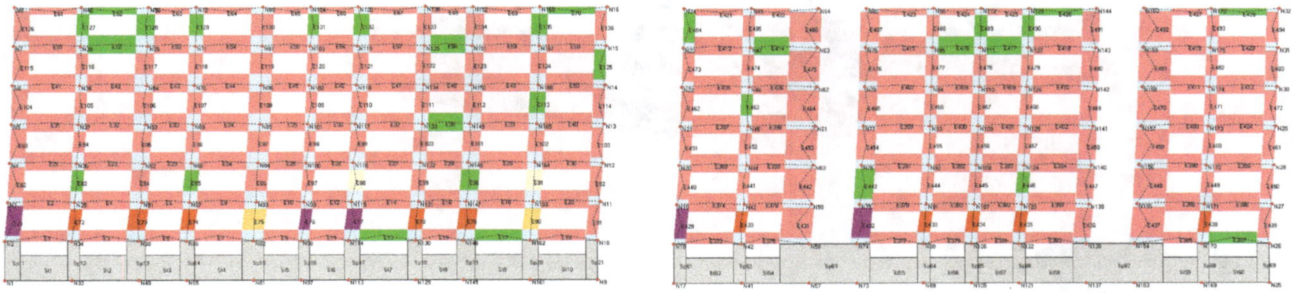

Figure 14. Façade walls damage of the original structure after exposure to PGA = 0.1 g.

4. Strengthening Intervention (EFM Model)

The proposed strengthening method followed here is a combination of traditional and modern strengthening techniques. This envisaged solution includes the construction of four new walls (Figure 15a) and strengthening the ground floor with fiber reinforced polymers (FRP) where concentration of damage was observed, see Figure 14. In this way, the structure moved from inadequate to the adequate positioning of the structural walls in plan [32]. The walls were positioned in such a way as to keep the symmetry of the structure and in that way circumvent possible undesirable torsional effects. The new walls are made of solid clay bricks connected with lime mortar in order to be compatible with the existing walls. The four new structural walls were built in place of existing separation walls in order to not affect the serviceability of the structure, besides being the most convenient location. The walls are founded on new continuous wall footings connected to the existing foundation. The two longitudinal walls surrounded with a red rectangle in Figure 15b were reinforced at the ground floor with carbon FRP bidirectional strips, having thickness of 2 mm and a width of 200 mm and spaced every 0.5 m, see S_d in Figure 15c. The used FRP material has the following mechanical properties: E = 240,000 MPa (modulus of elasticity), f_{fd} = 3182 MPa (tensile strength) and ε = 1.45% (ultimate elongation), area of the corresponding strips is $2 \times 200\ mm^2$.

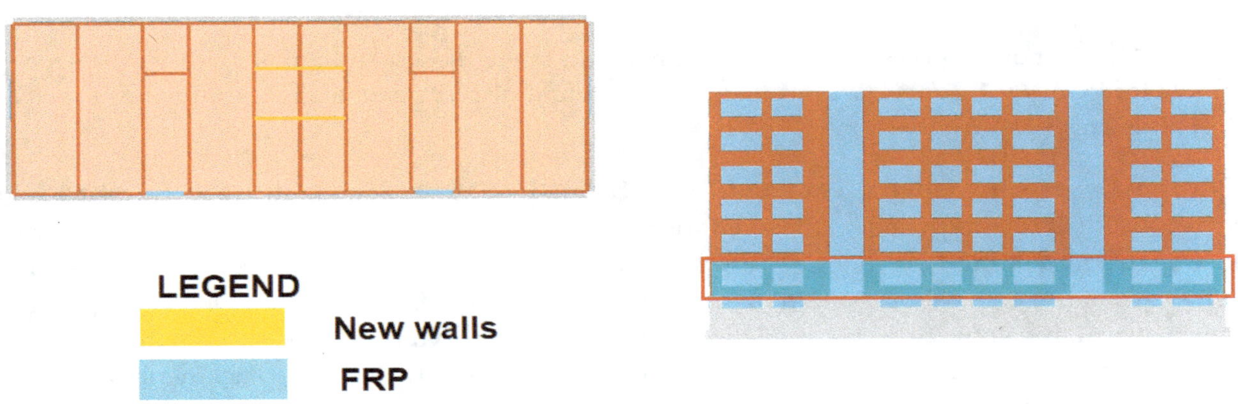

(a) (b)

Figure 15. *Cont.*

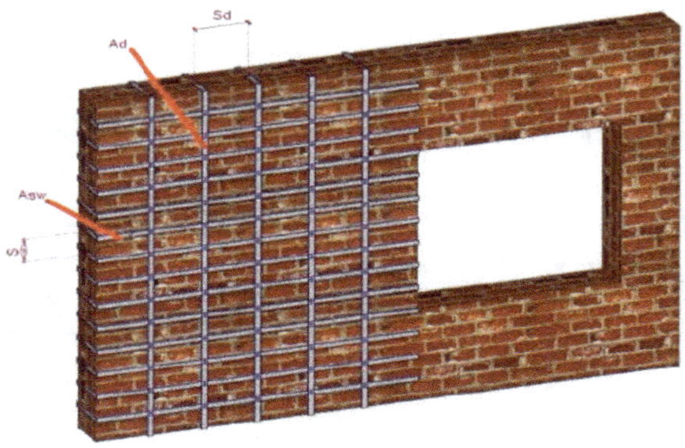

(c)

Figure 15. Strengthened structure (**a**) 4 additional walls; (**b**) FRP at the ground level; (**c**) FRP modeling in 3MURI [18].

The comparison of the performance of the unstrengthen and strengthened structure exposed to the elastic response spectra as defined in Eurocode 8 [37] for PGA equal to 0.1 g and 0.2 g and C soil type (Figure 16), shows that the strengthening procedure adopted could not overcome the problem of seismic vulnerability of this type of structures for larger earthquake motions. The structure has an adequate behavior if exposed to a smaller PGA, however bending failure occurs at the lower floors of the structure with the increase of PGA.

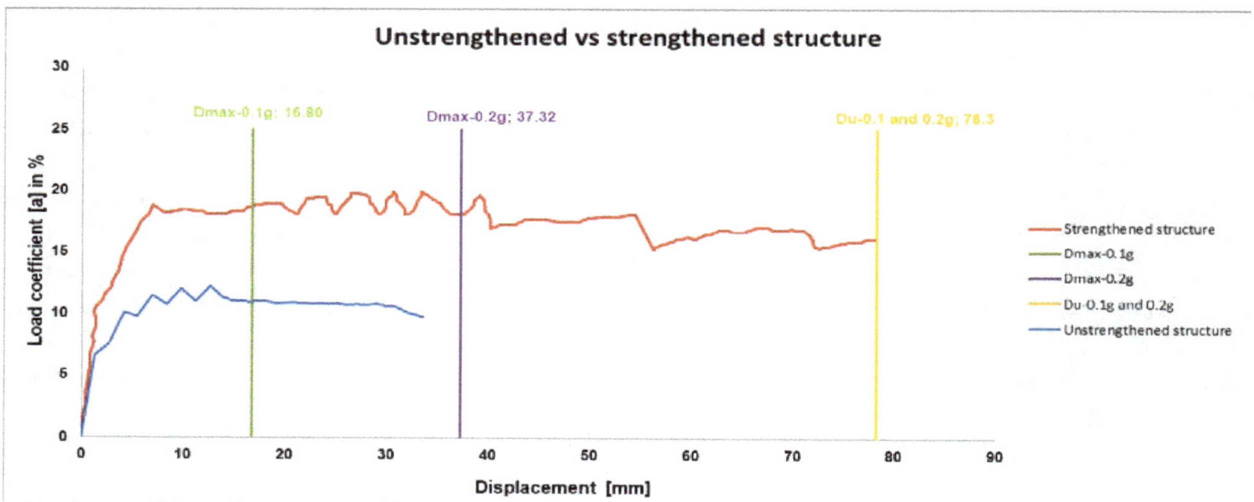

Figure 16. Pushover curves of the unstrengthen and strengthened structures, with performance points for PGA = 0.1 g and PGA = 0.2 g (D_{max}—demand displacement and D_u—ultimate capacity displacement).

Both analyses satisfied the Damage Limit State (DLS) check according to Eurocode 8 [37]. However, the structure exposed to the excitation of 0.2g did not pass the Ultimate Limit State (ULS) check, where the obtained value of q* (the relationship between the elastic response force and the yield strength of the equivalent system) as defined in reference [38] was 4.90, which is larger than the maximum acceptable value of 3 [38], while in the case of 0.1g this value was equal to 2.45. The available ductility (the ratio between the ultimate displacement and the elastic limit displacement) was 16.78 for the two analyzed cases. If the structure is to be exposed to an earthquake with PGA of 0.1g this intervention would be acceptable (Figure 17). However, this cannot be stated for an earthquake of a higher magnitude, where the concentration of damage due to bending is seen on the lower levels of the strengthened building (Figure 18) causing major damage on the masonry spandrels.

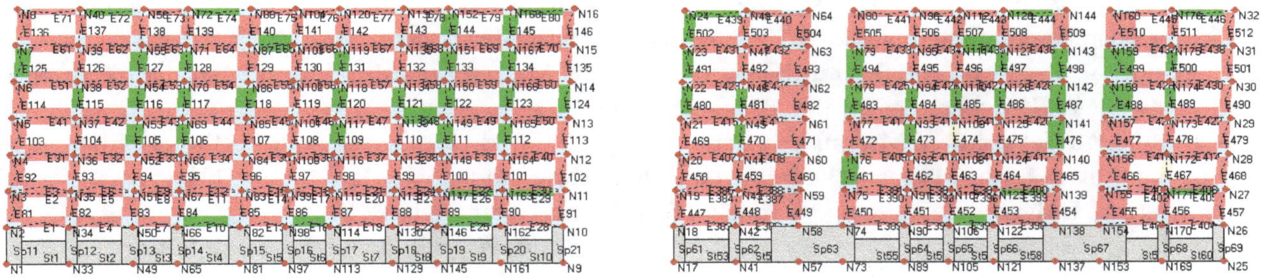

Figure 17. Wall damage after exposure to PGA = 0.1 g—satisfactory behavior of the structure.

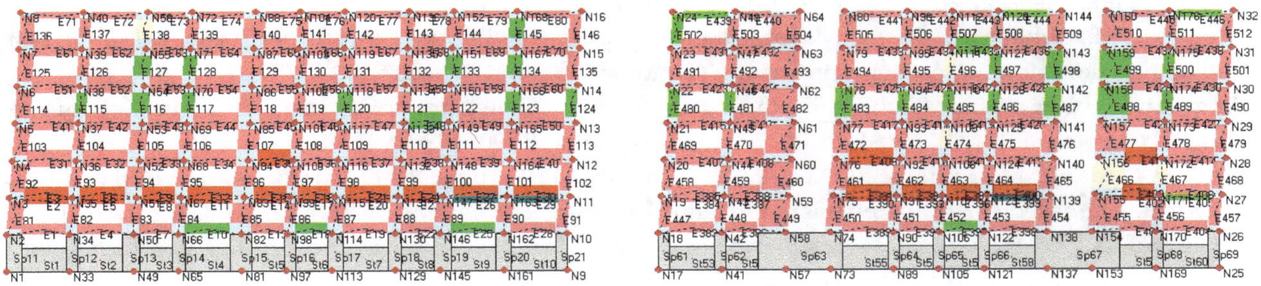

Figure 18. Wall damage after exposure to PGA = 0.2 g—inadequate behavior of the structure.

5. Conclusions

This article addresses the assessment of the seismic vulnerability of a typical URM building in Sarajevo, which is characteristic of the Balkan region, from Macedonia to Slovenia. Assessment of existing structures and determination of their vulnerability to seismic actions and possible intervention methods should be seen as a preventive measure for seismic risk mitigation. According to the hazard seismic map of Bosnia and Herzegovina, Sarajevo can expect earthquakes of up to 0.2g, giving a good basis and reason for assessment of structures that were built from 1920 to 1965 without the application of a seismic code, as such codes did not exist in this territory at that time. The vulnerability of these typologies is more than evident and adequate strengthening procedures should be planned and implemented in order to be in line with preventive measures. Focus should be put on earthquake awareness and mitigation of structural vulnerability so that in the case of future earthquakes, seismic losses are minimized and lives saved. It is an unfortunate fact that this type of structure has not been strengthened and, on the contrary, additional floors have been added on to it, which makes them even more vulnerable to future seismic actions.

Author Contributions: N.A. conducted the numerical simulations and analyzed the results. D.V.O. supervised the numerical process. P.B.L. revised the article and provided multiple suggestions for contents' change. All authors contributed in the writing of the article.

References

1. Bulajić, B.Đ.; Manić, M.I.; Lađinović, Đ. Effects of shallow and deep geology on seismic hazard estimates: A case study of pseudo-acceleration response spectra for the northwestern Balkans. *Nat. Hazards* **2013**, *69*, 573–588. [CrossRef]
2. Lee, V.W.; Herak, D.; Herak, M.; Trifunac, M.D. Uniform hazard spectra in western Balkan Peninsula. *Soil Dyn. Earthq. Eng.* **2013**, *55*, 1–20. [CrossRef]
3. BAS. *EN 1998-1:2018 Eurocode 8: Design of Structures for Earthquake Resistance—Part 1: General Rules, Seismic Actions and Rules for Buildings—National Annex*; Institute of Standardization of Bosnia and Herzegovina: Sarajevo, Bosnia and Herzegovina, 2018.

4. Ademović, N.; Kalman Šipoš, T.; Hadzima-Nyarko, M. Multidimensional Protocol for Evaluation of Seismic Risk for Bosnia and Herzegovina. *Bull. Earthq. Eng.* **2018**. under review.
5. Ademović, N. Structural and Seismic Behavior of Typical Masonry Buildings from Bosnia and Herzegovina. Master's Thesis, University of Minho, Braga, Portugal, 2011.
6. Ademović, N. Behavior of Masonry Structures in Bosnia and Herzegovina at the Effect of Earthquakes from the Viewpoint of Modern Theoretical and Experimental Knowledge. Ph.D. Thesis, Faculty of Civil Engineering, University of Sarajevo, Sarajevo, Bosnia and Herzegovina, 2012.
7. Lutman, M.; Tomazevic, M. *Unreinforced Brick Masonry Apartment Building*; Housing Report; Earthquake Engineering Research Institute and International Association for Earthquake Engineering: Ljubljana, Slovenia, 2002.
8. Bureau of Statistics CBS. *National Population and Housing Census 2013*; National Report; Agency for Statistics of Bosnia and Herzegovina: Sarajevo, Bosnia and Herzegovina, 2013.
9. Stojković, M.; Mihailov, V. *Seismic Microzonation Map of Banja Luka of the City Urban Area of Banja Luka*; Institute of Earthquake Engineering & Engineering Seismology: Skopje, Macedonia, 1972.
10. Regulations Regarding Technical Standards for the Construction of Buildings in Seismic Areas. Official Gazette SFRJ No. 31/81, 49.82, 29/83, 21/88 and 52/90. 1990.
11. Lee, V.W.; Manić, M.I.; Bulajić, B.Đ.; Herak, D.; Herak, M.; Trifunac, M.D. Microzonation of Banja Luka for performance-based earthquake-resistant design. *Soil Dyn. Earthq. Eng.* **2015**, *78*, 71–88. [CrossRef]
12. Grünthal, G. (Ed.) *European Macroseismic Scale 1998, EMS-98*; Center Europeen de Geodynamique et de Seismologie: Luxemburg, 1998; Volume 15, pp. 1–101.
13. JUS.B.D1.011. *Collection of Yugoslav Regulations and Standards for Construction Structures, Masonry Structures*; Book 5; Faculty of Civil Engineering of the University of Belgrade: Belgrade, Serbia, 1995.
14. JUS. U.M1.048. Regulations Regarding Technical Standards for Concrete and Reinforced Concrete. Official Gazette of SFRY 48/85. Belgrade. 1985.
15. CEN.Eurocode 2. *EN 1992-1-1:2004. Design of Concrete Structures—Part 1-1: General Rules and Rules for Buildings*; Committee European de Normalization: Bruxelles, Belgium, 2004.
16. CEN.Eurocode 6. *EN 1996-1-1:2005. Design of Masonry Structures. Part 1-1: General Rules for Reinforced and Unreinforced Masonry Structures*; Committee European de Normalization: Bruxelles, Belgium, 2005.
17. DIANA 9.4, TNO. Displacement method ANAlyser 9.4. Finite element analysis. User's Manual, release 9.4. Netherlands: S.n., Vol. Release 9.4. 2009. Available online: https://dianafea.com/ (accessed on 15 November 2018).
18. S.T.A. DATA 3 Muri. Manual, STA DATA srl, Torino. 2010. Available online: http://www.stadata.com/ (accessed on 15 November 2018).
19. Haach, V.G.; Vasconcelos, G.; Lourenço, P.B. Parametrical study of masonry walls subjected to in-plane loading through numerical modeling. *Eng. Struct.* **2011**, *33*, 1377–1389. [CrossRef]
20. Dolatshahi, K.M.; Aref, A.J. Two-dimensional computational framework of mesoscale rigid and line interface elements for masonry structures. *Eng. Struct.* **2011**, *33*, 3657–3667. [CrossRef]
21. Ghiassi, B.; Marcari, G.; Oliveira, D.V.; Lourenço, P.B. Numerical analysis of bond behavior between masonry bricks and composite materials. *Eng. Struct.* **2012**, *43*, 210–220. [CrossRef]
22. Marques, R.; Pereira, J.M.; Lourenço, P.B.; Parker, W.; Uno, M. Study of the Seismic Behavior of the "Old Municipal Chambers" Building in Christchurch, New Zealand. *J. Earthq. Eng.* **2013**, *17*, 350–377. [CrossRef]
23. Araujo, A.S.; Lourenço, P.B.; Oliveira, D.V.; Leite, J. Seismic Assessment of St James Church by Means of Pushover Analysis—Before and After the New Zealand Earthquake. *Open Civ. Eng. J.* **2012**, *6*, 160–172. [CrossRef]
24. Ademović, N.; Oliveira, D.V. Seismic Assessment of a Typical Masonry Residential Building in Bosnia and Herzegovina. In Proceedings of the 15th World Conference on Earthquake Engineering, Lisbon, Portugal, 24–28 September 2012; pp. 1–10.
25. Marques, R.; Lourenço, P.B. Unreinforced and confined masonry buildings in seismic regions: Validation of macro-element models and cost analysis. *Eng. Struct.* **2014**, *64*, 52–67. [CrossRef]
26. Formisano, A.; Iaquinandi, A.; Mazzolani, F.M. Seismic retrofitting by FRP of a school building damaged by Emilia-Romagna earthquake. *Key Eng. Mater.* **2015**, *624*, 106–113. [CrossRef]
27. Paparo, A.; Beyer, K. Modelling the seismic response of modern URM buildings retrofitted by adding RC walls. *J. Earthq. Eng.* **2015**, *20*, 587–610. [CrossRef]

28. Cattari, S.; Lagomarsino, S. Seismic assessment of mixed masonry-reinforced concrete buildings by non-linear static analyses. *Earthq. Struct.* **2013**, *4*, 241–264. [CrossRef]
29. Ademović, N.; Hrasnica, M.; Oliveira, D.V. Pushover analysis and failure pattern of a typical masonry residential building in Bosnia and Herzegovina. *Eng. Struct.* **2013**, *50*, 13–29. [CrossRef]
30. Maio, R.; Vicente, R.; Formisano, A.; Varum, H. Seismic vulnerability of building aggregates through hybrid and indirect assessment techniques. *Bull. Earthq. Eng.* **2015**, *13*, 2995–3014. [CrossRef]
31. Lourenço, P.B. Computational Strategies for Masonry Structures. Ph.D. Thesis, Technische Universiteit Delft, Delft, The Netherlands, 1996.
32. Tomaževič, M. *Earthquake-Resistant Design of Masonry Buildings*; Tom, I., Ed.; Imperial College Press: London, UK, 1999.
33. Cattari, S.; Lagomarsino, A. Strength criterion for the flexural behaviour of spandrel in un-reinforced masonry walls. In Proceedings of the 14th World Conference on Earthquake Engineering, Beijing, China, 12–17 October 2008; p. 41.
34. Tomaževič, M.; Iztok, K. Verification of Seismic resistance of confined masonry buildings. *Earthq. Eng. Struct. Dyn.* **1997**, *26*, 1073–1088. [CrossRef]
35. Tomaževič, M.; Bosiljkov, V.; Polona, W. Structural behavior factor for masonry structures. In Proceedings of the 13th World Conference on Earthquake Engineering, Vancouver, BC, Canada, 1–6 August 2004; p. 2642.
36. Seismosignal. Earthquake Engineering Software Solutions, Version 4.2.1. 2010. Available online: https://www.seismosoft.com/ (accessed on 15 November 2018).
37. CEN.Eurocode 8. *EN 1998-1-2004: Design of Structures for Earthquake Resistance—Part 1: General Rules, Seismic Actions and Rules for Buildings*; Committee European de Normalization: Bruxelles, Belgium, 2004.
38. OPCM 3274 Norme tecniche per Il progetto, la valutazione e l'adeguamento sismico degli edifici testo integrato dell'Allegato 2—edifici—all'ordinanza 3274 come modificato dall'OPCM 3431. 2005. Available online: http://zonesismiche.mi.ingv.it/pcm3274.html (accessed on 15 November 2018).

11

Evaluation of Repair Costs for Masonry Infills in RC Buildings from Observed Damage Data: The Case-Study of the 2009 L'Aquila Earthquake

Maria Teresa De Risi *, Carlo Del Gaudio and Gerardo Mario Verderame

Department of Structures for Engineering and Architecture, University of Naples Federico II, 80125 Naples, Italy; carlo.delgaudio@unina.it (C.D.G.); verderam@unina.it (G.M.V.)

* Correspondence: mariateresa.derisi@unina.it

Abstract: The estimation of direct and indirect losses due to earthquakes is a key issue in the Performance Based Earthquake Engineering framework. In commonly adopted loss computation tools, no specific data related to masonry infill panels, widespread in moment-resisting-frame residential buildings, are available to perform a probabilistic assessment of losses. To fill this gap, specific fragility and loss functions have been recently proposed in the last years. To assess their validity and estimate the relevance of the repair costs due to infills after earthquakes with respect to the total reconstruction process, the present work analyses the Reinforced Concrete residential buildings with masonry infills struck by the 2009 L'Aquila (Italy) earthquake, focusing on the dataset of "lightly" damaged buildings, where only damage to masonry infills occurred. Based on available data related to these buildings, the observed damage scenario after L'Aquila earthquake is first obtained. The repair costs for infills are estimated given this damage scenario. The resulting estimated repair costs are then compared with the actual repair costs presented in the available literature. The percentage influence of infills on the total repair costs due to earthquakes for residential buildings is lastly computed, resulting on average equal to the fifty percent.

Keywords: RC buildings; masonry infills and partitions; damage analysis; repair costs due to infills; post-earthquake surveys

---

1. Introduction

Earthquakes certainly represent a paramount kind of natural hazard due to the current impossibility to predict their occurrence and significant impact on civil structures worldwide in terms of social consequences, direct and indirect losses, and risk of casualties. FEMA P-58 [1] explicitly proposes a seismic performance assessment methodology for buildings according to the Performance-Based Earthquake Engineering (PBEE) concept [2], namely combining seismic hazard, structural response, damage analysis and associated consequences, the latter defined in terms of repair costs, repair time, and casualties. FEMA P-58 is also accompanied by a Performance Assessment Calculation Tool (PACT) [3], a user-friendly electronic calculation tool, including a repository of component-by-component (structural and non-structural) fragility and consequence data, which allows performing the probabilistic assessment of losses, as described in the FEMA P-58 methodology.

Several works over the last few years have carried out an estimation of the post-earthquake damage [4] and total repair costs (e.g., [5,6], among others). In particular, after the 2009 L'Aquila (Italy) earthquake, a systematic analysis of the actual costs for reconstruction was carried out, both for repairing and retrofitting activities [7–11]. Some of these studies also highlighted that no specific data related to masonry infill panels in typical moment-resisting-frame residential buildings are accounted for in the PACT tool [11,12]. This gap is particularly important for typical Mediterranean building stock, especially for Reinforced Concrete (RC) Moment Resisting Frames (MRF), generally characterized by masonry exterior infills and partitions. The analysis of post-earthquake damage data in [7,13–16] highlights the key role played by damage to the masonry infills and partitions. Therefore, seismic performance assessment of infilled RC frames needs to take into account also the contribution of these components in estimating properly the expected seismic performance of RC MRF both in terms of seismic response [17] and loss estimation [18]. To fill the above-mentioned gap, specific fragility and loss functions have been recently carried out in [12] and [18]. These proposals need to be compared with actual repair costs, specifically related to the masonry infills, obtained from real post-earthquake damage surveys, to assess their validity and estimate the relevance of the repair costs due to infills with respect to the total reconstruction process. This is the core of this work.

To this aim, the present work analyses the RC residential buildings with masonry infills struck by the 6 April 2009 L'Aquila earthquake, focusing on the dataset of "lightly" damaged buildings. "Lightly" damaged buildings are defined herein as those buildings where only damage to masonry infills occurred. The evaluation of repair costs can thus be made by neglecting the contribution due to repair activity to other structural components (e.g., vertical structures, horizontal structures, stairs, and roofs). All these buildings, in the Abruzzi region, after the 2009 earthquake, have been charged with a post-earthquake usability in-situ assessment procedure [4], providing for each building a very useful description of the damage severity and extent and its resulting usability rating. The observed damage scenario for the investigated building stock is obtained and analysed in detail.

The repair costs for infills are then estimated, starting from the cost analysis related to a single infill panel reported in Del Gaudio et al. [18], and by means of its extension to the whole analysed dataset of buildings. The resulting estimated repair costs are lastly compared with the actual repair costs properly analysed in Dolce and Manfredi [19], depending on the building's usability rating, to compute the percentage influence of infills on the total amount of all the repairing activities.

2. IN-SITU Post-Earthquake Damage and Safety Assessment

In-situ surveys are significantly important for a rapid response to earthquake emergency. Their primary goal is to judge the usability of inspected buildings, defining if they can still be used with a reasonable level of safety. Secondly, these data, properly elaborated, can also be used to provide statistics about damage to buildings after a seismic event, as recently published by the Italian Department of Civil Protection through the Da.D.O. platform ("Database di Danno Osservato" [20]).

A well-consolidated post-earthquake usability assessment procedure was proposed in the past by Baggio et al. [4], and it was widely used in Italy in the aftermath of recent events, from the 1997 Umbria-Marche earthquake to present day, thus representing the base of the above-mentioned Da.D.O. platform. These post-earthquake survey data are synthetically reported in the so-called AeDES (Agibilità e Danno nell'Emergenza Sismica, Usability and Damage in Post-Earthquake Emergency) form [4].

In general, the parameters collected by the AeDES form can be grouped into five macro-sections:

- Building identification: data related to the municipality and position of the building;
- Building description: number of stories, average storey height, average surface, construction and renovation age, use and utilisation (Figure 1);
- Building typology: information on vertical and horizontal (masonry or RC) structures, on the presence of tie rods or tie beams, of isolated columns, of mixed type structures among others, on plan and elevation regularity;
- Damage: (3 + 1) damage levels are considered ("Null"; "D1: Slight"; "D2–D3: Medium-Severe"; "D4–D5: Very heavy") based on European Macroseismic Scale [21] classification, with explicit indication about the damaged portion of the whole building (<1/3, 1/3–2/3, >2/3) for different structural components (vertical structures, floors, stairs, roofs, infills/partitions);

- Usability: six usability judgements are reported as a function of the risk conditions detected on the structure for structural and non-structural components (or external risk), and eventual short-term countermeasures are suggested.

Total number of stories		*Average storey height* [m]	*Average storey surface* [m²]		*Construction and renovation* [max 2]	*Use*
						A ☑ Residential
○ 1	○ 9	1 ○ ≤ 2.50	A ○ ≤ 50	I ○ 400 ÷ 500	1 ○ ≤ 1919	B ○ Production
○ 2	○ 10	2 ○ 2.50÷3.50	B ○ 50 ÷ 70	L ○ 500 ÷ 650	2 ○ 19 ÷ 45	C ○ Business
○ 3	○ 11	3 ○ 3.50÷5.0	C ○ 70 ÷ 100	M ○ 650 ÷ 900	3 ○ 46 ÷ 61	D ○ Offices
○ 4	○ 12	4 ○ > 5.0	D ○ 100 ÷ 130	N ○ 900 ÷ 1200	4 ○ 62 ÷ 71	E ○ Public services
○ 5	○ >12		E ○ 130 ÷ 170	O ○ 1200 ÷ 1600	5 ○ 72 ÷ 81	F ○ Warehouse
○ 6		*No. of basements*	F ○ 170 ÷ 230	P ○ 1600 ÷ 2200	6 ○ 82 ÷ 91	G ○ Strategic services
○ 7		A ○ 0 C ○ 2	G ○ 230 ÷ 300	Q ○ 2200 ÷ 3000	7 ○ 92 ÷ 01	H ○ Touristic
○ 8		B ○ 1 D ○ ≥3	H ○ 300 ÷ 400	R ○ > 3000	8 ○ ≥ 2002	

Figure 1. Extract of the AeDES form for building description—adapted from Baggio et al. [4].

Note that only residential buildings are analysed in the following (Figure 1).

The analysed damage can be related to vertical structures, floors, stairs, roofs, infills, or can be pre-existing damage. The "structural" damage can be also related to infills and partitions, thus recognising their primary role in structural responses. Possible damage grade and extent in infills is described in Section 2.1, as suggested in the AeDES user manual [4]. How this damage description can provide a usability judgment is the objective of Section 2.2.

2.1. Damage to Infills According to the Aedes Form

As for the other structural components, the damage grade to infills can be defined as "Null", "D1: Slight", "D2–D3: Medium-Severe", or "D4–D5: Very heavy". Damage extension can be reported in the following ranges: "<1/3", "1/3–2/3", ">2/3" of the whole building.

Damage grade D1 ("Slight") is a damage level that does not significantly change the behaviour of the building and does not affect the safety of the building' occupants. Particularly for infills, slight detachments (<1 mm) of the infill panels from the surrounding beams/columns can occur, with eventual cracks (<1 mm width) due to the participation of the infill to the total lateral strength of the building. This damage level for infills can contribute to the definition of a "low" total damage level in the building, unless there is a certain degree of risk of out-of-plane collapse due to the eventual absence of connection between the panel and other structural components.

A more severe damage level ("D2–D3: Medium-Severe") can significantly change the building's lateral strength, which will be lower in the case of a subsequent similar seismic shaking, even if without any collapse risk. The infill panel can present cracks (between 1 and 5 mm) due to detachment from the surrounding elements, diagonal cracks up to "some" millimetres, and quite evident corner crushing with some localized bricks expulsions. If a big number of infill panels (high extent) are affected by this damage level, the total structural risk can be "high"; otherwise, a lower damage level can be evaluated case-by-case for less significant damage.

Damage grade D4–D5, defined as "Very heavy" damage, significantly modifies the structural response leading to a possible partial or total collapse of the building. Cracks' width and extension on the infills are significantly more severe than for the previous damage level. Some examples of Medium-Severe and Very Heavy damage to infills are shown in Figure 2.

(a) (b) (c)

Figure 2. Examples of Medium-Severe and Very Heavy damage to infills: horizontal and vertical cracks between infill and beams/columns—widespread damage D2–D3 and locally D4–D5 (**a**); damage D4–D5 to infills (**b**); damage D4–D5 with out-of-plane collapse of one infill leaf (**c**)—adapted from [4].

2.2. Post-Earthquake Usability Rating According to the Aedes Form

The damage extent and severity for structural and non-structural elements are lastly converted in a usability judgment for the analysed building [4]. Six usability judgments can be selected for each building, from "A" to "F" (Figure 3). In particular:

- "A" ("Usable building") does not mean that the building has not suffered any damage, but that the repair of damage is not a necessary condition for the usability of the building;
- "B" ("Temporary Unusable") requires short-term countermeasures to reduce the risk to the occupants to an "acceptable" level; the building is unusable until these countermeasures are realized;
- The judgment "C" ("Partially Unusable") is like "B" but related to only a part of the building;
- "D" is a case in which a further or more expert investigation is required;
- "E" ("Unusable building") means that the building cannot be used at all and short-term countermeasures are not enough. Damages could be repaired, but the repairing activities must be considered as part of the "reconstruction process";
- "F" is related to external risk sources.

A	USABLE building	○
B	UNUSABLE building (totally or partially), but USABLE after short term countermeasures	○
C	PARTIALLY UNUSABLE building (1)	⊘
D	TEMPORARILY UNUSABLE building requiring a more detailed investigation	○
E	UNUSABLE building	○
F	UNUSABLE building due to external risk (1)	○

Figure 3. Usability judgments according to the AeDES form—adapted from Baggio et al. [4].

Only usability judgment "A", "B" and "E" will be considered in the following analyses reported herein, since the other usability judgements are related to exterior risk ("F"), or to not definitive analyses ("D"), or to a not-known unusable portion of the building ("C").

Note that the information about the usability judgment is particularly important since generally it is strictly related to the reconstruction refund level for each building [9,10,19]. In this work, the repair costs due to infills, estimated as explained in the following sections, will be compared with the total actual reconstruction cost (reported in [19]), depending on the usability judgment, to highlight the percentage incidence of the infills on the whole repair cost.

3. Post-L'Aquila 2009 Earthquake Observed Damage to RC Buildings

In this work, the attention is focused on the well- and sadly- known L'Aquila 2009 seismic event, whose main characteristics are briefly described in Section 3.1. After this event, an extensive post-earthquake survey campaign was performed, based on the AeDES form to evaluate the produced damage to residential buildings and to judge the usability of those buildings, as mentioned in Section 2. Thanks to this data collection, a subset of buildings are investigated herein, as explained and described in Section 3.2, to obtain an "observed" damage scenario (Section 3.3) allowing, in the end, to identify post-earthquake repair costs due to infills, one of the main aims of this work.

3.1. Seismic Input Description

On 6th April, 2009, an earthquake of magnitude Mw = 6.3 struck the Abruzzo region, heavily affecting the area in the proximity of L'Aquila city, and killing 308 people. The area near the epicentre, in the neighbourhood of L'Aquila Municipality, was seriously damaged, resulting in IX–X grade of MCS (Mercalli–Cancani–Sieberg) macro-seismic scale.

The related ShakeMap in terms of Peak Ground Acceleration (PGA) and spectral ordinates (PSA) (for periods of vibration, T, equal to 0.3, 1 and 3 sec) can be derived by means of the Italian National Institute of Geophysics and Volcanology (INGV) procedure [22]. The ShakeMap in terms of PGA is shown in Figure 4. The map is derived by means of the software package ShakeMap®by using different Ground Motion Prediction Equations and signals registered by Italian Strong Motion Network (Rete Accelerometrica Nazionale, RAN) and the Italian National Seismic Network (RSN).

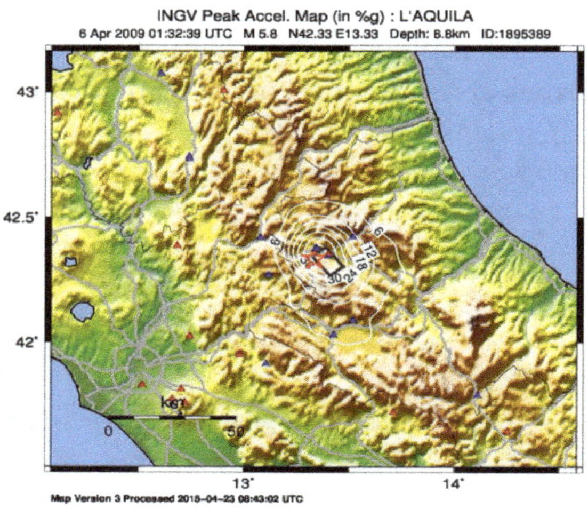

Figure 4. Shakemaps derived by the Italian National Institute of Geophysics and Volcanology (INGV) for PGA.

3.2. Investigated Database: "Lightly Damaged" Infilled RC Buildings

The dataset of buildings investigated in this work is made up of the MRF residential RC buildings located in the Abruzzi region, which after the 2009 earthquake have been involved in the post-earthquake usability assessment procedure by means of the AeDES form [4]. The focus herein has been first restrained from an original sample of 12223 to a subset of 7597 residential MRF RC buildings (for further detail, see [13]) from those collected in the Da.D.O. platform [20]. Among these data, only the buildings characterized exclusively by damage to infill panels are considered herein, since the aim of this work is the evaluation of repair costs due to infills in RC buildings, neglecting the contribution of repairing activity to other structural components (namely vertical structures, horizontal structures, stairs, roofs). An extract of the AeDES form is reported in Figure 5, to highlight the selection process among all the forms available in the Da.D.O platform. Therefore, only buildings for which the AeDES form reported damage to exterior infills and interior partitions and "Null" damage to all the other structural components are considered for the following analyses (as shown by the blue box in Figure 5). These buildings are defined herein as "lightly damaged buildings". The resulting database analysed herein is thus composed of 5095 RC buildings. The related frequency distribution of number of stories, year of construction, plan area (A), and suffered PGA during the main seismic shock are reported in Figures 6 and 7.

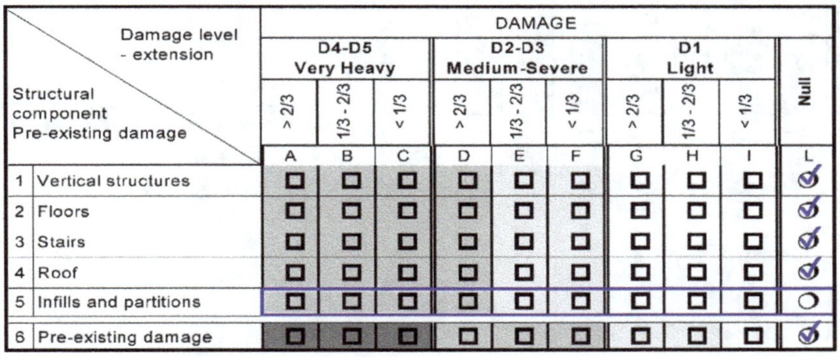

Figure 5. Damage description according to the AeDES forms and analysed damage—adapted from Baggio et al. [4].

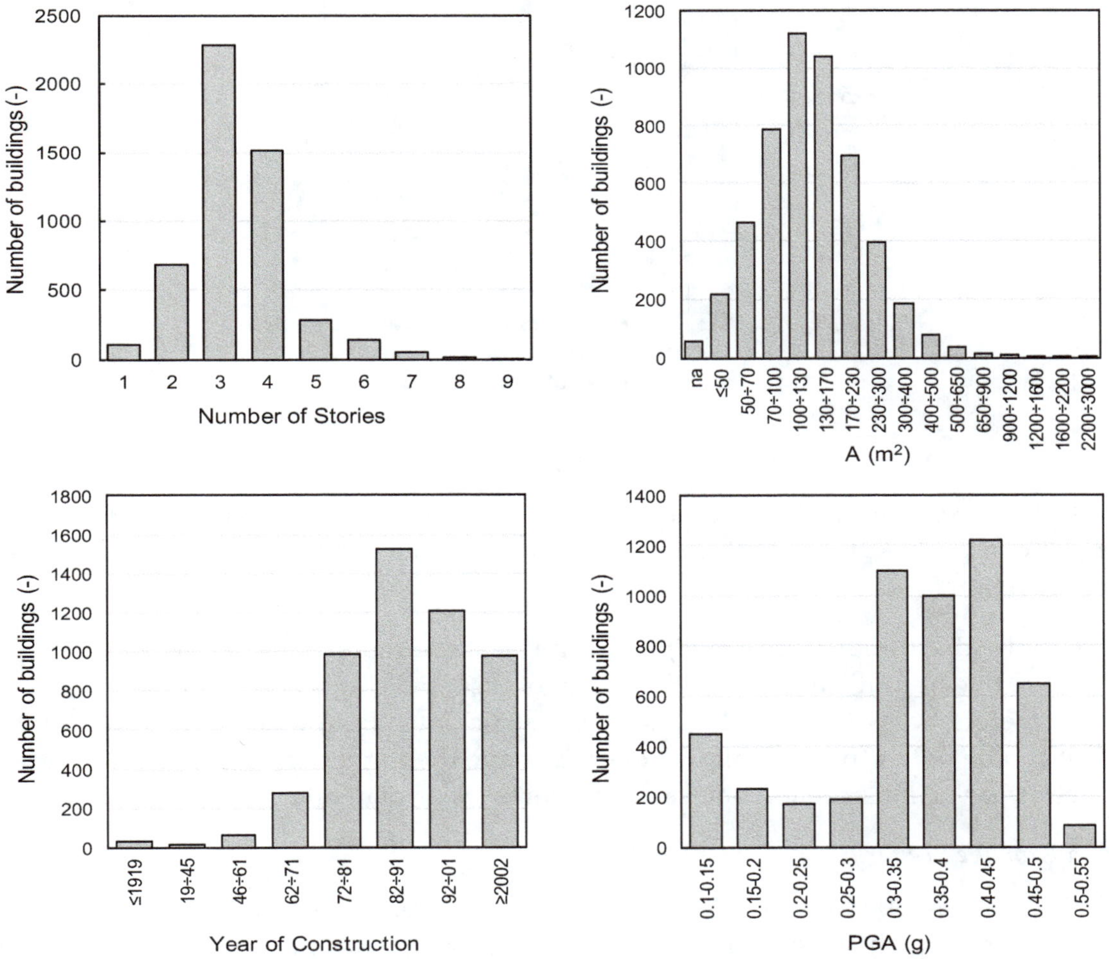

Figure 6. Frequency distributions of number of stories, plan area (A) ("na" = not available), year of construction, and PGA suffered during the L'Aquila 2009 earthquake for the analysed subset of buildings.

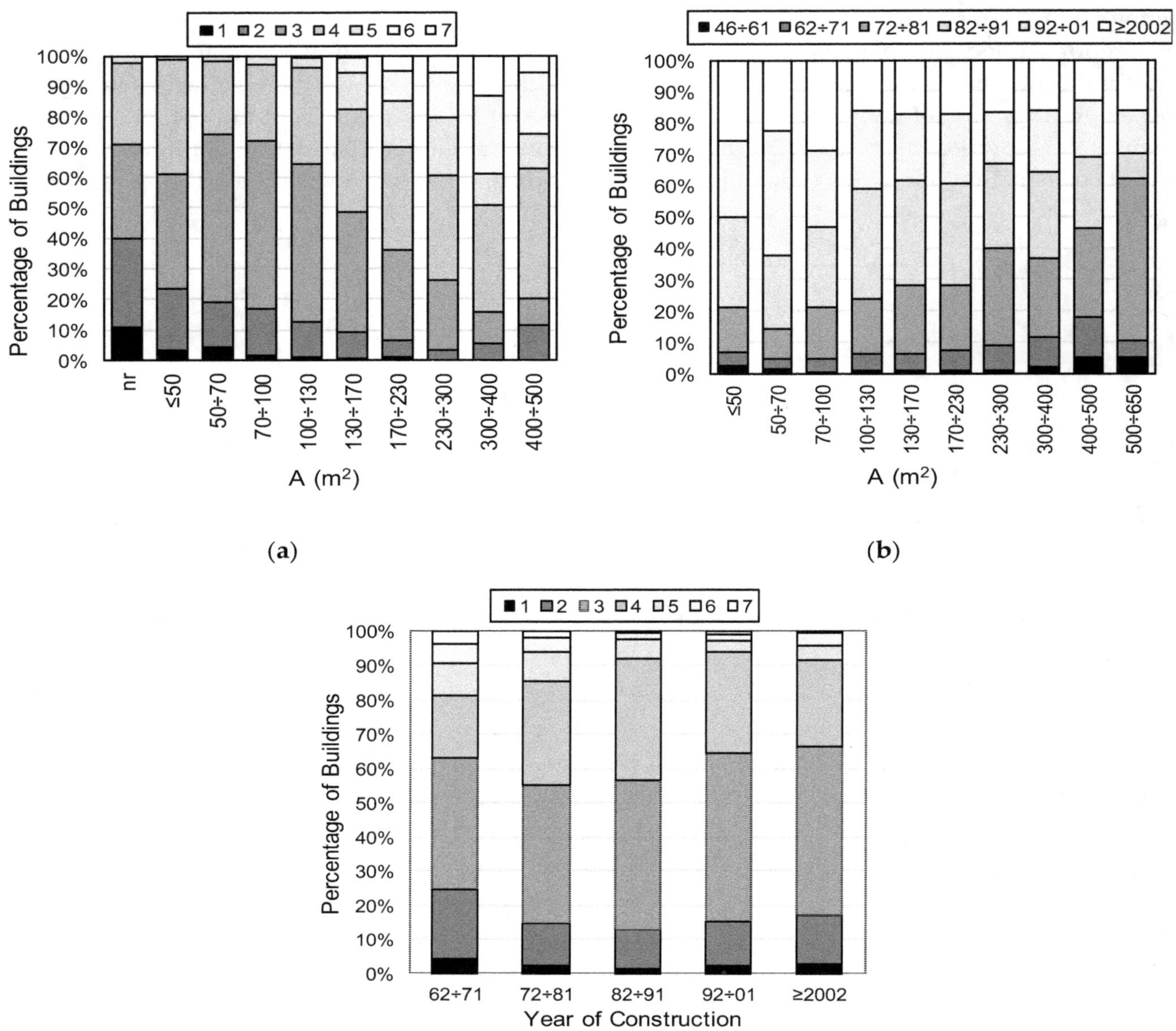

(c)

Figure 7. Cross-correlation in the frequency distributions for the analysed subset of buildings: number of storeys (from 1 to 7)—plan area (**a**), year of construction—plan area (**b**), number of storeys—year of construction (**c**).

Figure 6 shows that three-storey buildings are the most common; additionally, a relevant percentage of buildings have two or four stories. The major part of these buildings has a plan area varying between 50–300 square meters, was built after 1972, and was struck by a PGA level in the range of 0.30–0.50 g. In Figure 7, it can be noted that the tallest buildings have the highest plan surface. No other evident trends between year of construction and number of stories or plan area can be found (Figure 7).

3.3. Observed Damage Scenario

Based on data described in the previous section, the observed damage scenario is obtained and shown in this section. To obtain such a damage scenario, some assumptions are necessary, particularly in relation to the damage metric definition for infill panels.

Over the past years, authors have proposed definitions of different Damage States (DSs) through observation on the extent and severity of cracking patterns on the panels or the failure of brick units.

Some other research works also relate such damage levels to the achievement of the peak strength of the infilled frame or the achievement of given lateral strength reduction thresholds. Typically, three or four DSs have been defined in the literature, corresponding, for increasing damage level, to (i) the onset of cracking and first detachment between infill panel and the surrounding RC frame, (ii) the widening of previous damage pattern, (iii) the crushing and spalling of a significant portion of bricks and (iv) the partial/total collapse of the panel.

European Macroseismic Scale (EMS-98) [21], first, proposes three DSs specifically for infill panels in RC frames, as a function of a qualitative description of damage (Table 1): fine cracks (DS1), large cracks (DS2), collapse (DS3). DS4 and DS5 are also defined in the EMS-98 scale, but they are basically related to damage suffered by RC members (in the specific case, RC buildings). Therefore, due to the scope of the present work, these latter two DSs will be neglected in the following analyses.

Table 1. Correspondence of damage level according to EMS-98 and AeDES form (each background color represents the related DS in the following figures).

DS	EMS-98 [21]	AeDES form [4]	
	Damage Description	**Damage Severity**	**Damage Extent**
DS0	No Damage	D0—Null Damage	None
DS1	Negligible to Slight damage: Fine cracks in partitions and infills.	D1: Slight	<1/3
			1/3–2/3
			>2/3
DS2	Moderate damage: Cracks in partition and infill walls	D2–D3: Medium—Severe	<1/3
			1/3–2/3
			>2/3
DS3	Substantial to Heavy damage: Large cracks in partition and infill walls, failure of individual infill panels	D4–D5: Very Heavy	<1/3
			1/3–2/3
			>2/3

Similar to EMS-98, AeDES survey forms [4] define three DSs, as explained in Section 2, reporting a more accurate and detailed damage description with a quantitative indication of crack width for each one. A certain degree of correlation can be found between these two damage scales, as reported in Table 1.

Starting from the damage metric reported in Table 1, the collected buildings with damage to infills can be classified in DS1, DS2, or, DS3, depending on the information reported on the related AeDES form. The resulting "observed" damage scenario is shown in Figure 8a, reporting the number of buildings in each DS, where the DS of the whole building is assumed as the maximum observed damage level identified in the AeDES form for that building. In summary, 2406 buildings present no damage to infills and partitions (and no damage to vertical structures, roofs, stairs, etc.). A total of 1943 buildings fall within the damage level DS1, 555 are in DS2, and a smaller portion (191 buildings) presents a damage level DS3.

For each building, the maximum damage level shown in Figure 8a can be attained with different damage extents, as shown in Figure 5 and Table 1. Figure 8b additionally shows the distribution of the damage extent ("<1/3", "1/3–2/3", ">2/3") that defined the maximum achieved DS. Note that, for each given maximum DS, the prevalent damage extent is "<1/3", namely the maximum achieved damage is generally quite concentrated in a small portion of the building. Together with this maximum damage level, with its extent in a certain portion of the building, a less severe damage—distributed in other portions/stories of the buildings—can (co)exist. The information about this "co-existing" damage is shown in Figure 9, depending on the maximum achieved damage level. For all buildings with maximum damage level DS1 (1943 buildings), different damage extent can be present; their complement to the unity is assumed to be in DS0. About buildings with maximum damage level DS2 (555 buildings), a certain extent of less severe damage (DS1) or not damaged portions of the whole building (DS0) can be present, as shown in Figure 9. Similarly, for a building with a maximum damage level equal to DS3, some portions of the building with a damage level DS0, DS1 or DS2 can be present. Such information should be considered for a realistic repairing cost estimation, as explained in Section 4.

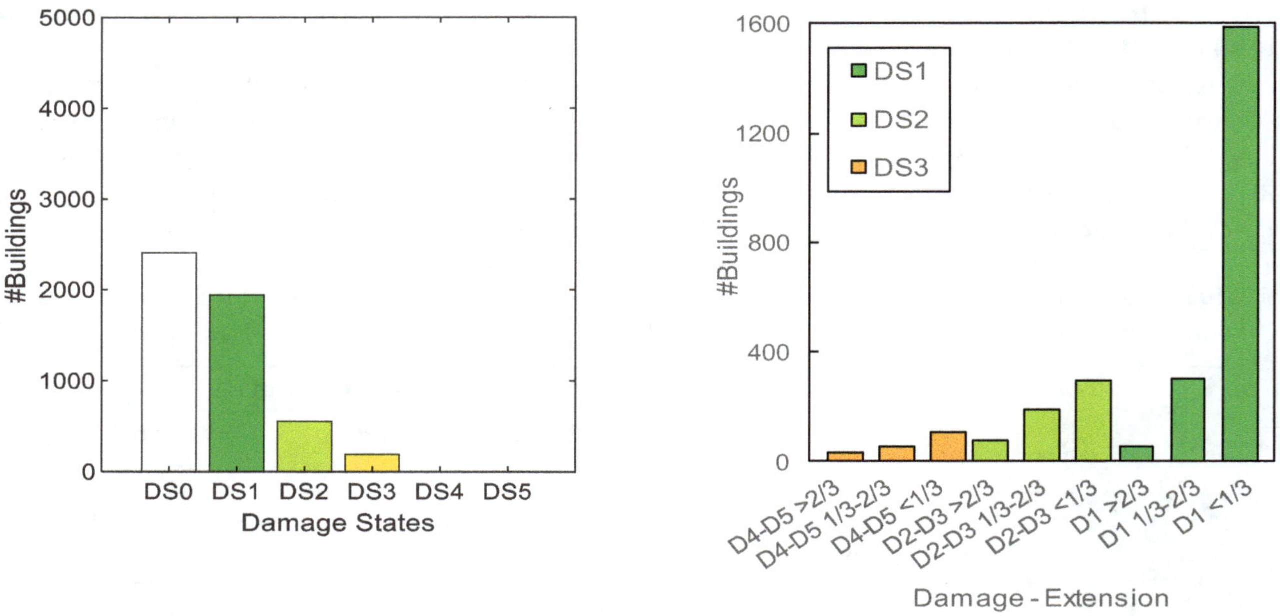

(a) (b)

Figure 8. "Observed" maximum damage scenario for "lightly damaged" RC buildings (**a**), and details about the extent of the maximum damage in each DS (**b**).

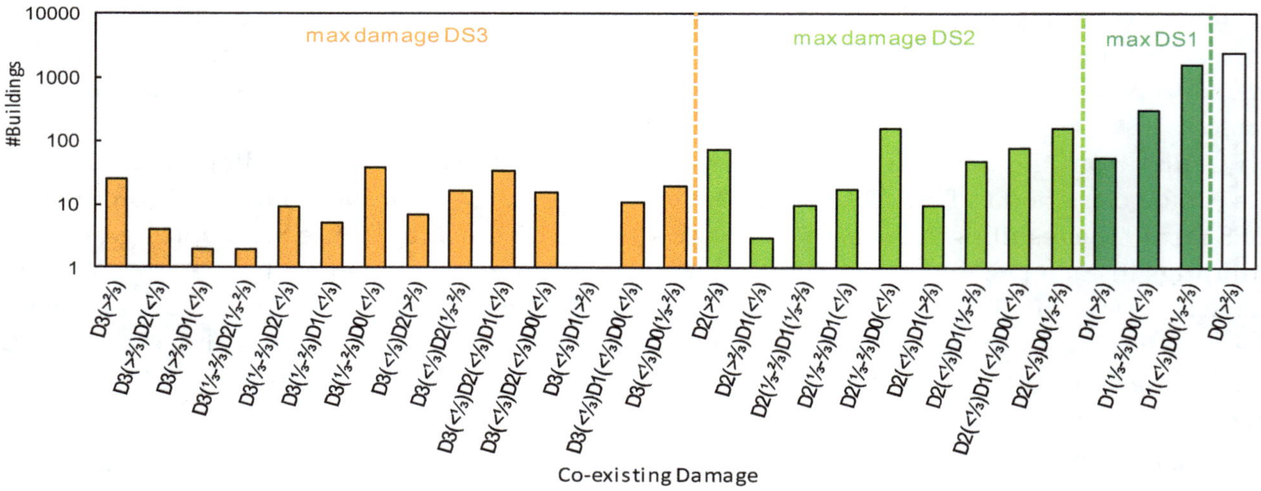

Figure 9. "Coexisting" damage given the maximum damage level (Number (#) of buildings is in logarithmic scale).

Further analysis of the observed damage scenario reported in Figure 8a can be carried out by means of the decomposition of the whole scenario depending on plan area, number of storeys, year of construction of the analysed building stock, or depending on the suffered PGA during the seismic event, as shown in Figure 10. It can be noted that:

- For high-rise buildings, a more severe damage level (i.e., higher percentage of DS2 and DS3) is observed, likely due to the generally higher seismic vulnerability of higher buildings, (all the other main characteristics being the same);

- Construction age is not a significant parameter in the definition of damage trend;
- A higher plan area is associated with more severe damage levels;
- Damage severity increases for higher values of PGA, as expected.

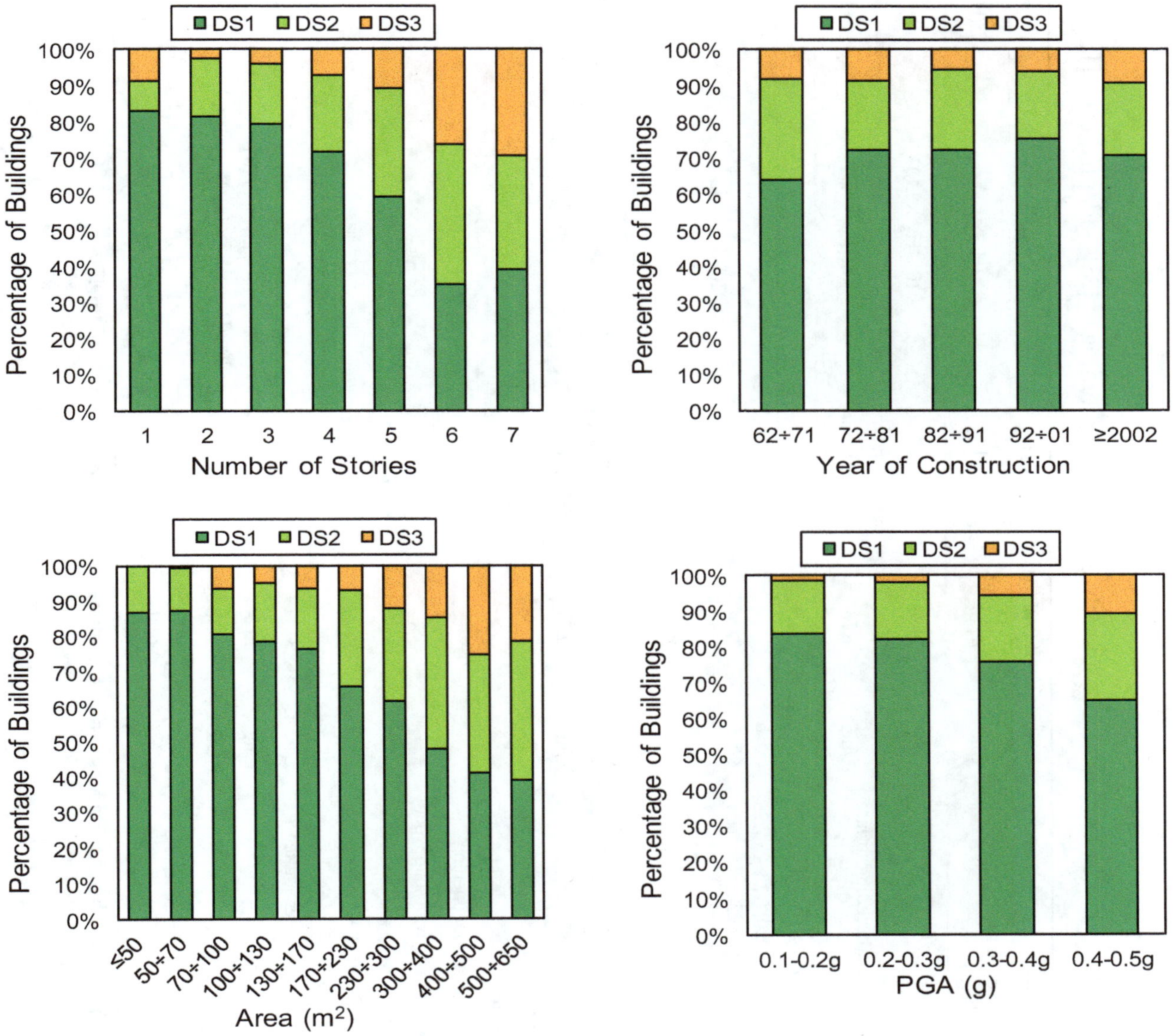

Figure 10. Analysis of the observed damage scenario for "lightly-damaged" RC buildings depending on the PGA of the event, year of construction, number of storeys and plan area of the buildings.

4. Repair Costs Estimation for Masonry Infills

The repair cost estimation performed herein belongs to "component-level" loss prediction methodologies [1,3,5]. Only repair costs due to infills are analysed and presented, to show their percentage incidence on the total repair costs and to provide a lower bound for the actual repair costs for infilled RC buildings.

The repair costs estimation provided in this section is directly derived from: (i) the cost analysis related to a single infill panel and (ii) the "observed" damage scenario reported and commented in the previous section, as explained in Sections 4.1 and 4.2.

4.1. Repair Cost Evaluation for A Single Masonry Infill Panel

First, a list of considered macro-activities is needed. To this aim, the main operations in repairing a single infill panel damaged during a seismic event has to be defined, as reported in Table 2.

Table 2. Extension of intervention for each elementary actions and related unit cost according to [23]—adapted from [18].

Activity Group	Elementary Actions	Unit	Unit Cost (c_j)	$A_{j,DS1}$	$A_{j,DS2}$	$A_{j,DS3}$
a. Preliminary operations	Install scaffolding with steel scaffolding and multi-directional ringlock rosettes […].	€/m^2	25.72	$(H_w \times L_w)$	$(H_w \times L_w)$	$(H_w \times L_w)$
b. Demolition activities	Demolition of single leaf masonry brick with mechanical equipment	€/m^2	13.84	-	$10\%(H_w \times L_w)$	$(H_w \times L_w)$
	Render or plaster removal up to 5 cm thick (including surface brushing of affected area)	€/m^2	8.58	$(0.20\ m) \times d$	$30\%(H_w \times L_w)$	$(H_w \times L_w)$
c. Construction activities	Construction of infill panels:					
	Construction of double leaf cavity masonry wall (12 × 25 × 25 hollow clay brick for exterior leaf and 8 × 25 × 25 hollow clay brick for interior leaf)	€/m^2	72.58	-	$10\%(H_w \times L_w)$	$(H_w \times L_w)$
	Cavity thermal insulation with Glass mineral wool	€/m^2	22.78			
d. Finishings	Coat rendering or plastering (3 layers) with cement mortar	€/m^2	23.86	$(0.20\ m) \times d$	$(30 + 10)\%$ $(H_w \times L_w)$	$(H_w \times L_w)$
	Water-based primer application	€/m^2	2.53	$(H_w \times L_w)$	$(H_w \times L_w)$	$(H_w \times L_w)$
	Coating painting (3 layers) for each side with water-based paints	€/m^2	13.19	$(H_w \times L_w)$	$(H_w \times L_w)$	$(H_w \times L_w)$
e. Windows or door frame installation	Installation of new or old window or door opening					
	Installation of new window or door opening with solid wood frame with douglas […]	€/m^2	542.6	-	-	$(H_{op} \times L_{op})$ *
	Installation of old window or door opening with solid wood frame, glazing and rolling shutter	€/m^2	62.36	-	-	-
	Removal of solid wood frame, box and roller shutter	€/m^2	34.12	-	-	$(H_{op} \times L_{op})$ *
f. Landfill	Landfill transportation	€/m^3	26.24	$(0.20\ m) \times d \times (0.1\ 0m)$	$30\%(H_w \times L_w) \times (0.10\ m) + 10\%(H_w \times L_w) \times (s_w\ ^{**} + 0.10\ m)$	$(H_w \times L_w) \times (s_w + 0.10\ m)$
	Landfill disposal *	€/t	15.18			
g. Technical cost	Technical cost	%	8.00	$\sum_{j}^{\{a,b,c,d,e,f\}} c_j A_{j,DS1}$	$\sum_{j}^{\{a,b,c,d,e,f\}} c_j A_{j,DS2}$	$\sum_{j}^{\{a,b,c,d,e,f\}} c_j A_{j,DS4}$

* "Hop" and "Lop" are the height and length of the opening, respectively; ** s_w = infill thickness.

Each group of activities—from a) to g)—is made of a list of elementary actions, established on engineering judgement to restore the infill panel to its undamaged state, according to Del Gaudio et al. [18], where the corresponding unit cost (c_j) has been evaluated from the Price List of Public Works in Abruzzi Region [23].

Summing up the product of the unit cost (c_j) and the area of intervention ($A_{j,DSi}$) for all the activity groups, the total cost of restoration C_{DSi}^{TOT} of an infill panel damaged during a seismic event for a given damage state (DS_i, with $i = 1, \ldots, 3$) can be evaluated as reported in Equation (1):

$$C_{DSi}^{TOT} = \sum_{j}^{\{a,b,c,d,e,f,g\}} c_j A_{j,DSi} \tag{1}$$

Note that Del Gaudio et al. [18] reported the repair costs related to four conditions: the first one was related to a *light cracking damage*; the second one to an *extensive cracking damage*; the third one was assumed as the *"economically convenience" limit to repair without demolish/reconstruct*; the fourth one corresponds to the *total infill collapse/failure condition* [18]. Therefore, the cost related to the *light cracking damage* has been used for DS1; that related to the *extensive cracking damage* for DS2; that related to the *"economically convenience" limit to repair without demolish/reconstruct* is neglected in what follows; lastly the one related the *total infill collapse/failure condition* has been used herein (Figure 11) to determine the repair cost of buildings characterized by the DS3 according to the definition of EMS-98 (in which the total failure of the infill is assumed).

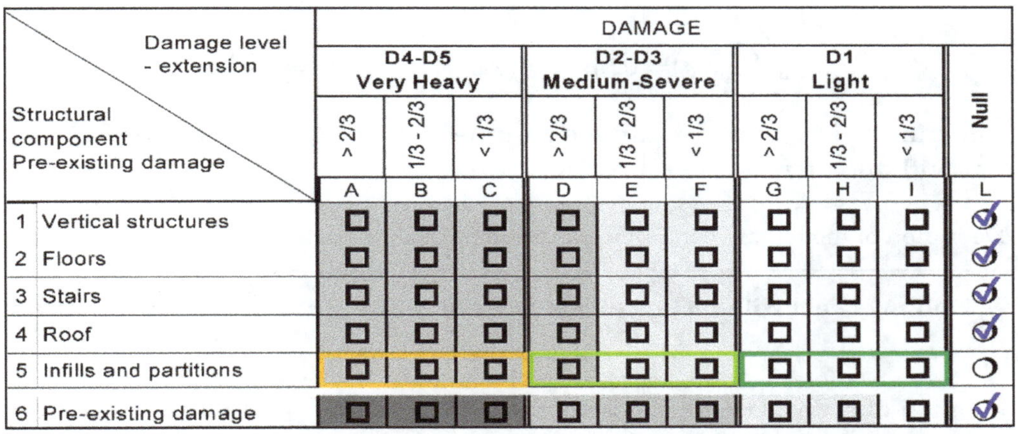

Hollow + Hollow Clay bricks (€/m²)	C_{DS3}	C_{DS2}	C_{DS1}	C_0
Solid panel	285.8	105.3	77.0	0
Panel with window	331.4	101.3	73.0	0
Panel with door	374.9	97.4	69.2	0

Figure 11. Repair costs (C_{DSi}^{TOT}) for double leaf hollow clay bricks.

Hence, C_{DSi}^{TOT} has been evaluated as suggested in [18] for three panel typologies (panel without openings, panel with window opening and panel with door opening), by means of a Monte Carlo simulation technique, considering a number of 1000 simulations varying the length dimension of the infill panel from 4.00 m to 5.00 m and assuming the panel height equal to 2.75 m. It is assumed herein that the clay infill panel is realized with a double leaf cavity masonry wall with (hollow + hollow) panel, constituted by (12 × 25 × 25) cm hollow clay brick (void percentage > 55%) for exterior leaf and (8 × 25 × 25) cm hollow clay brick (void percentage > 55%) for interior leaf, with thermal insulation, as generally widespread in the L'Aquila region [24]. Openings are assumed as constituted by wood frames with plan dimensions of (1.20 × 2.20) m^2 or (0.90 × 1.50) m^2, for door or window opening, respectively. The resulting C_{DSi}^{TOT} for each panel typology at DS_i (with i = 1, ... ,3) are reported in Figure 11.

Note that the values in Figure 11 can be considered as expected (mean) values of economic losses for restoring a damaged infill partition after an earthquake. A dispersion value around them may be considered due to variability related to different professional practices or considering uncertainty in contractor pricing strategies. However, this aspect is not investigated herein.

The above reported remarks and resulting data reported in Table 2 are related to exterior infill panels, namely to infills belonging to the external skin of the building. To obtain a realistic repair cost prediction for a whole building, the interior partitions should also be considered. Therefore, some hypotheses are assumed herein to evaluate the repair cost of interior infill partitions with cement mortar and (8 × 25 × 25) cm hollow clay bricks, starting from the costs obtained for the corresponding (hollow + hollow) double leaf exterior infill panel. Such assumptions and their results are summarized in Table 3, where the activities to be subtracted/added to the total cost of an exterior infill are listed and quantified. Additionally, the equivalent length of interior infill panels ($L_{int,x}$ and $L_{int,y}$) along the two main orthogonal directions (x and y) is determined by assuming that the geometric percentage of interior infills (with thickness $s_{w,int}$) was equal to the 50% [25] of the geometric percentage of exterior infills (with thickness $s_{w,ext}$), as shown in Equations (2) and (3):

$$s_{w,int}(L_{int,x}) = 0.5[s_{w,ext}(2L_x)] \rightarrow L_{int,x} = \frac{s_{w,ext}}{s_{w,int}} L_x \quad (2)$$

$$s_{w,int}\left(L_{int,y}\right) = 0.5\left[s_{w,ext}(2L_y)\right] \rightarrow L_{int,y} = \frac{s_{w,ext}}{s_{w,int}} L_y \quad (3)$$

where $s_{w,int}$ = 80 mm and $s_{w,ext}$ = 200 mm are assumed. As a consequence, the $L_{int,x}$ – to – $L_{int,y}$ ratio results coherent with the plan aspect ratio (PR = L_x/L_y).

Table 3. Definition of the repair cost for interior infill partitions starting from the costs of exterior infills.

(Hollow + Hollow) Panel without Openings: Exterior Infill		C^{TOT}_{DS1}	C^{TOT}_{DS2}	C^{TOT}_{DS3}
		77	**105.3**	**285.8**
subtract	Scaffolding with steel scaffolding and multi-directional ring-lock rosettes, with fiberglass monofilament netting for scaffolding enclosure	25.7	25.7	25.7
	Demolition of single leaf masonry brick with mechanical equipment	-	1.4	13.8
	Construction of double leaf cavity masonry wall (12 × 25 × 25) cm hollow clay brick for exterior leaf and (8 × 25 × 25) cm hollow clay brick for interior leaf	-	7.3	72.6
add	Interior infill partition with cement mortar and (8 × 25 × 25) cm hollow clay bricks	-	2.6	26.3
Costs for interior partition with hollow clay brick		51.3	73.5	199.9

4.2. *Probabilistic Evaluation of Repair Costs Due to Infill Panels for a Whole RC Building*

The total repair cost estimation due to infills in a RC building obviously starts from the repair cost related to a single infill panel, but additionally requires the definition of some Random Variables (RVs) to identify the complete configuration of the damaged infills throughout the whole building, belonging to a building stock. The necessary RVs, assumed here with uniform probability density functions, are listed below:

- Plan Area (*A*), assumed as a continuous RV within the ranges presented in Figure 6;

- Plan aspect Ratio (PR), assumed as a continuous RV within the range [1; 2.5] (as suggested in [16] for a building stock in the same geographic area)
- Damage Extent (DE), assumed as a continuous RV within the ranges [0; 1/3], [1/3; 2/3], [2/3; 1] (as in the AeDES form)
- Presence of Openings (OP), assumed as a discrete RV among the cases: "no opening", "window", "door"

Therefore, for each building belonging to the collected database, starting from its own plan area range and damage extent range from the related AeDES form, 1000 random samples are generated in a Monte Carlo simulation approach, thus assuming A_j, PR_j, DE_j, OP_j with $j = 1, \ldots, 1000$. Then, the following cascading quantities can be defined for each building and for each sample *j*:

- Longitudinal (L_{xj}) and Transversal (L_{yj}) plan length:

$$L_{xj} = A_j/PR_j;\ L_{yj} = A_j/L_{xj} \tag{4}$$

- Exposed infills area (S): Plan perimeter (P) × Building height (H) (the latter defined as the number of stories (n_s) multiplied by the inter-story height, h, assumed equal to 3 meters), as in Equation (4):

$$S_j = P_j \times H = 2(L_{xj} + L_{yj}) \times n_s \times 3\ \text{m} \tag{5}$$

- Damaged infills area at DS_i (S_{DSij}): Exposed infills area × Damage Extent at the damage level DS_i, as in Equation (5):

$$S_{DSij} = S_j \times DE_{ij} \tag{6}$$

- Repair cost at a given DS_i (RC_{DSij}): Damaged infills area at DS_i × Repair cost at that DS_i, as shown in Equation (6):

$$RC_{DSij} = S_{DSij} \times C^{TOT}{}_{DSi} \tag{7}$$

- Mean total repair cost TRC, as the sum of RC_{DSi}, for $i = 1, \ldots, 3$, averaged among all the samples.

5. Resulting Repair Costs for Infills and Discussion

The result of the procedure described in Section 4 is summarized in Figure 12, in terms of TRC per the total plan area unit (average plan surface × number of stories) in Euro (€)/m^2 and depending on the maximum observed DS. Mean and median values, 16th and 84th percentiles are shown in Figure 12 and reported in Table 4 for each maximum DS.

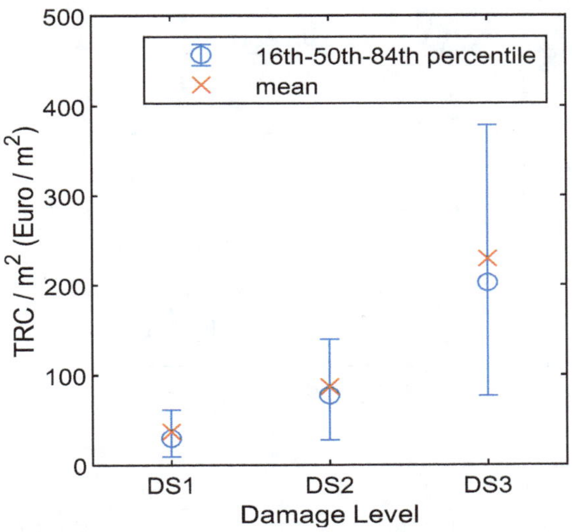

Figure 12. TRC due to infills depending on the maximum attained DS.

Table 4. Median, 16th and 84th percentiles for TRC depending on the maximum DS.

TRC (€/m^2)	Mean	Median	16th Percentile	84th Percentile
DS1	36.96	29.63	9.45	61.3
DS2	87.05	77.43	27.72	139.23
DS3	228.86	202.7	77.1	378.5

A further de-composition of these costs is performed and explained below. The TRC reported in Figure 12 can be disaggregated as shown in Figure 13, depending on the maximum damage level to investigate about the influence, if any, of the parameters number of stories, age of construction, and average plan surface on the outcome TRC/m^2.

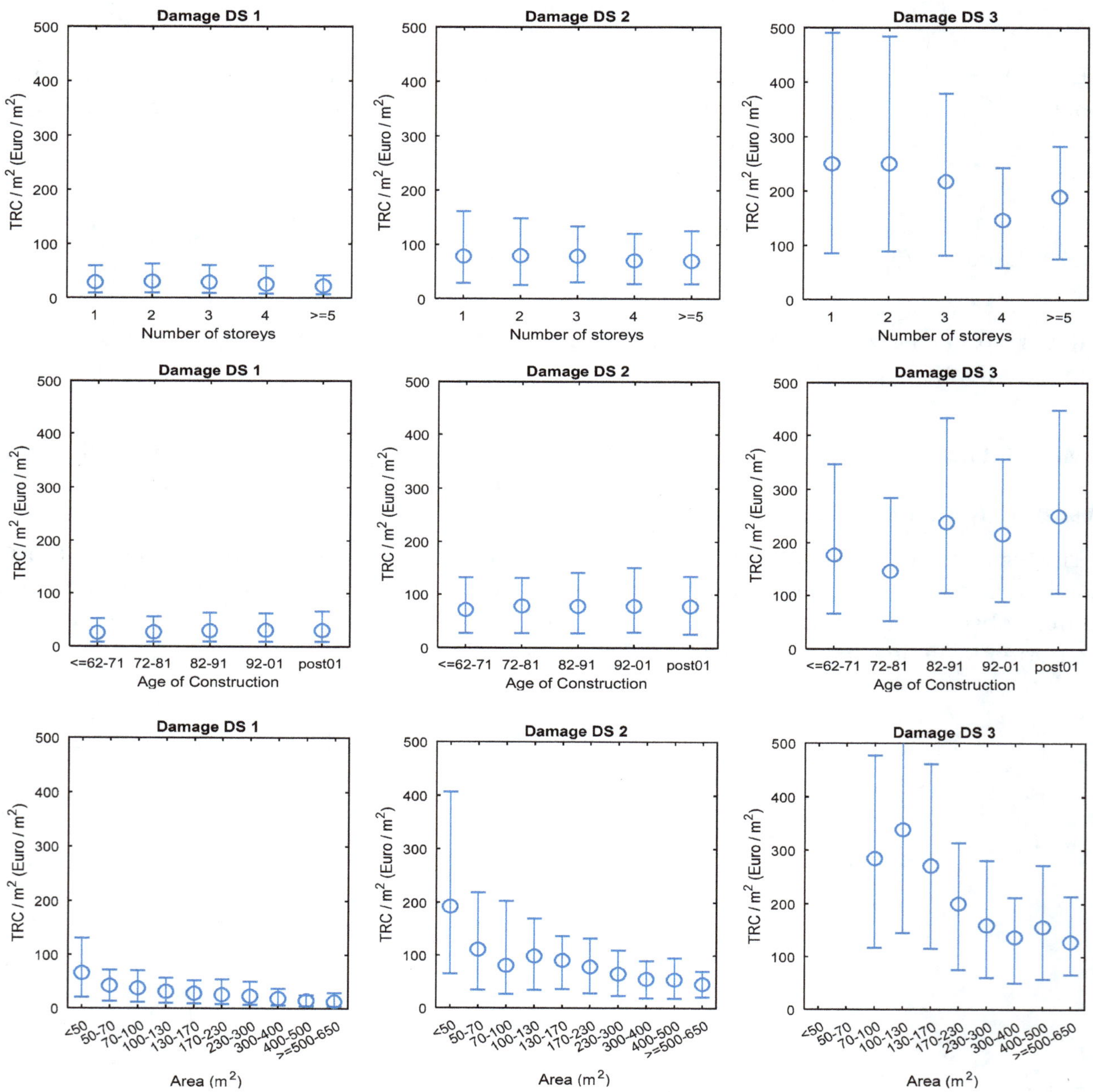

Figure 13. Trends of TRC/m2 with number of storeys, age of construction, and average plan surface depending on the maximum achieved DS (the figure shows 50th, 16th and 84th percentiles).

Figure 13 highlights that the age of construction is generally not significantly influent on the total repair cost per square meter. On the contrary, the TRC/m^2 very slightly decreases for increasing number of storeys and more significantly decreases with the average plan area. Such a trend seems to be counter-intuitive, due to the (opposite) trends of damage distribution, shown in Figure 10. Actually, DSs distributions are related to the maximum observed damage level, which is generally quite concentrated in a small portion ("<1/3") of the building (Figure 8b). Therefore, when TRC is divided by the whole plan surface (average plan surface x number of stories), a decreasing trend is observed with increasing average plan surface or number of stories. Additionally, the decreasing trend is more evident depending on the average surface, likely because for increasing plan area, the percentage incidence of the (at least) exterior infills decreases, thus leading to a reduction of the TRC per surface unit.

Comparison with the Actual Repair Costs

The usability judgments carried out in the aftermath of the seismic event have been strictly related to the (mean) total actual repair cost and, consequently, to the reconstruction refund level, as reported in [19] and [9]. Therefore, the mean actual costs depending on the usability judgments can be a useful term of comparison for the mean repair costs due to infills estimated herein, to highlight the percentage incidence of the infills on the total actual repair cost. To this aim, first a correlation between the usability judgment and the damage extent and severity should be assumed, in compliance with the AeDES form. According to the AeDES form (as explained in more detail in Section 2), the damage level D1, whatever the extent, generally leads to the definition of a "low" structural risk for the building, namely to a usability judgment "A" (i.e., "Usable building"). Therefore, DS1 as defined above, is completely included within the usability judgment "A", together with the damage level DS0. AeDES manual [4] also suggests that, if a big number of infill panels (high extent) are interested by damage level D2–D3, the total structural risk can be "high"; otherwise, a lower structural risk level can be evaluated for less significant damage. Therefore, when D2–D3 has an extent "<2/3", usability judgment "B" is obtained. When the extent of the damage level D2–D3 is ">2/3" or the building reaches the damage level D4–D5, its usability rating is "E". In summary, the complete assumed correlation between DSs by EMS-98, AeDES damage grade and extent, and AeDES usability judgments is reported in Table 5.

Table 5. Assumed correspondence between usability judgments and damage levels according to EMS-98 and AeDES form (background colors are used in the figures to tag each DS or usability judgment).

DS	EMS-98 [21]	AeDES form [4]		
	Damage Description	**Damage Severity**	**Damage Extent**	**Usability Judgement**
DS0	No Damage	D0—Null Damage	None	A
DS1	Negligible to Slight damage: Fine cracks in partitions and infills.	D1: Slight	<1/3	
			1/3–2/3	
			>2/3	
DS2	Moderate damage: Cracks in partition and infill walls	D2–D3: Medium—Severe	<1/3	B
			1/3–2/3	
			>2/3	E
DS3	Substantial to Heavy damage: Large cracks in partition and infill walls, failure of individual infill panels	D4–D5: Very Heavy	<1/3	
			1/3–2/3	
			>2/3	

Starting from this definition of usability rating, the percentage of buildings with "A", "B", and "E" judgements can be obtained, as reported in Figure 14. Note that the main difference between the DSs distribution (shown in Figure 8) and the usability judgments distribution is the attribution of damage level D2-D3 for an extent ">2/3" to the most severe usability rating ("E").

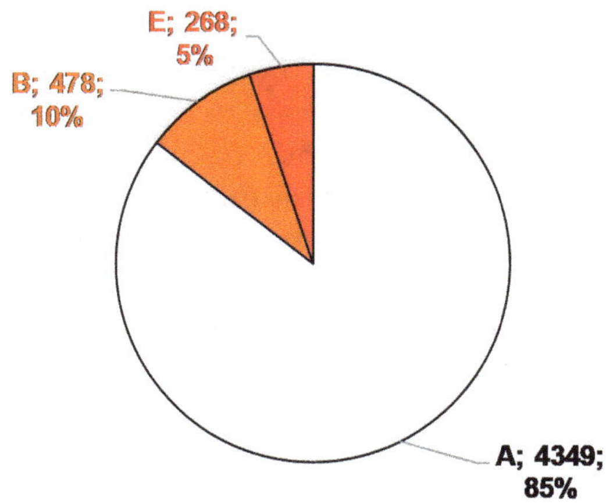

Figure 14. Usability judgments frequency and related damage extent.

In summary, a total of 4349 buildings result in "A" rating; 478 buildings have judgement "B" and 268 buildings "E". Note that in the following analyses, to evaluate the repair costs for "A" buildings, only those actually damaged are considered (namely only buildings in DS1), since buildings in DS0 clearly do not require any repairing activities.

The estimated repair costs are the same already evaluated in the previous Section, but now TRC/m^2 is plotted depending on the usability rating, as shown in Figure 15 (and in Table 6), where 16th, 50th and 84th percentiles are reported together with the mean values (red crosses).

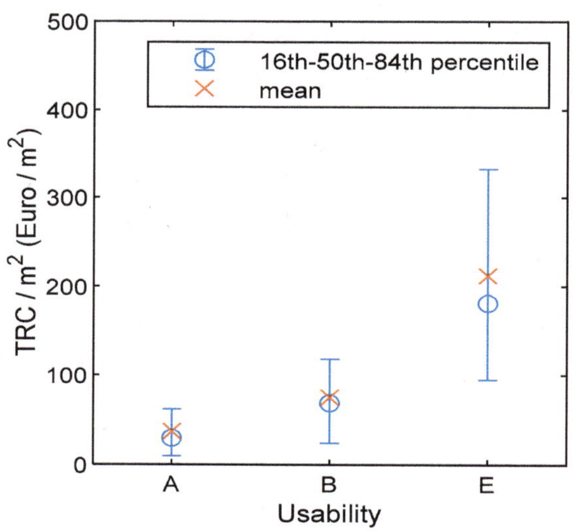

Figure 15. TRC depending on the usability judgment (costs for judgment "A" only consider buildings in DS1).

Table 6. Mean, median, 16th and 84th percentiles for the estimated TRC depending on the usability judgment (costs for judgment "A" only consider buildings in DS1).

TRC (€/m^2)	Mean	Median	16th Percentile	84th Percentile
A	36.94	29.63	9.43	61.28
B	74.33	68.17	24.07	117.97
E	211.77	180.78	94.36	332.20

The actual repair costs can be obtained from Dolce and Manfredi [19], depending on the usability judgments, specifically for those buildings that, after the post-L'Aquila earthquake in-situ surveys, have been classified:

- As "B" rating, or
- In usability judgment "E" with "high risk due to damage to infills and slight structural damages" (defined as "E according to O.P.C.M. 3779" in [19]), the latter representing the subset of data with "E" rating more similar to the dataset analysed in this work ("lightly-damaged buildings").

The actual repair costs include building safety measures, eventual demolition, transportation and landfill disposal, repair and finishing activities, services (sanitary and electrical equipment, plumbing), other non-structural components (floors, roof and chimney, lights ...), technical works, charges for design and technical assistance [26]. These actual repair costs are shown in Table 7 and compared with the estimated TRC reported above. Note that no actual costs can be obtained for "A" usability rate, because they were not explicitly reported in [19].

Table 7. Comparison between mean estimated TRC and mean actual repair cost.

Usability Judgement	Mean Estimated TRC (€/m^2)	Mean Actual Cost (Dolce and Manfredi [19]) (€/m^2)	Mean Estimated TRC/ Mean Actual Cost (-)
A	36.94	not available	-
B	74.33	183.76	0.40
E	211.77	342.35 *	0.62

* "E according to O.P.C.M. 3779".

Table 7 clearly shows that the predicted repair costs due to masonry infills represent a lower bound for the total actual repair cost for the analysed buildings dataset, as expected. The mean estimated TRC does not include the repairing of the structural components (even if very slight damage for the considered buildings dataset), services (sanitary and electrical equipment, plumbing) and other non-structural components (floors, roof and chimney, lights ...). The reparability analysis of these additional components would require the exact knowledge of the disposal of the services with respect to the infill panels distribution or some arbitrarily assumptions on the floor replacement, for example. Such "additional activities" can represent up to a maximum of 50% of the total repair cost depending on the damage grade and extent on the vertical structural components ([26,27]). Additionally, the mean TRC estimated herein consider the minimum damaged infill area to be repaired/replaced, without any conversion in a minimum effective (integer) number of infill panels, conversion that would be quite arbitrary. This circumstance is much more emphasized for "B" Usability judgements, where a predicted-versus-actual TRC ratio equal to 0.4 is observed. As a matter of fact, for these buildings, the predicted repair activities concerning the demolition/construction of masonry infill panels and plaster removal/application are restricted to only 10% and 30%, respectively, of the exposed area. These percentages have to be considered as the minimum strictly necessary in compliance with the damage classification reported by the AeDES form. Likely, the actual area of intervention could be affected by a stronger level of invasiveness, well beyond this conventional limit, producing a higher actual TRC.

In conclusion, the repair cost due to masonry infills only, including all the activities listed in Table 2, represent an average of 50% of the total actual repair cost. This outcome is in good agreement with a recent study by [26], which investigated a specific case-study building among those collected after the post-L'Aquila earthquake survey defined as "E according to OPCM 3779". It shows that infills represent about 55% of the sum of costs related to services (sanitary and electrical equipment, plumbing), structural components, infills and partitions, other non-structural components (floors, roof and chimney, lights ...), in tune with the outcome obtained herein.

It could be stated that repair costs (as per the evaluation) should be multiplied by a factor equal to 2, on average, to obtain the actual total repair cost, for lightly-damaged RC residential buildings. Further studies are certainly necessary to confirm the outcome obtained and commented herein.

6. Conclusions

In commonly adopted loss computation tools (e.g., PACT tool [3] proposed by FEMA P-58 [1]), no specific data related to masonry infill panels, as widespread in moment-resisting-frame residential buildings, are available to perform a reliable probabilistic assessment of losses. To fill this gap, specific fragility and loss functions have been recently carried out in the literature, e.g., by Del Gaudio et al. [18], as adopted in the present study. This proposal has been compared herein with actual repair costs, specifically related to masonry infills, obtained from real post-earthquake damage surveys to estimate the relevance of the repair costs due to infills with respect to the total reconstruction process.

To this aim, Reinforced Concrete (RC) residential buildings with masonry infills struck by the 2009 L'Aquila earthquake have been analysed, focusing on the dataset of 5095 "lightly" damaged buildings, where only damage to masonry infills (or no damage at all) occurred. All these buildings have been assessed with a post-earthquake usability in-situ assessment procedure (Baggio et al., [4]), which provided a comprehensive description of the damage severity and extent. The observed damage scenario for this building stock has been obtained and analysed, starting from the Damage States (DSs) definition (from DS1 to DS3, increasing damage severity) by the EMS-98 (Grunthal [21]), and AeDES survey forms (Baggio et al. [4]). In summary, 2406 buildings presented no damage to infills and partitions (and no damage to vertical structures, roofs, stairs, etc.). A total of 1943 buildings were in damage level DS1, 555 were in DS2, and a smaller portion (191 buildings) presented a damage level of DS3.

The repair costs for infills have been estimated given the observed damage scenario, starting from the cost analysis related to a single infill panel reported in Del Gaudio et al. [18] and extending to the whole analysed dataset of buildings. Mean values of these costs resulted in a range from about 37 €/m^2 to about 230 €/m^2, passing from DS1 to DS3. They were not affected by the age of construction of the buildings; only slightly decreasing for an increasing number of storeys and more significantly decreasing with the average plan area. The predicted repair costs due to masonry infills have been lastly compared with the related actual repair costs reported in Dolce and Manfredi [19], specifically related to buildings with damage to infills and only slight damage to structural components. The predicted repair costs due to masonry infills resulted equal, on average, to 50% of the total actual repair cost.

It is worth noting that the predicted costs estimated in this work refer to a specific infill typology, commonly adopted in the L'Aquila region. Nevertheless, this study can be extended in future research to other infill typologies, also considering different kinds of doors/windows technology and materials, or variability of the estimated costs. Therefore, further studies are certainly necessary to support the outcomes obtained herein.

Author Contributions: Conceptualization, M.T.D.R., C.D.G., G.M.V.; methodology, M.T.D.R., C.D.G.; software, C.D.G.; validation, M.T.D.R., C.D.G.; formal analysis, M.T.D.R.; investigation, M.T.D.R., C.D.G., G.M.V.; data curation, C.D.G.; writing—original draft preparation, M.T.D.R.; writing—review and editing, C.D.G., G.M.; supervision, G.M.V.; project administration, M.T.D.R., G.M.V.; funding acquisition, M.T.D.R., G.M.V.

References

1. FEMA. *Seismic Performance Assessment of Buildings*; Fema: Washington, DC, USA, 2012.
2. Moehle, J.; Deierlein, G.G. A framework methodology for performance-based earthquake engineering. In Proceedings of the 13th World Conference on Earthquake Engineering, Vancouver, BC, Canada, 1–6 August 2004. Article number 13.
3. FEMA. Performance Assessment Calculation Tool (PACT). Available online: http://www.fema.gov/medialibrary/assets/documents/90380 (accessed on 15 October 2018).
4. Baggio, C.; Bernardini, A.; Colozza, R.; Coppari, S.; Corazza, L.; Della Bella, M.; Di Pasquale, G.; Dolce, M.; Goretti, A.; Martinelli, A.; et al. Field manual for post-earthquake damage and safety assessment and short term countermeasures. *JRC Sci. Tech. Rep.* **2007**.

5. Cremen, G.; Baker, J.W. A Methodology for Evaluating Component-Level Loss Predictions of the FEMA P-58 Seismic Performance Assessment Procedure. *Earthq. Spectra* **2019**, *35*, 193–210. [CrossRef]
6. Cardone, D.; Gesualdi, G.; Perrone, G. Cost-benefit analysis of alternative retrofit strategies for RC frame buildings. *J. Earthq. Eng.* **2017**, *23*, 208–241. [CrossRef]
7. Dolce, M.; Goretti, A. Building damage assessment after the 2009 Abruzzi earthquake. *Bull. Earthq. Eng.* **2015**, *13*, 2241–2264. [CrossRef]
8. De Martino, G.; Di Ludovico, M.; Prota, A.; Moroni, C.; Manfredi, G.; Dolce, M. Estimation of repair costs for RC and masonry residential buildings based on damage data collected by post-earthquake visual inspection. *Bull. Earthq. Eng.* **2017**, *15*, 1681–1706. [CrossRef]
9. Di Ludovico, M.; Prota, A.; Moroni, C.; Manfredi, G.; Dolce, M. Reconstruction process of damaged residential buildings outside historical centres after the L'Aquila earthquake: Part I—"Light damage" reconstruction. *Bull. Earthq. Eng.* **2017**, *15*, 667–692. [CrossRef]
10. Di Ludovico, M.; Prota, A.; Moroni, C.; Manfredi, G.; Dolce, M. Reconstruction process of damaged residential buildings outside historical centres after the L'Aquila earthquake: Part II—"Heavy damage" reconstruction. *Bull. Earthq. Eng.* **2017**, *15*, 693–729. [CrossRef]
11. Del Vecchio, C.; Di Ludovico, M.; Pampanin, S.; Prota, A. Repair costs of existing RC buildings damaged by the L'Aquila earthquake and comparison with FEMA P-58 predictions. *Earthq. Spectra* **2018**, *34*, 237–263. [CrossRef]
12. Cardone, D.; Perrone, G. Developing fragility curves and loss functions for masonry infill walls. *Earthq. Struct.* **2015**, *9.1*, 257–279. [CrossRef]
13. Del Gaudio, C.; De Martino, G.; Di Ludovico, M.; Manfredi, G.; Prota, A.; Ricci, P.; Verderame, G.M. Empirical fragility curves from damage data on RC buildings after the 2009 L'Aquila earthquake. *Bull. Earthq. Eng.* **2017**, *15*, 1425–1450. [CrossRef]
14. Del Gaudio, C.; Ricci, P.; Verderame, G.M.; Manfredi, G. Development and urban-scale application of a simplified method for seismic fragility assessment of RC buildings. *Eng. Struct.* **2015**, *91*, 40–57. [CrossRef]
15. Del Gaudio, C.; Ricci, P.; Verderame, G.M.; Manfredi, G. Observed and predicted earthquake damage scenarios: The case study of Pettino (L'Aquila) after the 6th April 2009 event. *Bull. Earthq. Eng.* **2016**, *14*, 2643–2678. [CrossRef]
16. Del Gaudio, C.; Ricci, P.; Verderame, G.M. A class-oriented mechanical approach for seismic damage assessment of RC buildings subjected to the 2009 L'Aquila earthquake. *Bull. Earthq. Eng.* **2018**, *16*, 1–25. [CrossRef]
17. De Risi, M.T.; Del Gaudio, C.; Ricci, P.; Verderame, G.M. In-plane behaviour and damage assessment of masonry infills with hollow clay bricks in RC frames. *Eng. Struct.* **2018**, *168*, 257–275. [CrossRef]
18. Del Gaudio, C.; De Risi, M.T.; Ricci, P.; Verderame, G.M. Empirical drift-fragility functions and loss estimation for infills in reinforced concrete frames under seismic loading. *Bull. Earthq. Eng.* **2019**, *17*, 1285. [CrossRef]
19. Dolce, M.; Manfredi, G. *Libro Bianco Sulla Ricostruzione Privata Fuori dai Centri Storici nei Comuni Colpiti dal Sisma dell'Abruzzo del 6 Aprile 2009*; Doppiavoce Editore: Napoli, Italy, 2015. (In Italian)
20. Dolce, M.; Speranza, E.; Giordano, F.; Borzi, B.; Bocchi, F.; Conte, C.; Pascale, V. Da. DO—A web-based tool for analyzing and comparing post-earthquake damage database relevant to national seismic events since 1976. In *Atti del XVII Convegno ANIDIS L'ingegneria Sismica in Italia*; Pisa University Press: Pisa, Italy, 2017; pp. 347–357.
21. Grunthal, G. *Cahiers du Centre Europeen de Geodynamique et de Seismologie: Volume 15—European Macroseismic Scale*; European Center for Geodynamics and Seismology: Luxembourg, 1998.
22. Michelini, A.; Faenza, L.; Lauciani, V.; Malagnini, L. ShakeMap implementation in Italy. *Seismol. Res. Lett.* **2008**, *79*, 688–697. [CrossRef]
23. B.U.R.A. *Price List of Public Works in Abruzzi Region*; Regione Abbruzzo: L'Aquila, Italy, 2017. (In Italian)
24. Ricci, P.; De Luca, F.; Verderame, G.M. 6th April 2009 L'Aquila earthquake, Italy: Reinforced concrete building performance. *Bull. Earthq. Eng.* **2011**, *9*, 285–305. [CrossRef]
25. Del Gaudio, C.; Ricci, P.; Verderame, G.M.; Manfredi, G. Urban-scale seismic fragility assessment of RC buildings subjected to L'Aquila earthquake. *Soil Dyn. Earthq. Eng.* **2017**, *96*, 49–63. [CrossRef]
26. Del Vecchio, C.; Di Ludovico, M.; Prota, A.; Pampanin, S. Repair costs analysis for case study buildings damaged in the 2009 L'Aquila earthquake. In Proceedings of the NZSEE Conference, Christchurch, New Zealand, 1–3 April 2016.

12

Parametric Study on Seismic Rehabilitation of Masonry Buildings using FRP based upon 3D Non-Linear Dynamic Analysis

Eissa Fathalla and Hamed Salem *

Structural Engineering Department, Faculty of Engineering, Cairo University, Giza 12613, Egypt; eissa.tokyo.concrete@gmail.com
* Correspondence: hamedhadhoud@yahoo.com

Abstract: Unreinforced load-bearing masonry (URM) buildings represent a significant portion of the non-engineered old buildings in many developing countries aiming to reduce the construction cost. The walls of those buildings are developed to resist gravity loads. Lateral loads induced by earthquakes or wind may cause severe their damage. In the current study, a numerical investigation is carried out for a seismic assessment of a typical four-story, load-bearing building in Giza, Egypt. The full 3D nonlinear dynamic analysis is carried out using the Applied Element Method (AEM), which proved to be efficient in such case where partial or total collapse is expected. The study includes two earthquake zones in Egypt called zone (3) and zone (5B), which are the actual studied building seismic zone and the highest seismic activity zone in Egypt, respectively. Carbon fiber reinforced polymers (CFRP) laminates with different thicknesses and different configurations are used in strengthening unreinforced masonry walls to study the efficiency of the proposed rehabilitation technique on a realistic structure.

Keywords: AEM; load-bearing masonry walls; seismic rehabilitation

---

1. Introduction

Load-bearing masonry buildings represent a significant portion of the old buildings in Egypt. Those buildings are mostly non-engineered and constructed without engineering supervision. Their walls are mainly used to resist gravity loads. Lateral loads induced by earthquakes (EQ) or the wind are not taken into account; therefore they may cause severe structural damages. Therefore, seismic rehabilitation of such buildings is believed to be crucial for upgrading their lateral capacity. They also may need to be upgraded to meet more extreme design seismic requirements.

Currently, many researchers focus on the use of innovative strengthening techniques involving fiber reinforced polymers (FRP) materials [1]. FRP can be used for strengthening structural members in the form of laminates or sheets. It appears that the use of FRP composites to strengthen unreinforced masonry walls might be a powerful technique to enhance both in-plane and out of plane behavior of walls where it was investigated in previous research studies both experimentally and numerically [2–12].

In the current study, a numerical investigation is carried out for seismic rehabilitation of a four-story, load-bearing building located in Giza, Egypt by using carbon a fiber reinforced polymer (CFRP). The magnitude of earthquakes used in the current study is according to the Egyptian design code for loads [13]. The study includes two earthquake zones in Egypt known as zone (3) and zone (5B) with design magnitudes of 15% and 30% of the gravitational acceleration, respectively. The full 3D nonlinear dynamic analysis is carried out using the Applied Element Method (AEM) [14]. The AEM is based on a discrete crack approach and is capable of predicting the nonlinear structural behavior as well as local damage and total collapse. Therefore, it is believed to be efficient for the current study where partial or total collapse is expected. CFRP laminates with different thicknesses and different configurations are used in strengthening the URM walls to study the efficiency of the proposed strengthening technique.

2. The Applied Element Method

The AEM simulates the structure by virtually dividing it into small elements that are connected by normal and shear springs positioned at specific contact points around the surface of the elements [14]. Each assembly of springs represent the deformations and stresses of a particular volume. The AEM has been considered a reliable method to track the collapse of structures passing through all the phases of the application of loads, elastic stage, cracking initiation, reinforcement yield, rupture, elements separation, and collision with ground and adjacent structures. Maekawa's compression model [15] is used for concrete modeling under compression, which is shown in Figure 1a. For concrete shear springs, the linear relation of shear stress and shear strain is assumed until the cracking of concrete occurs. Then the shear stresses drop, as shown in Figure 1b. The level of the drop depends on aggregate interlocking and friction at the crack surface. For reinforcement springs, Figure 2 shows the model, which is previously presented for cyclic loading of reinforcing steel bars [16] and is used in AEM. The authors utilize the constitutive models of concrete to simulate the masonry structure in which they both have the same general trend [17]. Moreover, we validate our method by previous experimental program, where the validation results are shown in the following section.

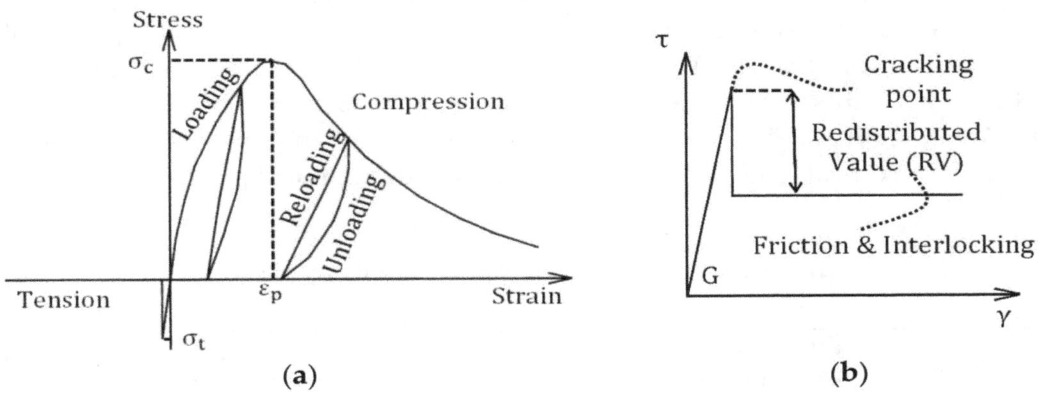

Figure 1. (**a**) Axial stresses in concrete springs due to relative displacement. (**b**) Shear stresses in concrete springs due to relative displacement.

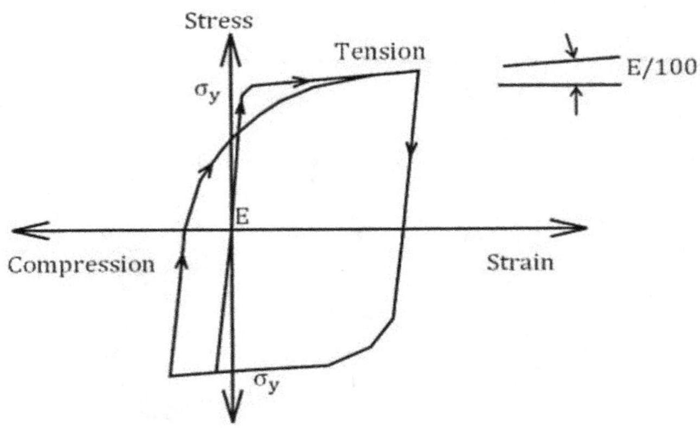

Figure 2. Stresses in steel springs due to relative displacement.

Although the Finite Element Method (FEM) is a well-established and robust structural analysis method, it may not be the optimum solution for the scope of the present study. Many drawbacks are associated with the FEM progressive collapse analysis. The elements' separation, falling, and collision with each other are complicated to simulate. Previous research studies [18–20] showed that the computations associated with the simulation of collapses of real structures based on conventional FEM are very costly; therefore they followed another approach based on multibody models. Even though the progressive collapse analysis is possible in the explicit-integration FEM [21,22], it still has drawbacks that the element separation is introduced through erosion of highly stressed elements and the element's size should be as small as possible to enable such erosion acceptably, which causes the computations to be very costly and impractical when analyzing real structures. Additionally, cracks are introduced due to these elements' erosion, where the crack width equals to the removed elements' size. Thus, no shear transfer is possible across the crack surface and there is no opportunity for the crack closure, which does not allow correct modelling of members' seismic behavior. Consequently, in the current study, the numerical analysis was carried out using the AEM, where it is widely validated and has shown considerable agreement with real cases. It also covers many cases such as static, dynamic, and collapse cases [22–34]. The software used in the analysis is Extreme Loading for Structures (ELS®, Charlotte, NC, USA) [35], which is based on the AEM.

3. Validation of the Analysis Method

The experimental results of Santa Maria et al. [8] are used for validating the AEM results. Santa Maria et al. [8] carried out an experimental program to study the behavior of masonry walls externally reinforced with CFRP and subjected to in-plane cyclic loading. This experiment was carried out on six full-scale masonry walls. Two of the walls were not retrofitted (control) and the rest were retrofitted with CFRP strips that have different configurations and reinforcement areas, which are shown in Table 1 and Figure 3. The specimens were subjected to an in-plane cyclic load under a constant uniformly distributed vertical load of 98 kN. The loading setup of the experiment is shown in Figure 4.

Table 1. CFRP reinforcement of the tested walls [8].

Specimen Name	Reinforcement Configuration	Strip Width (mm)	Total Area of Reinforcement (m^2)	Ratio of FRP Reinforcement (%)
MLC-00-CA-SF-01	-	-	-	-
MLC-00-CA-SF-02	-	-	-	-
MLC-00-CA-FX-01	Diagonal	300	3.37	0.2
MLC-00-CA-FX-03	Diagonal	200	2.25	0.13
MLC-00-CA-FH-02	Horizontal	150	1.78	0.42
MLC-00-CA-FH-04	Horizontal	100	1.19	0.28

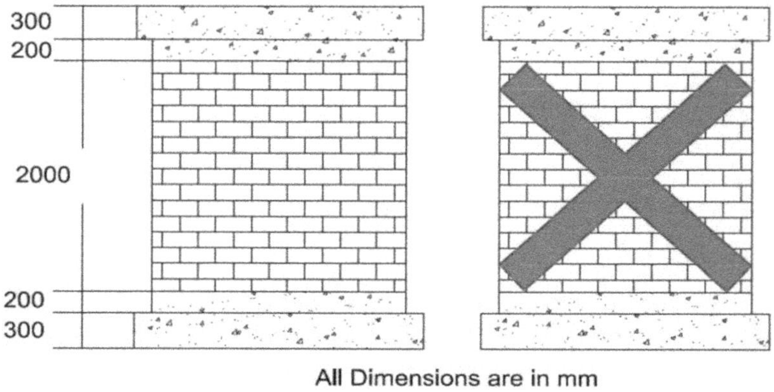

Figure 3. Configuration of exterior reinforcement for the experimental program [8].

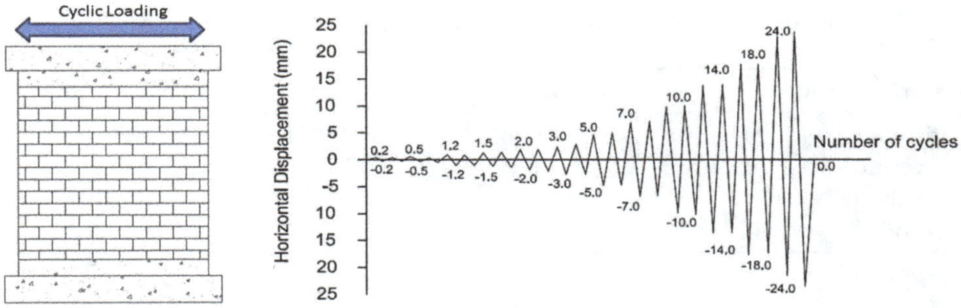

Figure 4. Horizontal displacement program for the cyclic shear tests [8].

The tested specimens were modeled using ELS software (version 3.1, Applied Science International, Morrisville, NC, USA). Two models are adopted for each tested wall, which are described below.

1. Walls are modeled with real brick configurations and connected with mortar joints (Masonry mesh), which is shown in Figure 5.
2. Walls are modeled with simplified quadrilateral mesh including average properties of brick and mortar, which is shown in Figure 6 where the masonry wall is divided in the wall plane to 40×40 discretized elements.

Figure 5. Walls modeled as masonry mesh.

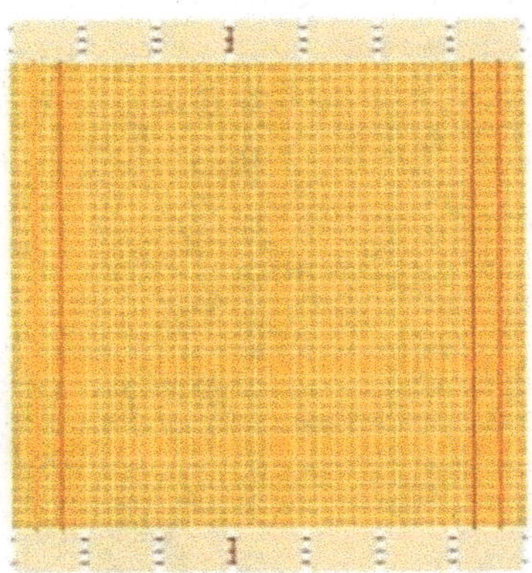

Figure 6. Walls modeled as simplified mesh 40×40.

The Material properties of the masonry are shown in Table 2. Since the CFRP laminates are not expected to carry compressive forces, the compressive strength is chosen to be a small value (0.35 N/mm^2) in order to permit simulation's computations. Interface elements are utilized in the simulation program for modeling the epoxy material in which the de-bonding of the CFRP laminates can be simulated if it occurs.

Table 2. Material properties of the experimental program.

Material Type		Bricks	Mortar	CFRP (0.13 mm)	Epoxy
Young's Modulus	N/mm^2	6618	10,000	230,000	3000
Compressive Strength	N/mm^2	11	25	-	80
Tensile Strength	N/mm^2	0.85	5	3500	50
Specific Weight	kN/m^3	18	22	19	12

For the masonry mesh, two materials were used including one for bricks and the other for mortar. The stiffness of the material used in the case of modeling walls as simplified mesh was an average of mortar and bricks. Since bricks were weaker than mortar, cracking was governed by bricks. Figure 7 shows the numerical model for the retrofitting scheme of the walls using CFRP, where the CFRP are attached to the walls using an epoxy adhesive material.

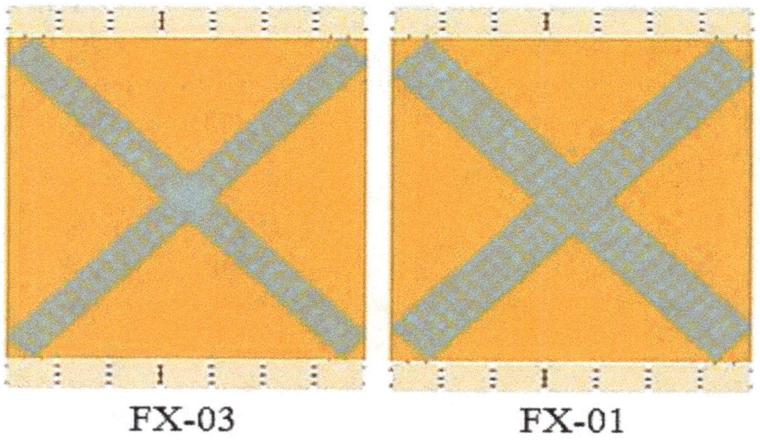

Figure 7. Configuration of exterior reinforcement for the numerical model.

A comparison of hysteresis loops of tested walls for both the experimental model and the numerical model for the masonry panel (SF-01 or SF-02), (FX-01), and (FX-03) are shown in Figures 8–10, respectively. The obtained results show that the AEM with utilizing the concrete constitutive laws is an acceptable and reasonable accurate method for the analysis of the masonry walls. It is evident that the experimental and the numerical model are in close agreement for both the masonry mesh and the simplified mesh (40 × 40). Therefore, it is decided to use the simplified mesh for the modeling of masonry walls to reduce the problem size for the full-scale model analysis of the multi-story building.

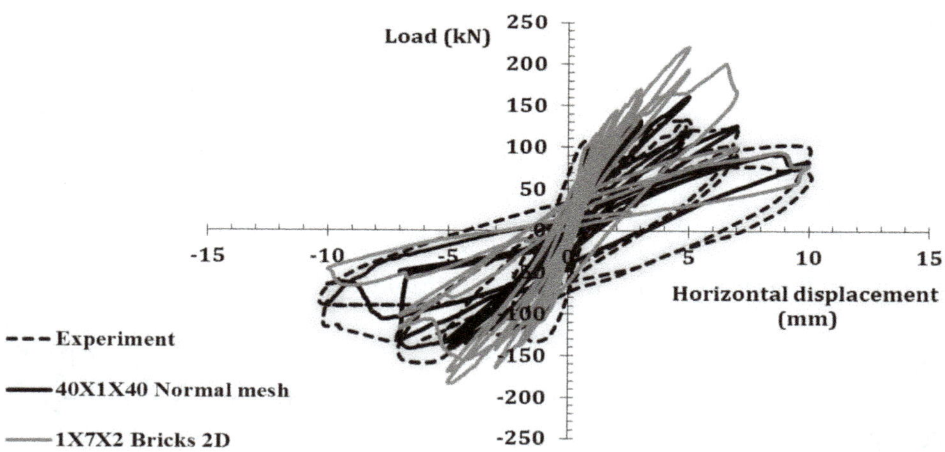

Figure 8. Comparison of experimental and analytical results for the URM wall (SF-01 or SF-02).

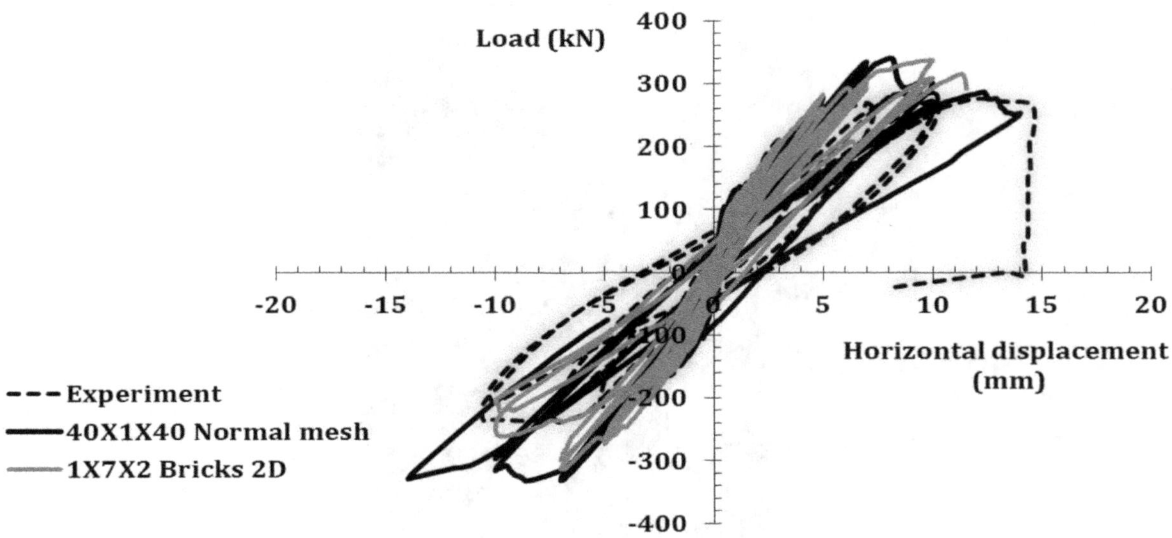

Figure 9. Comparison of experimental and analytical results for the FX-01 wall panel.

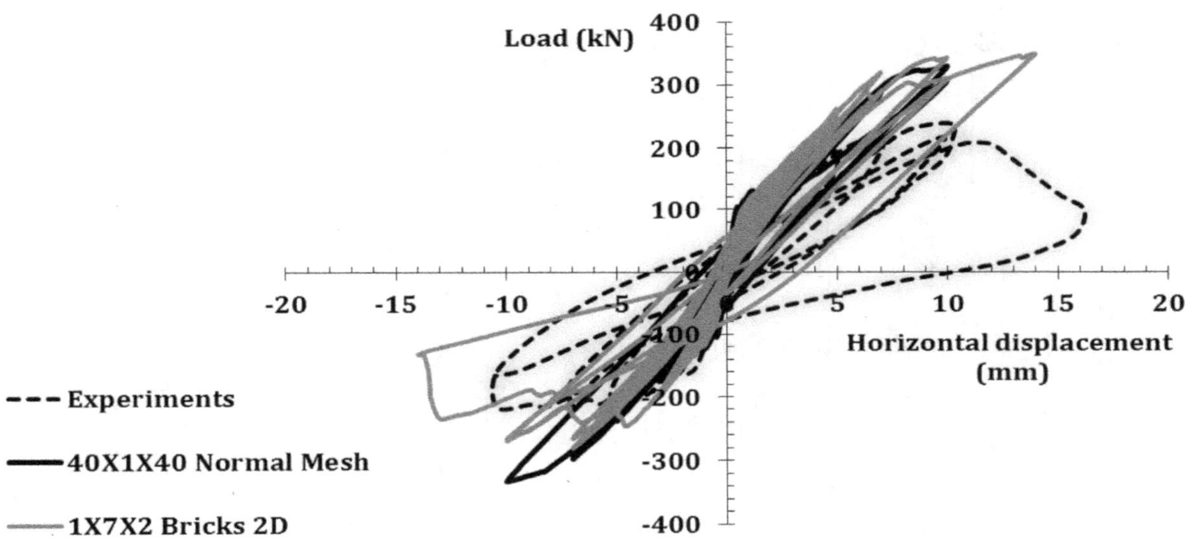

Figure 10. Comparison of experimental and analytical results for the FX-03 wall panel.

4. Case Study

4.1. Description of the Studied Structure

The investigated case study is an existing four-story residential building located in the Faisal district in Giza, Egypt and is constructed with load-bearing masonry walls. Figure 11 shows a picture of the building while Figure 12 shows the plan of the typical floor showing dimensions of walls (W), doors (D), and openings (O). All floors are 3 m high. Since no data is available for the reinforced concrete (RC) slab reinforcement, bottom reinforcement is reasonably assumed with a diameter of 8 mm and 200 mm spacing in both directions along with an additional top reinforcement with a diameter of 8 mm and 200 mm spacing, as shown in Figure 13. It should be noted that the staircase is omitted in the model for simplicity. The RC slabs are supported on RC beams at the walls locations, where these beams are directly supported on the masonry bearing walls, which is similar to a prior construction practice in Egypt. A detailed three-dimensional model is built using ELS by taking into consideration all structural components. Figure 14 shows different views of the numerical model.

Figure 11. Case study building.

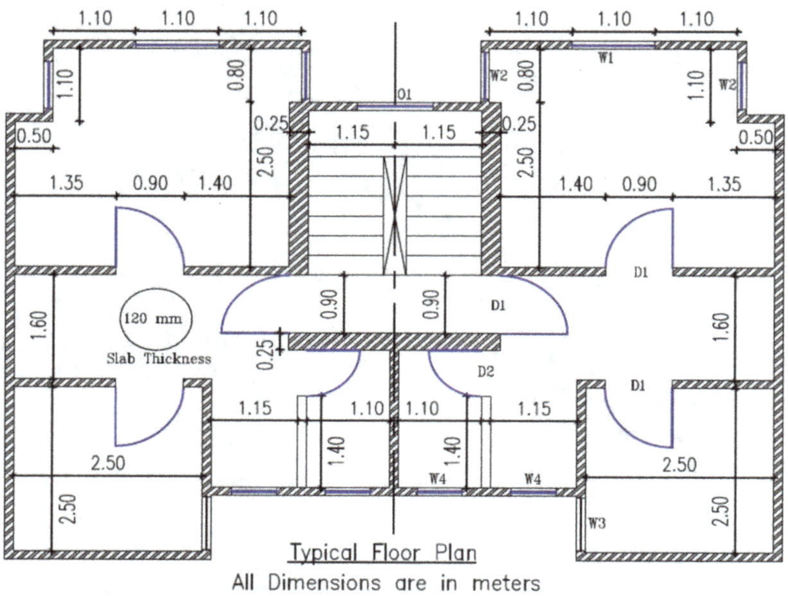

Note: All Masonry walls are 120 mm except mentioned

Type	Dimensions (mm)		
	Length	Width in plan	Height from floor
D1	2200	900	---
D2	2200	700	---
W1	1200	1100	1100
W2	1200	700	1100
W3	1200	800	900
W4	1000	600	1100
O1	1000	1000	Above floor level

Figure 12. Typical floor plan of the studied case.

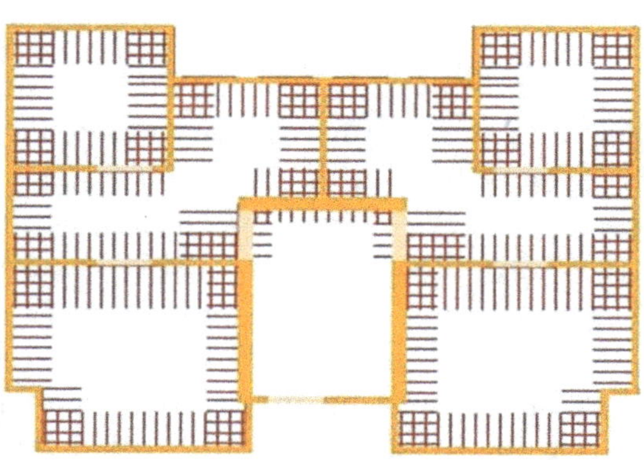

Bottom Reinforcement
D8@ 200 mm

Additional
Top Reinforcement
As shown in figure
D8@ 200 mm

D: Rebar's diameter

Figure 13. Reinforcement details of the reinforced concrete slabs.

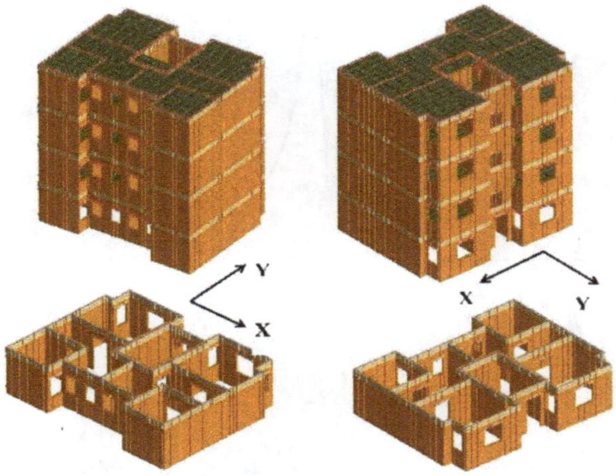

Figure 14. Different views of the numerical model.

4.2. Material Properties

Properties of concrete, reinforcement, bricks, CFRP (width = 0.5 m and thickness = 0.5 mm), and epoxy are listed in Table 3.

Table 3. Material properties of the studied case.

Material Type		Concrete	Bricks	CFRP	Epoxy	Steel Reinforcement
Young's Modulus	N/mm^2	22,000	6618	230,000	3000	200,000
Compressive Strength	N/mm^2	25	10	0.35	80	-
Tensile Strength	N/mm^2	2.5	1	3500	50	360
Specific Weight	kN/m^3	25	19	18	12	78
Ultimate Strength	N/mm^2	-	-	-	-	520

4.3. Loads

The self-weight of the structure, the floor weight, the live loads, and the earthquake load are applied to the studied structure. The floor and live loads on the slabs are considered equal to 2.0 kN/m^2 and 2.5 kN/m^2, respectively. The ultimate load combination given by the ECP [13] is used (1.12 D.L + 0.25 L.L + S) where D.L is the dead load, L.L is the live load, and S is the seismic load. The coupled orthogonality effect is not considered in this paper in order to study the effect of strengthening of each direction separately. This load combination is applied to the building for 20 s, which is the duration of the applied earthquake.

4.4. Earthquake (EQ) Characteristics

In the current research, two earthquake magnitudes are considered.

1. First case. Design earthquake magnitude according to the actual location of the building in Giza (Zone 3), where the peak ground acceleration equals 0.15 g.
2. Second case. Design earthquake magnitude according to the most active seismic location in Egypt (zone 5B), where the peak ground acceleration equals 0.3 g.

Artificial acceleration time history records are created from the response spectrum using SIMQKE software (version 2.1, Massachusetts Institute of Technology, Cambridge, MA, USA) [36], as shown in Figure 15. The basis of SIMQKE software is creating random amplitudes and phase angles derived from a stationary power spectral density function of the motion. An envelope function of the form, shown in Figure 16, [36] is used to simulate the transient character of a real earthquake.

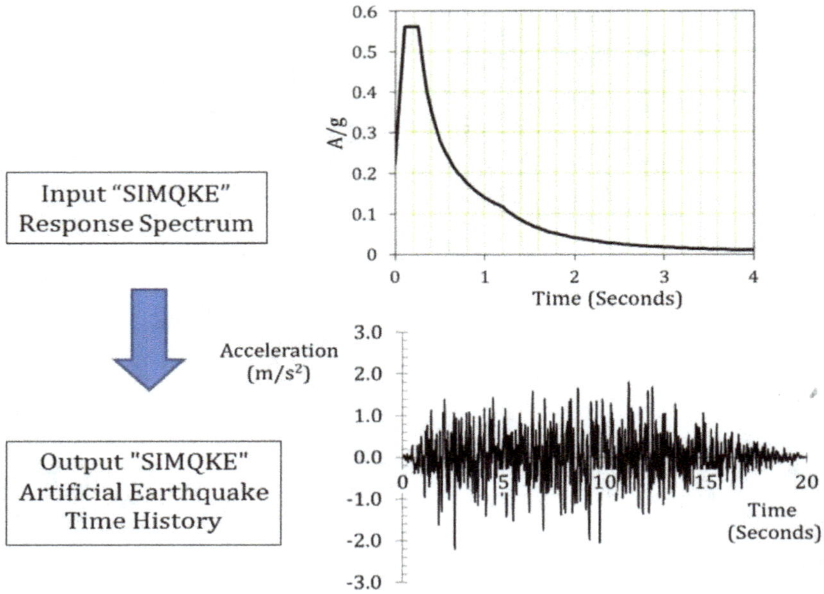

Figure 15. (SIMQKE) software conversion from response spectrum to earthquake time history acceleration [36].

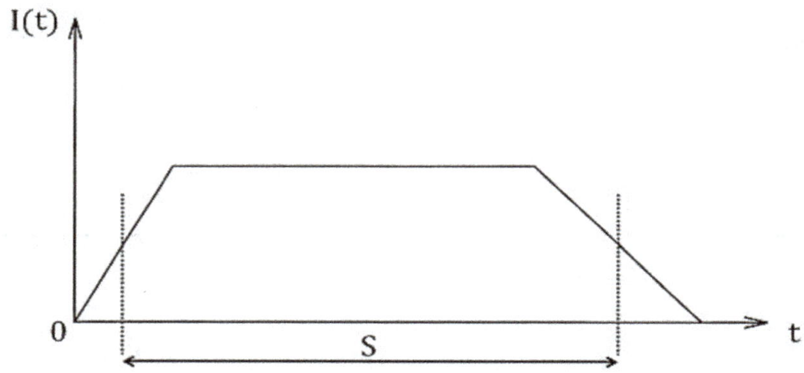

Figure 16. Intensity envelope function used to simulate the transient character of real earthquakes [36].

The simulated final motion z(t) is shown Equation (1).

$$z(t) = I(t) \sum_{k=0}^{n} \sin(\omega_n t + \Phi_n) \quad (1)$$

z(t): Simulated final motion; I(t): Deterministic envelope function; n: The nth contributing sinusoid function; ω_n: Frequency content of the nth contributing sinusoid function; Φ_n: Phase angle of the nth contributing sinusoid function.

The final motion represented by Equation (1) represents a motion stationary in frequency content with maximum acceleration approximately equal to the target maximum acceleration. The software uses a frequency range bounded by 0.5 ω_{low} and 2.0 ω_{high}, where ω_{low} and ω_{high} define the range for the required values of the response spectrum. The frequency step is $\omega_{n+1} = 0.005\ \omega_n$, where it is purposed to be a fraction of the smallest value of half-bandwidth.

The elastic response spectra (RS) of the ECP design code and the utilized artificial earthquakes (produced by SIMQKE) of zone 3 and zone 5B [13] are shown in Figures 17 and 18, respectively. Figures 19 and 20 show the acceleration time history of earthquakes for both studied cases.

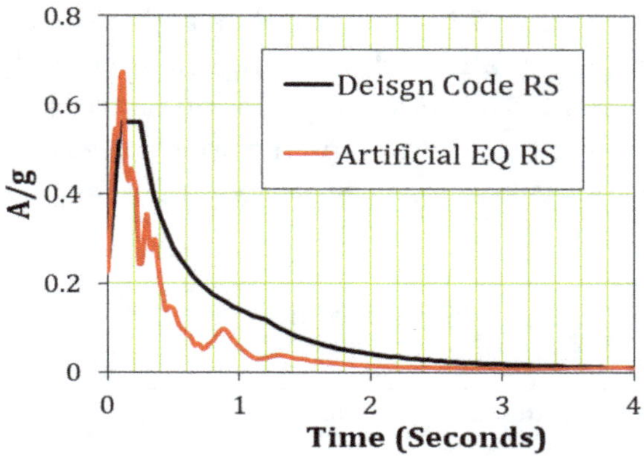

Figure 17. Design code and utilized artificial earthquake response spectrum of case (1).

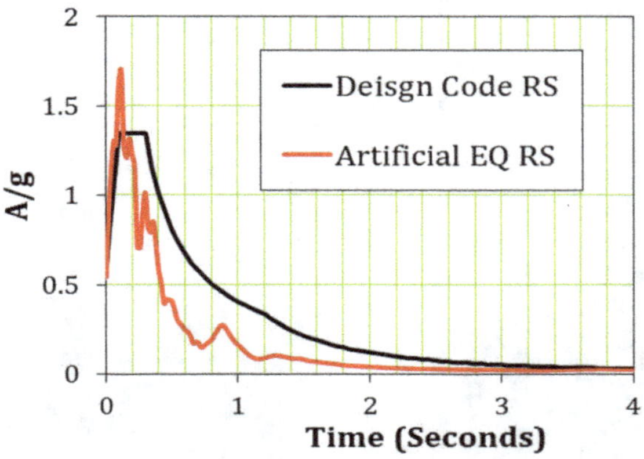

Figure 18. Design code and utilized artificial earthquake response spectrum of case (2).

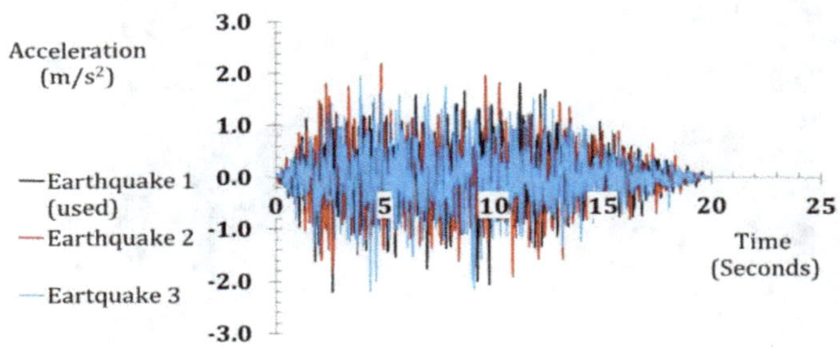

Figure 19. Acceleration time history of earthquakes in case (1).

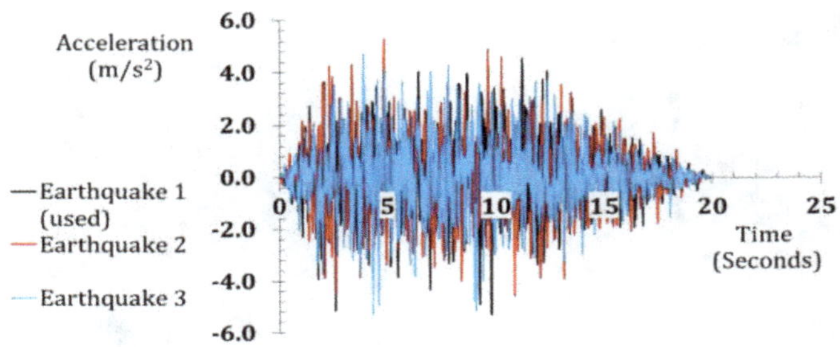

Figure 20. Acceleration time history of earthquakes in case (2).

4.5. Case Studies

Table 4 shows the studied cases where the building with non-retrofitted walls represents the reference case. In the retrofitted cases, the walls are retrofitted with CFRP laminates of width 0.5 m and thickness 0.5 mm. Figures 21 and 22 show the retrofitting scheme for the walls in both cases including solid walls only and solid walls with openings, respectively. For solid walls, the CFRP is attached to the walls in the form of two intersecting diagonal strips. Since CFRP are not expected to resist compression forces, the two diagonal implementations are essential to resist the reversed cyclic loading during the earthquake application. As for the openings, the CFRP is attached in parallel to the sides of the openings.

Table 4. Studied cases.

Studied Cases (Legend)	Earthquake Direction	Earthquake Zone	Retrofitting Scheme
E.Q. X- ZONE 3	X	3	-
E.Q. Y- ZONE 3	Y	3	-
E.Q. X- ZONE 5B	X	5B	-
E.Q. Y- ZONE 5B	Y	5B	-
E.Q. X- ZONE 5B- SW	X	5B	Solid walls
E.Q. Y- ZONE 5B- SW	Y	5B	Solid walls
E.Q. X- ZONE 5B- SW & OP	X	5B	Solid walls & openings
E.Q. Y- ZONE 5B- SW & OP	Y	5B	Solid walls & openings

Figure 21. Retrofitting scheme (for solid walls only).

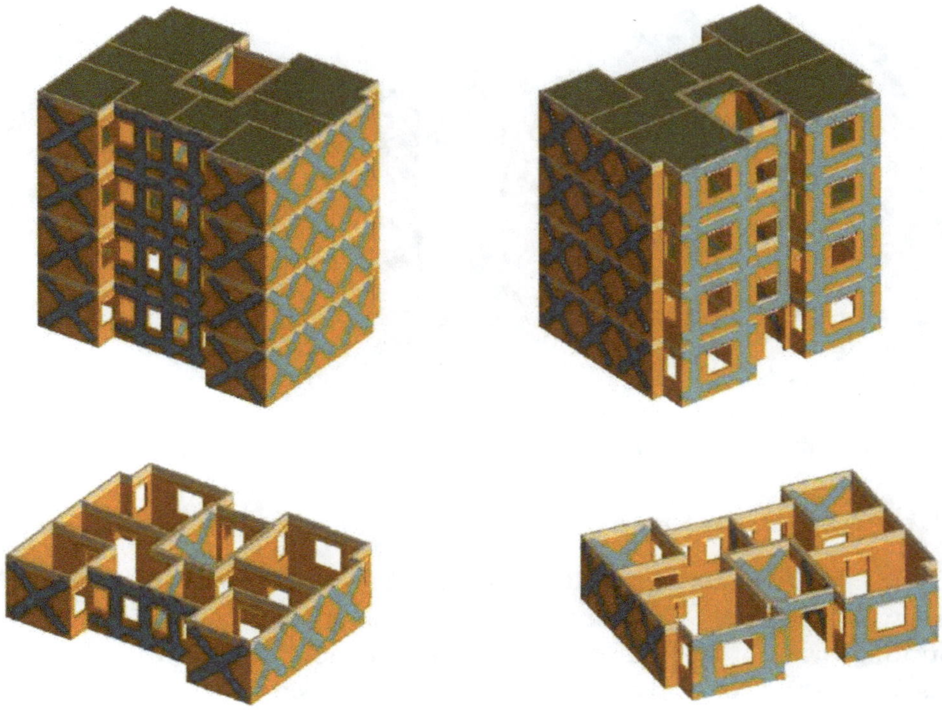

Figure 22. Retrofitting scheme for both solid walls and openings.

5. Numerical Results

The behavior of the studied building in resisting seismic loads and the efficiency of retrofitting using CFRP sheets are investigated with regard to the overall structural integrity, stability, and the damage level induced in the building due to the seismic action. In the following subsections, the deformed shapes of the studied building have a magnification factor of 20.

5.1. *Reference (Non-Retrofitted) Case in Zone 3*

As shown in Figures 23 and 24, the building is proven to maintain its global stability with minimal local damage during earthquake loading in the x-direction and the y-direction, respectively. The studied mid-rise masonry load-bearing building shows good behavior in resisting earthquakes in zone 3.

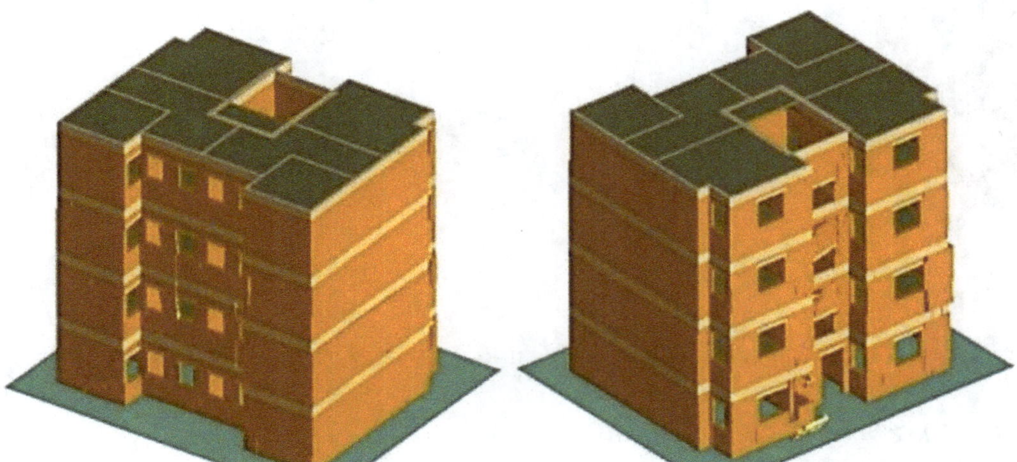

Figure 23. Deformed shape for the case (E.Q. X- ZONE 3).

Figure 24. Deformed shape for the case (E.Q.Y- ZONE 3).

5.2. Reference (Non-Retrofitted) Case in Zone 5B

In this case, the building collapses entirely. A progressive collapse is observed for seismic loading in both directions, which is shown in Figures 25 and 26. The collapse starts by the failure of the bearing walls at the ground floor. In other words, the mid-rise masonry load-bearing building cannot resist earthquake loads for seismic zone 5B and retrofitting of the masonry walls would be necessary.

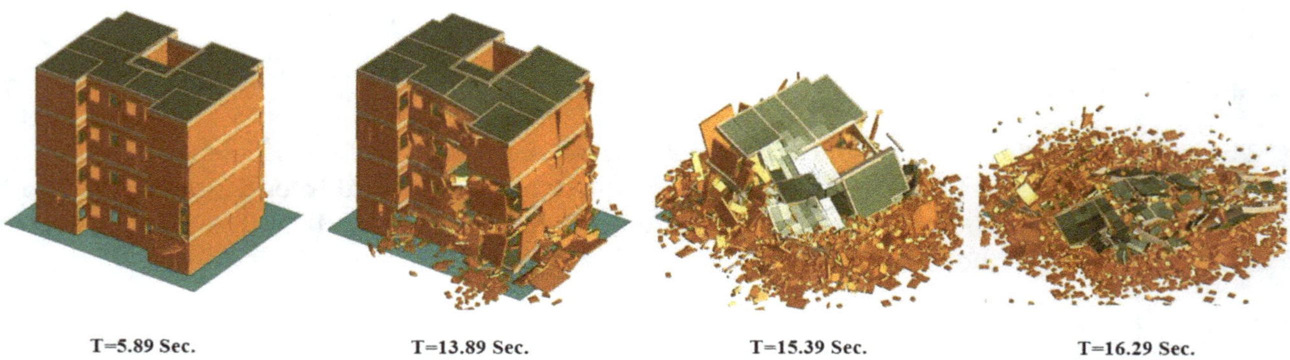

Figure 25. Progressive collapse of the building for the case study (E.Q. X- ZONE 5B).

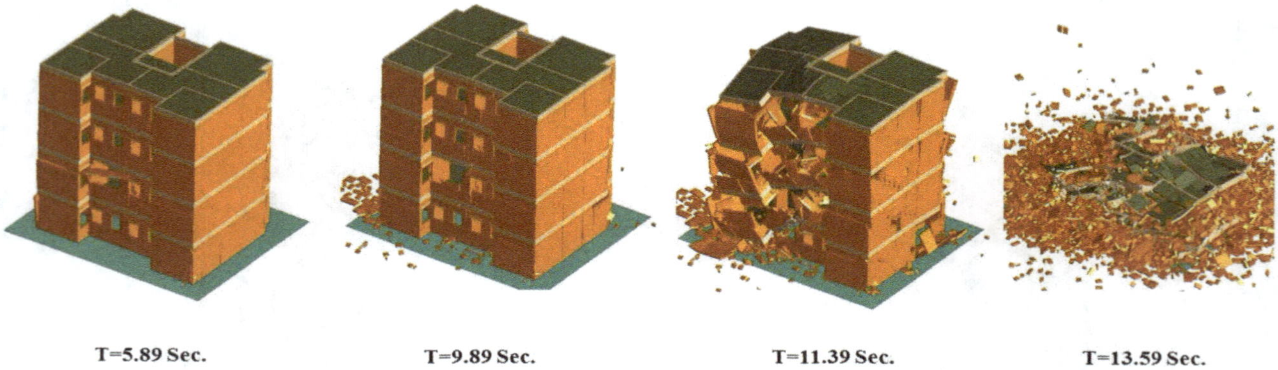

Figure 26. Progressive collapse of the building for the case study (E.Q.Y- ZONE 5B).

5.3. Case of Retrofitting Solid Walls in Zone 5B

Figures 27 and 28 show the final damage after the application of the earthquakes loading in the x-direction and the y-direction, respectively. The building is proven to maintain its global stability with the high local damage that occurs around the wall's opening. The externally bonded CFRP laminates are proven to be an efficient technique for strengthening the masonry mid-rise walls bearing building subjected to seismic loads.

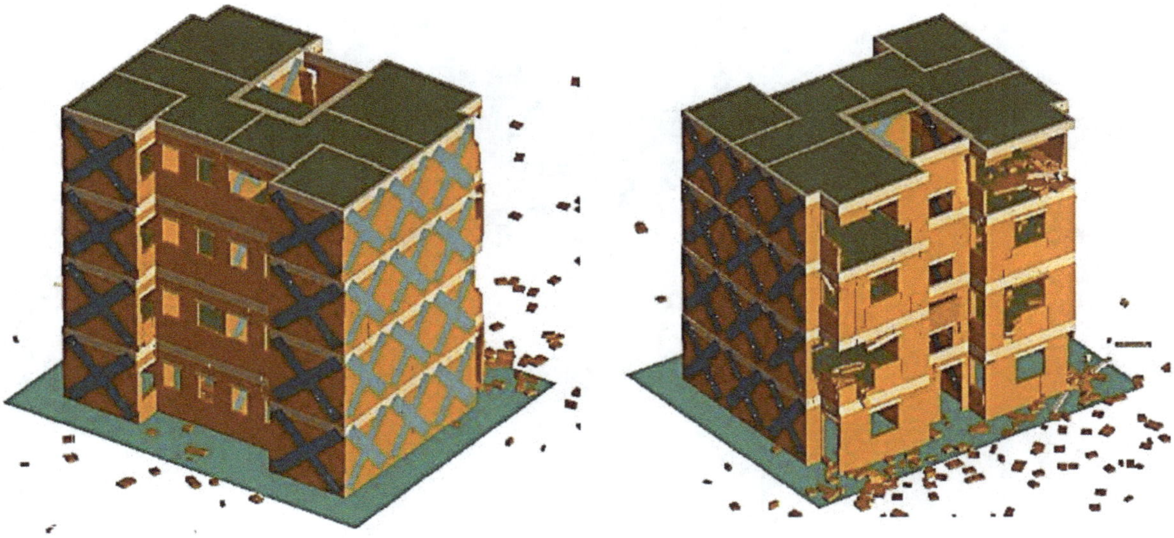

Figure 27. Damage obtained for the case study (E.Q. X- ZONE 5B- SW).

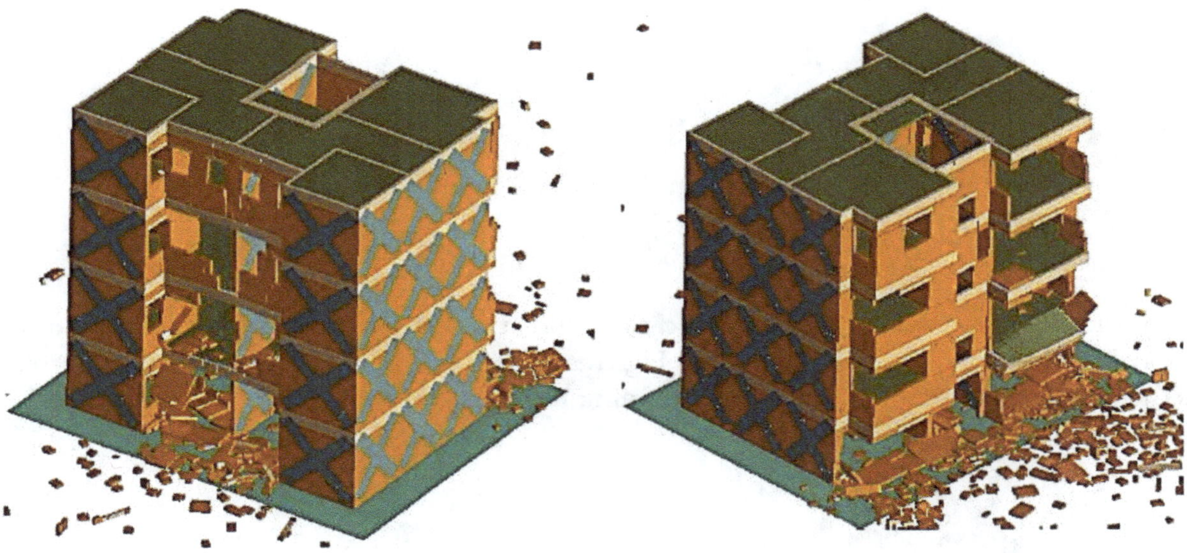

Figure 28. Damage obtained for the case study (E.Q. Y- ZONE 5B- SW).

Figure 29 shows the major principal strain contours at the ground floor walls for the case of the earthquake loading in the y-direction. The principal strain contours give a good indication of a cracking pattern with the lighter colors at the location of crack localization. Figure 30 shows the normal stresses versus time in the CFRP attached to one of the ground floor walls. As seen in Figure 30, the stresses in the CFRP do not reach its tensile strength (3500 N/mm^2) and the CFRP, therefore, do not rupture in the analysis. A sudden increase in the CFRP stresses is evident at time equals to 2.0 s. The sudden increase of the FRP stresses is explained by the crack occurrence in the walls at the location of the CFRP laminate, which could successfully bridge the crack and eventually maintain the wall stability.

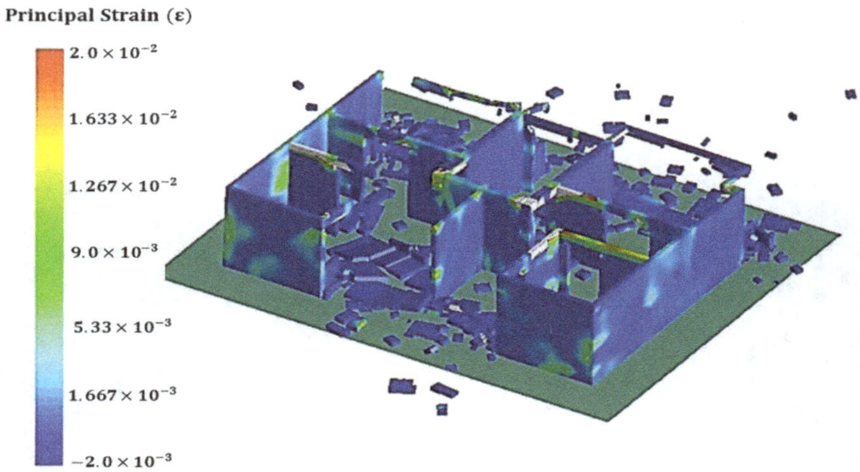

Figure 29. Principal strain contours of walls at the ground floor.

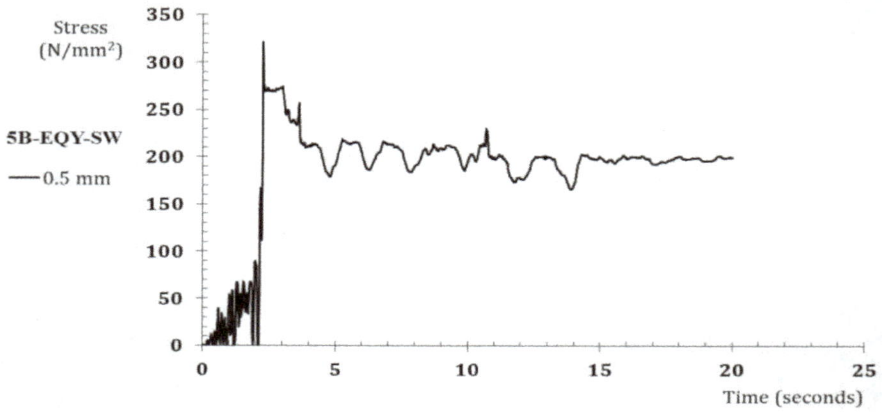

Figure 30. Normal stresses versus time for CFRP laminates at the ground floor.

5.4. *Case of Retrofitting Solid Walls and Openings in Zone 5B*

Numerical results verified that the building could successfully maintain its global stability with minor local damage, which is illustrated in Figures 31 and 32. The building's overall high stability proves the efficiency of the proposed strengthening technique of the current case study.

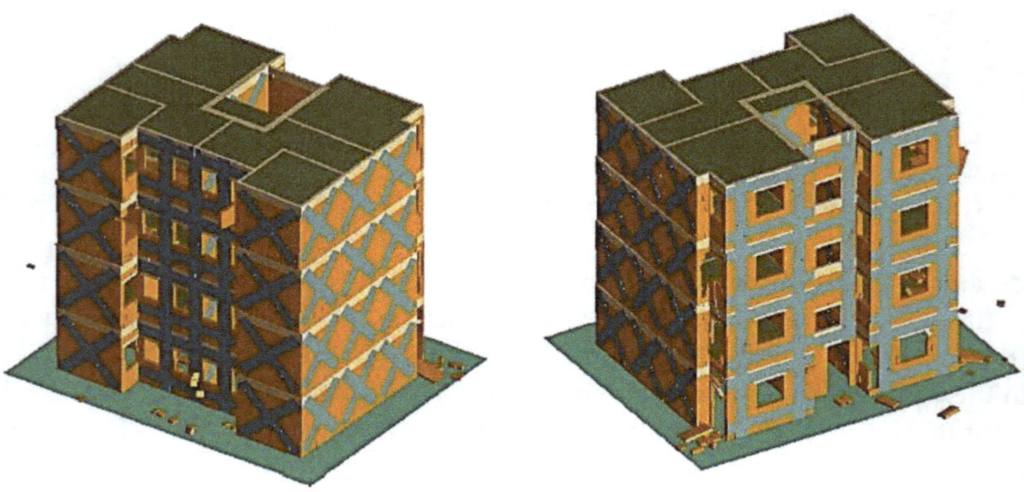

Figure 31. Deformed shape for the case study (E.Q. X- ZONE 5B- SW & OP).

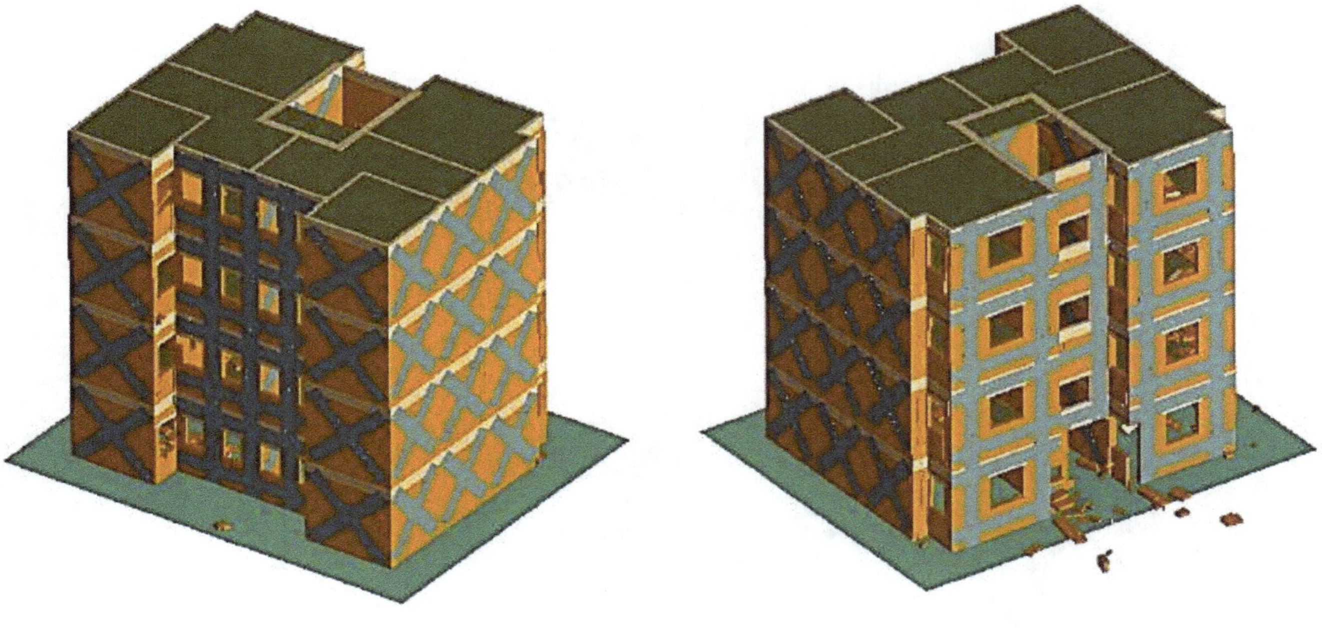

Figure 32. Damage obtained for the case study (E.Q. Y- ZONE 5B- SW & OP).

5.5. Optimization of the Amount of CFRP

Different thicknesses of CFRP (0.25, 1, and 1.5 mm) are used in the analysis of the retrofitted solid walls case to determine the minimum thickness of the CFRP laminates. Figures 33–36 show the sample of analysis results for the case (E.Q. X- ZONE 5B- SW). As seen in Figure 33, the lateral displacement at the top of the building decreases to almost half when CFRP thickness increases from 0.25 to 1.5 mm. The reason for such a reduction in displacement is that the CFRP reduces the induced cracks from the earthquake motion especially at the locations of the CFRP laminates where it is clearly seen in the principal strain contours in Figure 29, which was shown previously in Section 5.3. However, the base shear does not show a remarkable change, as shown in Figure 34. Figure 35 illustrates the hysteresis base shear-displacement curves for the building regarding different thicknesses of the CFRP where it can be recognized that the behavior is the same with higher deformability in the 0.25 mm thickness case. Figure 36 illustrates the stress versus time for the different CFRP thicknesses where the stresses in the CFRP of thickness 0.25 mm reach maximum value of 230.7 MPa, which is very much below the ultimate strength of the CFRP.

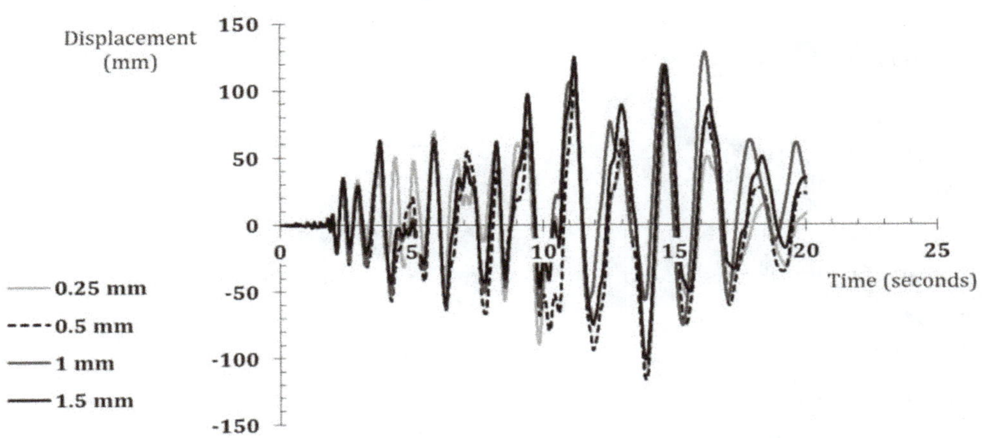

Figure 33. Effect of CFRP thickness. Top displacement vs. time for case (E.Q. X- ZONE 5B- SW).

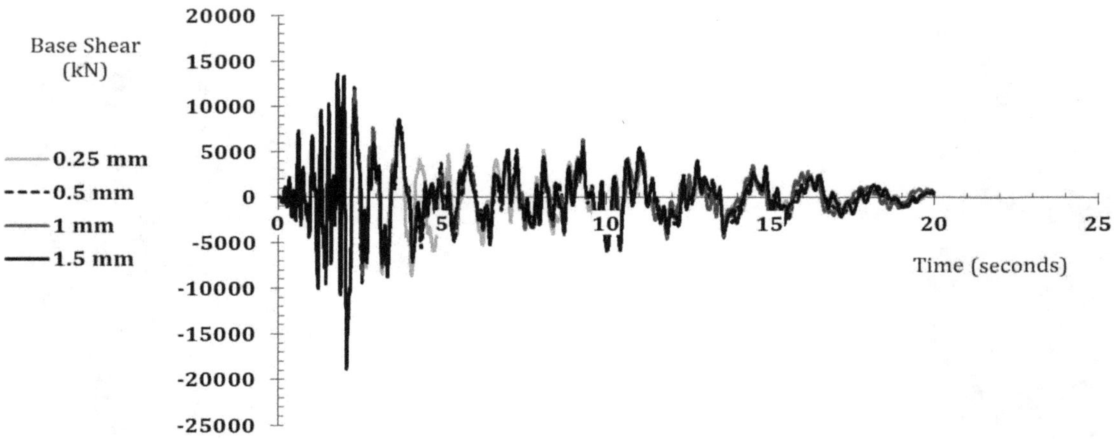

Figure 34. Effect of CFRP thickness. Base shear vs. time for case (E.Q. X- ZONE 5B- SW).

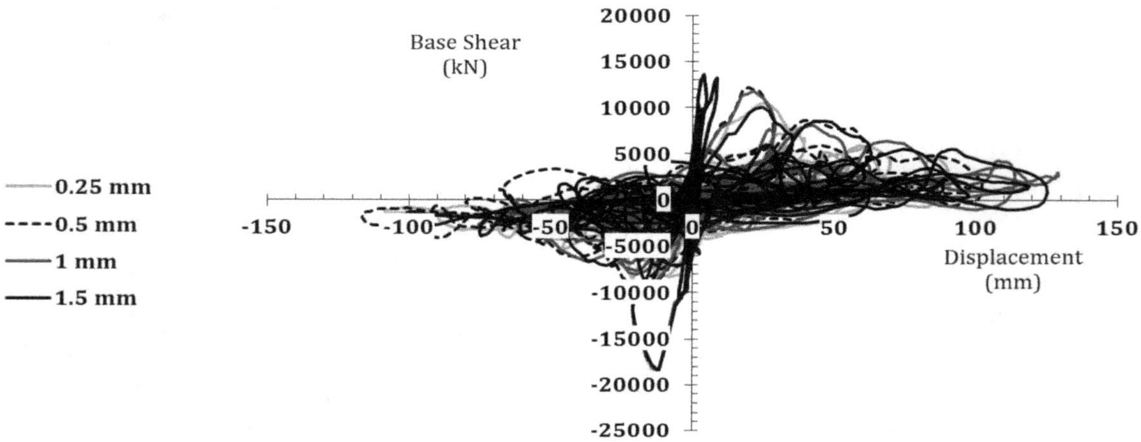

Figure 35. Effect of CFRP thickness. Base shear vs. top displacement for case (E.Q. X- ZONE 5B-SW).

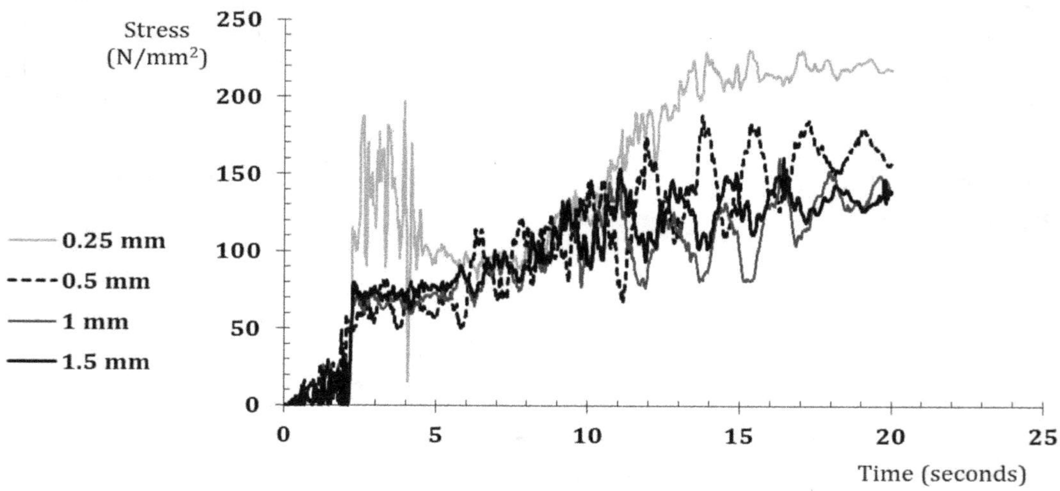

Figure 36. Effect of CFRP thickness. Normal stresses in CFRP versus time for case (E.Q. X- ZONE 5B- SW).

6. Conclusions

First, a validation is done for the experimental program of masonry walls under lateral loading by using the Applied Element Method simulation program. Afterward, a three-dimensional nonlinear dynamic analysis is carried out for numerically investigating the seismic rehabilitation of a typical masonry load-bearing residential building in Giza, Egypt by using carbon fiber reinforced polymer

(CFRP). The study included two earthquake zones in Egypt. Based on the results obtained from the studied cases, the following conclusions can be drawn.

(1) On the basis of the validation results of the previous experimental program, the constitutive models of concrete can be used for simulating the masonry as simplification where the simulation results show good agreement with the experimental results.
(2) The studied mid-rise masonry bearing walls building shows good behavior in resisting earthquakes for seismic zone 3 (peak ground acceleration of 0.15 g) while it shows a complete collapse in the seismic zone 5B (peak ground acceleration of 0.3 g).
(3) The externally bonded CFRP laminates are proven to be efficient for seismically strengthening the masonry mid-rise bearing wall buildings. It shows a good capability of preventing structural collapse with minor local damage. For the current case study, strengthening with CFRP laminates of width 0.5 m and thickness 0.25 mm is proven to be sufficient for preventing the total collapse of the building.

Author Contributions: H.S. built the research scheme. E.F. collected the required information for the targeted existing structure. E.F. conducted the numerical simulations and analyzed the results. H.S. supervised the numerical process. E.F. and H.S. wrote the paper.

Acknowledgments: The authors gratefully acknowledge Applied Science International, LLC, for granting the license of the simulation software (ELS®) to be used in the current research.

References

1. American Concrete Institution. *Abstract of: State of the Art Report on Fiber Reinforced Plastic (FRP) for Concrete Structures*; (ACI 440 R-95); American Concrete Institution: Farmington Hills, MI, USA, 1995; Volume 92, p. 5.
2. Stratford, T.; Pascale, G.; Manfroni, O.; Bonfiglioli, B. Shear Strengthening Masonry Panels with Sheet Glass-Fiber Reinforced Polymer. *J. Compos. Constr.* **2004**, *8*, 434–443. [CrossRef]
3. Petersen, R.B. In-Plane Shear Behavior of Unreinforced Masonry Panels Strengthened with Fiber Reinforced Polymer Strips. Ph.D. Thesis, The University of Newcastle, Callaghan, Australia, 2009.
4. Chuang, S.; Zhuge, Y.; Wong, T.; Peters, L. Seismic retrofitting of unreinforced masonry walls by FRP strips. In Proceedings of the Pacific Conference on Earthquake Engineering, University of Canterbury, Christchurch, New Zealand, 13–15 February 2003.
5. Tumialan, J.G.; Morbin, A.; Micelli, F.; Nanni, A. Flexural strengthening of URM walls with FRP laminates. In Proceedings of the Third International Conference on Composites in Infrastructure (ICCI 2002), San Francisco, CA, USA, 10–12 June 2002.
6. Li, T.; Galati, N.; Tumialan, J.G.; Nanni, A. FRP strengthening of URM walls with openings-Experimental results. *ACI Struct. J.* **2005**, *4*, 569–577.
7. Silva, P.F.; Belarbi, A.; Li, T. In-plane performance assessment of URM walls retrofitted with FRP. *TMS J.* **2006**, *24*, 57–70.
8. Maria, H.S.; Alcaino, P.; Luders, C. Experimental Response of Masonry Walls Externally Reinforced with Carbon Fiber Fabrics. In Proceedings of the 8th U.S. National Conference on Earthquake Engineering, San Francisco, CA, USA, 18–22 April 2006.
9. ElGawady, M.A.; Lestuzzi, P.; Badoux, M. In-plane seismic response of URM walls upgraded with FRP. *J. Compos. Constr.* **2005**, *9*, 524–535. [CrossRef]
10. Lignola, G.P.; Prota, A.; Manfredi, G. Numerical investigation on the influence of FRP retrofit layout and geometry on the in-plane behavior of masonry walls. *J. Compos. Constr.* **2012**, *16*, 712–723. [CrossRef]
11. Pantò, B.; Cannizzaro, F.; Caddemi, S.; Caliò, I.; Chácara, C.; Lourenço, P.B. Nonlinear modelling of curved masonry structures after seismic retrofit through FRP reinforcing. *Buildings* **2017**, *7*, 79. [CrossRef]
12. Ramaglia, G.; Lignola, G.P.; Fabbrocino, F.; Prota, A. Numerical investigation of masonry strengthened with composites. *Polymers* **2018**, *10*, 334. [CrossRef]
13. Housing and Building Research Center, Egypt. *The Egyptian Code for Calculating Loads and Actions in Structural and Block Works*; ECP (201); Housing and Building Research Center: Giza, Egypt, 2012.

14. Tagel-Din, H.; Meguro, K. Applied Element Method for Dynamic Large Deformation Analysis of Structures. *Int. J. Jpn. Soc. Civ. Eng.* **2010**, *17*, 215s–224s. [CrossRef]
15. Maekawa, K.; Pimanmas, A.; Okamura, H. *Nonlinear Mechanics of Reinforced Concrete*; Spon Press: London, UK, 2003.
16. Menegotto, M.; Pinto, P. Method of analysis for cyclically loaded reinforced concrete plane frames including changes in geometry and non-elastic behavior of elements under combined normal force and bending. In *IABSE Symposium of Resistance and Ultimate Deformability of Structure acted on by Well Defined Repeated Loads*; International Association for Bridge and Structural Engineering: Zurich, Switzerland, 1973.
17. Angelillo, M.; Lourenço, P.B.; Milani, G. *Masonry Behaviour and Modelling*; Springer: Vienna, Austria, 2014; Volume 551.
18. Hartmann, D.; Breidt, M.; Nguyen, V.; Stangenberg, F.; Höhle, S.; Schweizerhof, K.; Mattern, S.; Blankenhorn, G.; Möller, B.; Liebscher, M. Structural Collapse Simulation under Consideration of Uncertainty—Fundamental Concept and Results. *Comput. Struct.* **2008**, *86*, 2064–2078. [CrossRef]
19. Giamundo, V.; Sarhosis, V.; Lignola, G.; Sheng, Y.; Manfredi, G. Evaluation of different computational modelling strategies for the analysis of low strength masonry structures. *Eng. Struct.* **2014**, *73*, 160–169. [CrossRef]
20. Pantò, B.; Cannizzaro, F.; Caliò, I.; Lourenço, P. Numerical and Experimental Validation of a 3D Macro-Model for the In-Plane and Out-Of-Plane Behavior of Unreinforced Masonry Walls. *Int. J. Arch. Heritage* **2017**, *11*, 946–964.
21. Helmy, H.; Salem, H.; Mourad, S. Computer-Aided Assessment of Progressive Collapse of Reinforced Concrete Structures according to GSA Code. *J. Perform. Constr. Facil.* **2013**, *27*, 529–539. [CrossRef]
22. Helmy, H.; Salem, H.; Mourad, S. Progressive collapse assessment of framed reinforced concrete structures according to UFC guidelines for alternative path method. *Eng. Struct.* **2014**, *42*, 127–141. [CrossRef]
23. Helmy, H.; Elfouly, A.; Salem, H. Numerical simulation of demolition of Perna Seca Hospital using the Applied Element Method. In Proceedings of the Structures Congress, Chicago, IL, USA, 29–31 March 2012.
24. Meguro, K.; Tagel-Din, H. AEM Used for Large Displacement Structure Analysis. *J. Nat. Disaster Sci.* **2002**, *24*, 65–82.
25. Park, H.; Suk, C.G.; Kim, S.K. Collapse Modeling of model RC Structure using Applied Element Method. *J. Korean Soc. Rock Mech.* **2009**, *19*, 43–51.
26. Salem, H. Computer-aided design of framed reinforced concrete structures subjected to flood scouring. *J. Am. Sci.* **2011**, *7*, 191–2000.
27. Salem, H.; Helmy, H. Numerical investigation of collapse of the Minnesota I-35W bridge. *Eng. Struct.* **2014**, *59*, 635–645. [CrossRef]
28. Salem, H.; Ehab, M.; Mostafa, H.; Yehia, N. Earthquake Pounding Effect on Adjacent Reinforced Concrete Buildings. *Int. J. Comput. Appl.* **2014**, *106*. [CrossRef]
29. Salem, H.; Mohssen, S.; Kasa, K.; Hosoda, A. Collapse Analysis of Utatsu Ohashi Bridge Damaged by Tohoku Tsunami using Applied Element Method. *J. Adv. Concr. Technol.* **2014**, *12*, 388–402. [CrossRef]
30. Tagel-Din, H. Collision of Structures during Earthquakes. In Proceedings of the 12th European Conference on Earthquake Engineering, London, UK, 9–13 September 2002.
31. Asprone, D.; Nanni, A.; Tagel-Din, H. Applied Element Method Analysis of Porous GFRP Barrier Subjected to Blast. *Adv. Struct. Eng.* **2010**, *13*, 153–169. [CrossRef]
32. Almusallam, T.H.; Elsanadedy, H.M.; Abbas, H.; Ngo, T.; Mendis, P. Numerical analysis for progressive collapse potential of a typical-framed concrete building. *Int. J. Civ. Environ. Eng.* **2010**, *10*, 36–42.
33. Kang, H.; Kim, J. Progressive Collapse of Steel Moment Frames Subjected to Vehicle Impact. *J. Perform. Constr. Facil.* **2015**, *29*, 6. [CrossRef]
34. Griffin, J.W. Experimental and Analytical Investigation of Progressive Collapse through Demolition Scenarios and Computer Modeling. Master's Thesis, Civil Engineering, Raleigh, NC, USA, 2008.
35. Extreme Loading of Structures (ELS). Available online: http://www.extremeloading.com (accessed on 10 January 2018).
36. Gasprini, D.A.; Vanmarcke, E.H. *SIMQKE User's Manual and Documentation*; Massachusetts Institute of Technology: Cambridge, MA, USA, 1976.

Permissions

All chapters in this book were first published in MDPI; hereby published with permission under the Creative Commons Attribution License or equivalent. Every chapter published in this book has been scrutinized by our experts. Their significance has been extensively debated. The topics covered herein carry significant findings which will fuel the growth of the discipline. They may even be implemented as practical applications or may be referred to as a beginning point for another development.

The contributors of this book come from diverse backgrounds, making this book a truly international effort. This book will bring forth new frontiers with its revolutionizing research information and detailed analysis of the nascent developments around the world.

We would like to thank all the contributing authors for lending their expertise to make the book truly unique. They have played a crucial role in the development of this book. Without their invaluable contributions this book wouldn't have been possible. They have made vital efforts to compile up to date information on the varied aspects of this subject to make this book a valuable addition to the collection of many professionals and students.

This book was conceptualized with the vision of imparting up-to-date information and advanced data in this field. To ensure the same, a matchless editorial board was set up. Every individual on the board went through rigorous rounds of assessment to prove their worth. After which they invested a large part of their time researching and compiling the most relevant data for our readers.

The editorial board has been involved in producing this book since its inception. They have spent rigorous hours researching and exploring the diverse topics which have resulted in the successful publishing of this book. They have passed on their knowledge of decades through this book. To expedite this challenging task, the publisher supported the team at every step. A small team of assistant editors was also appointed to further simplify the editing procedure and attain best results for the readers.

Apart from the editorial board, the designing team has also invested a significant amount of their time in understanding the subject and creating the most relevant covers. They scrutinized every image to scout for the most suitable representation of the subject and create an appropriate cover for the book.

The publishing team has been an ardent support to the editorial, designing and production team. Their endless efforts to recruit the best for this project, has resulted in the accomplishment of this book. They are a veteran in the field of academics and their pool of knowledge is as vast as their experience in printing. Their expertise and guidance has proved useful at every step. Their uncompromising quality standards have made this book an exceptional effort. Their encouragement from time to time has been an inspiration for everyone.

The publisher and the editorial board hope that this book will prove to be a valuable piece of knowledge for researchers, students, practitioners and scholars across the globe.

List of Contributors

Jiří Pazderka
Department of Building Structures, Faculty of Civil Engineering, Czech Technical University in Prague, Thakurova 7, 160 00 Prague, Czech Republic

Tereza Pavlů, Jakub Řepka and Diana Mariaková
Department of Building Structures, Faculty of Civil Engineering, Czech Technical University in Prague, Thakurova 7, 160 00 Prague, Czech Republic
Research Team Architecture and the Environment, University Centre for Energy Ecient Buildings, Czech Technical University in Prague, Trinecka 1024, 273 43 Bustehrad, Czech Republic

Kristina Fořtová
Research Team Architecture and the Environment, University Centre for Energy Ecient Buildings, Czech Technical University in Prague, Trinecka 1024, 273 43 Bustehrad, Czech Republic

Radu Claudiu Fierascu, Mihaela Doni and Irina Fierascu
National Institute for Research & Development in Chemistry and Petrochemistry—ICECHIM Bucharest, Emerging Nanotechnologies Group, 202 Spl. Independentei, 060021 Bucharest, Romania

Mojmir Uranjek and Matjaž Skrinar
Chair of Structural Mechanics, Faculty of Civil Engineering, Transportation Engineering and Arhitecture University of Maribor, Maribor 2000, Slovenia

Tadej Lorenci
Department of Health and Safety, Abrasiv Muta, Muta 2366, Slovenia

José Carlos Domingues
Department of Civil Engineering, University of Coimbra, 3030-788 Coimbra, Portugal

Tiago Miguel Ferreira
ISISE, Institute of Science and Innovation for Bio-Sustainability (IB-S), Department of Civil Engineering, University of Minho, 4800-058 Guimarães, Portugal

Romeu Vicente
RISCO, Department of Civil Engineering, University of Aveiro, 3810-193 Aveiro, Portugal

João Negrão
Department of Civil Engineering, University of Coimbra, 3030-788 Coimbra, Portugal

Tanja Kalman Šipoš and Kristina Strukar
Faculty of Civil Engineering and Architecture Osijek, Department for Technical Mechanics, 31000 Osijek, Croatia

Rabindra Adhikari
Department of Civil Engineering, Cosmos College of Management and Technology, Lalitpur 44600, Nepal
Interdisciplinary Research Institute for Sustainability, IRIS, Kathmandu 44600, Nepal

Pratyush Jha
Interdisciplinary Research Institute for Sustainability, IRIS, Kathmandu 44600, Nepal
Digicon Engineering Consult, Lalitpur 44600, Nepal

Dipendra Gautam
Structural and Geotechnical Dynamics Laboratory, StreGa, DiBT, University of Molise, Via F. de Sanctis, CB 86100 Campobasso, Italy

Giovanni Fabbrocino
Structural and Geotechnical Dynamics Laboratory, StreGa, DiBT, University of Molise, Via F. de Sanctis, CB 86100 Campobasso, Italy
ITC-CNR, Construction Technologies Institute, National Research Council of Italy (CNR), L'Aquila Branch, Via Carducci, AQ 67100 L'Aquila, Italy

Ridho Bayuaji
Department of Civil Infrastructure Engineering, Faculty of Vocations, Institut Teknologi Sepuluh Nopember, Surabaya 60116, Indonesia

Totok Ruki Biyanto
Department of Engineering Physic, Faculty of Industrial Technologi, Institut Teknologi Sepuluh Nopember, Surabaya 60116, Indonesia

André Furtado, Nelson Vila-Pouca, Humberto Varum and António Arêde
Construct-Lese, Departamento de Engenharia Civil, Faculdade de Engenharia da Universidade do Porto, 4200-001 Porto, Portugal

George C. Manos, Lambros Kotoulas and Evangelos Kozikopoulos
Lab. Strength of Materials and Structures, Aristotle University, 541 24 Thessaloniki, Greece

Naida Ademović
Faculty of Civil Engineering in Sarajevo, University of Sarajevo, Sarajevo 71 000, Bosnia and Herzegovina

Daniel V. Oliveira and Paulo B. Lourenço
ISISE, Institute of Science and Innovation for Bio-Sustainability (IB-S), Dept. of Civil Engineering, University of Minho, Guimarães 4800-058, Portugal

Maria Teresa De Risi, Carlo Del Gaudio and Gerardo Mario Verderame
Department of Structures for Engineering and Architecture, University of Naples Federico II, 80125 Naples, Italy

Eissa Fathalla and Hamed Salem
Structural Engineering Department, Faculty of Engineering, Cairo University, Giza 12613, Egypt

Index

www.ingramcontent.com/pod-product-compliance
Lightning Source LLC
LaVergne TN
LVHW081039210926
848485LV00004B/1057